CW00766207

NEW IDEAS IN ASTRONOMY

Participants in Venice

New Ideas in Astronomy

Proceedings of a Conference
held in honor of the 60th birthday of Halton C. Arp
Venice Italy, May 5–7 1987

Edited by

F. BERTOLA
Instituto di Astronomia, Universitá di Padova

J. W. SULENTIC
University of Alabama, Tuscaloosa

B. F. MADORE
California Institute of Technology
David Dunlap Observatory, University of Toronto

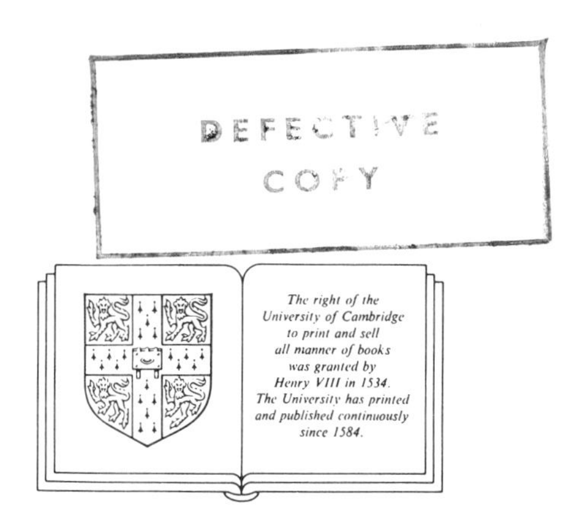

CAMBRIDGE UNIVERSITY PRESS

Cambridge

New York New Rochelle

Melbourne Sydney

Published by the Press Syndicate of the University of Cambridge
The Pitt Building, Trumpington Street, Cambridge CB2 1RP
32 East 57th Street, New York NY 10022, USA
10 Stamford Road, Oakleigh, Melbourne 3166, Australia

First published 1988

Printed in Great Britain at the University Press, Cambridge

British Library cataloguing in publication data

New ideas in astronomy
1. Astronomy
I. Bertola, F. II. Sulentic, J. W.
III. Madore, Barry F. (Barry Francis)
1948– IV. Arp, Halton C.
520

Library of Congress cataloguing in publication data

ISBN 0 521 345626 hard covers
ISBN 0 521 paperback

TABLE OF CONTENTS

Extragalactic Theory and Cosmology

FOREWORD

The conference on "New Ideas in Astronomy" was held May 5-7, 1987 to honor Dr. Halton C. Arp on the occasion of his 60th birthday. When the idea for a birthday conference was first suggested to Dr. Arp, his response was to express a wish that any such event not be simply a tribute to his past accomplishments. Instead, he suggested that such an event should represent a forum for new ideas in astronomy. Hence the evolution of our conference into a forum for classical reviews, new observational results and for ideas, old and new, in a wide range of astronomical fields. It is our belief that the conference represented an affirmation of the value for new and sometimes unpopular ideas. The setting for the meeting was Venice and the location in Venice was Palazzo Loredan, seat of the Istituto Veneto di Scienze Lettere ed Arti. The effect of Venice and of Palazzo Loredan upon the conference participants produced a "serenity" that complemented a stimulating and open exchange of ideas on subjects for which there is frequently little agreement. We were fortunate to have both invited and contributed speakers whose presentations spanned fields ranging from bioastronomy to cosmology. We are grateful to all of them for contributing to the success of the meeting.

ACKNOWLEDGEMENTS

The Istituto Veneto di Scienze, Lettere ed Arti not only hosted the meeting at its seat in Palazzo Loredan, but also co-sponsored the conference. We were also sponsored by the Istituto Italiano per gli Studi Filosofici and the local astronomical institutions: Dipartimento di Astronomia dell'Università di Padova and Osservatorio Astronomico di Padova. The cultural institutions "Palazzo Grassi" and "La Biennale" both graciously hosted the participants at receptions and exhibitions. Our conference was followed by another gathering, entitled "Kosmos", which involved an exchange of ideas between scientists and philosophers. This meeting was sponsored by Istituto Gramsci Veneto, which provided support for a number of speakers common to both conferences. We would also like to express our thanks to J. Boyd, G. Frezza, C. Sulentic and M. Zanon as well as G. Piotto and A. Rifatto for assistance with the successful organization and running of the conference. Finally, we thank Drs M. Calvani, D. DeSmet and R. Rampazzo for assistance with the software aspects of preparing these proceedings.

CONFERENCE PARTICIPANTS

T. van Albada
C. Barbieri
R. Barbon
L. Benacchio
F. Berefelt
P. Bernacca
G. Bertin
F. Bertola
D. Bettoni
A. Bianchini
C. Bonoli
S. Bonometto
W. Brinkmann
G. Borner
L. Buson
G. Burbidge
M. Burbidge
M. Calvani
M. Capaccioli
G. Chincarini
C. Chiosi
J. Clavel
U. Curi
S.V.M. Clube
S. Cristiani
F. De Felice
M. Della Valle
G. De Zotti
S. Di Serego Alighieri
R. J. Dickens
S. D'Odorico
R. Falomo
G. Favero
M.G. Fracastoro
A. Franceschini
D. Galletto
P. Giannone
E. Giraud
G. Giuricin
L. Gratton
J. Heidmann
E. Held
F. Hoyle
A. Hjalmarson
R. Keys
E. Khachikian
T. Kiang
R. Kraft
F. Lucchin
C. Maccagni
L. Malagnini
J. Manoussaynaki
A. Mammano
F. Mardirossian
M. Mezzetti
W. Napier
J. Narlikar
P. Padovani
J.C. Pecker
G. Piotto
P. Rafanelli
A. Renzini
M. Roberts
L. Rosino

K. Rudnicki
R. Ruffini
G. Romano
R. Sancisi
R. Sanders
P. Shaver
W. Saslaw
G.F.O. Schnur
N. Sharp
L. Secco
D. Sciama
R. Stalio
J.W. Sulentic
W.G. Tifft
M. Toffolatti
M. Tosi
A. Treves
M. Turatto
A. Vallenari
A. Vaesterberg
J.P. Vigier
J. Wampler
M. van Woerden
R. Wolstencroft
L. Woltjer
M.H. Ulrich
G. Zamorani

HISTORY, BIOASTRONOMY
AND
OUR GALAXY

THE ASTRONOMICAL TRADITION IN THE VENETO

Carlo Maccagni
Università di Genova
and
"Domus Galilæana" Pisa, Italy

SUMMARY

The science of the heavens has had diverse forms over the centuries, each of which has always impacted upon the way of life and on the culture of man. To follow the formation and development of astronomical thought means to reconstruct historically one of the most significant lines of evolution of civilization. In the case of the Veneto, it is a civilization that assumes a mediterranean and continental dimension, with the expansion of the political and economic power of the Republic of Venice, especially from the 12th century, and with the prestige acquired from the University of Padua, already among the European universities in the 14th century.

Man in prehistoric times observed the heavens in order to better arrange his own activities according to the seasons and read, in the movement of the stars, mythological stories full of symbolic and religious significance. Archeological testimony of this initial period remains in the "motte" and "castellieri" scattered in good number throughout the Veneto.

The renewal of western Latin civilization in medieval Europe, thanks above all to the mediation of the Arab culture, inherited the knowledge elaborated in the classical epoch. A science of the heavens, in particular, was a unique discipline constructed from the inextricable combination of mathematical astronomy and astrology. The latter became a normal component of every aspect of life. It also constituted the frame of reference that permitted the rational ordering of the empirical and dispersed knowledge of physicians. Astrology, for this reason, was an essential discipline of the faculty of medicine: originating from this teaching at Padua, astronomy and mathematics acquired in three centuries their proper autonomy.

Pietro d'Abano, professor of medicine and astronomy at Padua (1306-15), founding on the Almagest of Ptolemy the teaching of astrological medicine, sets down the basis for this future development. He also inspired the cycle of Giottesque frescoes in the Palazzo delle Ragione of Padua (1306) while the same Giotto was contemporaneously painting Halley's Comet in the Scrovegni Chapel.

Another professor of the University, Giovanni Dondi dall'Orologio, physician and astronomer, constructed in the middle of the 13th century a famous and complex astronomical clock, the "Astrarium", of which he also leaves an accurate description. Towards the middle of the following century, Giovanni Fontana, physician of Udine, dedicated various writings on astronomical instruments.

The University of Padua was also illuminated by foreign astronomers of European importance, such as Georg Purbach (between 1450-53), Johannes Müller Regiomontanus (1463-64), and Paul of Middleburg (1479-80). The first were connected with Cardinal Bessarione, who left to Venice (1468) his library that was rich in Greek manuscripts. Paul of Middleburg elaborated a first plan of calendar reform for the Fifth Lateran Ecumenical Council (1512-17) that did not come into effect.

At Venice, researchers such as Giorgio Valla made known the ancient texts of astronomy, which were published in large numbers by typographers of Venice and the Veneto. These included the first edition of the works of Ptolemeus (1515) and other medieval and contemporaneous works of astronomy and astrology. Meanwhile, also at Venice, noble and cultivated persons - Marcantonio Michiel, Daniele Barbaro, Jacopo Contarini – and academics – Accademia della Fama – cultivated studies of astrology-astronomy, gnomonics, cartography, and navigation. They also collected scientific instruments. At Padua, such a tradition continued with the Accademia Delia and with Gianvincenzo Pinelli, who was later in contact with Henry Savile and Galileo Galilei.

From 1501-03, Copernicus studied at Padua in the same period and same environment in which astronomers such as Giovanni Battista Amico, physician- astronomers such as Alessandro Achillini, Giovanni Battista della Torre, and Girolamo Fracastoro and philosophers such as Celio Calcagnini did. They expressed, with works of diverse character, the same dissatisfaction for the geocentricism either philosophised by Aristotle or "mathematically developed" by Ptolemy.

Galileo Galilei, professor at Padua from 1592 to 1610, represents the culmination of this tradition. The discoveries announced in *Sidereus Nuncius* are for several reasons connected also to the atmosphere between Padua and Venice in which he passed "the 18 best years of his life". The condemnation of Galileo slackened, in general, the development of science, above all in Italy, while the contemporaneous beginning of the political and economic decadence of Venice was reflected in the life of the University.

Thus, while the previous union of mathematics and astronomy resolved itself in favor of mathematics, the Republic favored the applied disciplines: hydraulics, cartography – in which excelled Vincenzo Maria Coronelli, from 1685 official state cartographer –civil and military architecture, and naval engineering. The reform of the University in 1739 also went in this direction and, while it furnished a laboratory for the teaching of mechanics, did not foresee an astronomical observatory. The construction of the observatory was decided only in 1765, and the realization was entrusted to Giuseppe Toaldo. It is the final act of the Republic of Venice as a sovereign state in favor of science.

(Translation provided by J.Sulentic)

IS THE UNIVERSE FUNDAMENTALLY BIOLOGICAL

Fred Hoyle
University of Cambridge, England

My first point has very far reaching implications if it is true. No process known to me produces organic material in quantity from inorganic materials alone. Always in every example I have studied a material of biological origin is brought in by the back door or by a side door. In the well known Urey-Miller experiment the ammonia was obtained by a process involving hydrogen of bioorigin, while the methane used in the experiment was overtly of biological origin. Biological processes for producing organic materials are far more powerful than non-biological processes for two reasons. First, biological catalysts are much more efficient than non-biological ones. Second, and under cosmic conditions, non-biological catalysts would be poisoned almost instantaneously by sulphur gases.

The first overwhelming indication that the universe must be fundamentally biological has come in recent years, with the realization that most of the material of the interstellar grains must be organic in composition. This is shown by the absorption spectrum of the grains in the 3-4 μm waveband. There is no deep absorption due to water ice near 3.1 μm but there is a broad absorption near 3.4 μm due to groupings of C atoms with hydrogen, together with a broad absorption near 3 μm due to OH groups. The spectrum obtained for all the grains along the line of sight from the galactic centre to the Earth, obtained in the background radiation of an M2 supergiant star near the galactic centre, is very like the absorption spectrum of dry bacteria measured in the laboratory. Either the grains are bacteria or they are composed of organic material similar to within a close margin to the organic materials present in bacteria, which is to say amino acids, nucleic acids,lipids and polysaccharides in much the proportions found in microorganisms.

The hypothesis that the grains are actually bacteria can be tested from the visual extinction of starlight. When dry, bacteria are hollow, with a low value for their mean refractive index. Taking the known degree of hollowness and the known refractive index of biomaterial, together with the known size distribution of spore-forming bacteria, an entirely parameter - free calculation of the visual extinction of starlight can be made. The result is excellent, in contrast to the very many inorganic models Professor Chandra Wickramasinghe and I have examined. A many-parameter (about ten parameters) model involving a combination of graphite and silicon grains is quite markedly worse.

Of the astronomical objects known to us, the most plausible source of biomaterial are comets. They have the correct overall composition with respect to the main life-forming elements. If all stars have cometary systems similar to the solar system, then in quantity they could supply most of the grains in the interstellar medium. And from recent observations in the 3-4 μm band we know that the million or more tons of small particles ejected by Comet Halley on 30-31 March 1986 showed the characteristic bacteria spectrum in emission.

The Earth is perpetually embedded in a halo of cometary material ejected from short-period comets and perhaps a 1000 tons of this material is swept up by the terrestrial atmosphere annually. If this material possesses a biological component that interacts pathogenically with terrestrial plants and animals then we have an exceedingly sensitive test for biomaterial arriving at the Earth from outside. This is because pathogenic interaction would involve multiplying an incident microorganism billions of times within the cells of a host plant or animal. The situation would be logically analogous to the detection of a high energy particle with a Geiger counter, with a single incident particle causing a large cascade of secondaries. This is a much more sensitive test in principle than attempting to detect incoming microorganisms individually.

Viruses taken individually are too small to descend the stratosphere under gravity alone. They descend due to large scale stratospheric air motions which are most effective in the later months of winter, essentially because the temperature difference between pole and equator is greatest in these months. This period coincides exactly with the familiar late winter period of respiratory diseases. It also predicts a 6-month phase difference between the northern and southern geographic hemispheres, which is also born out by experience.

For many so-called infectious diseases the old notion of viruses or bacteria being transferred from person to person does not stand up to experimental test. To make an epidemic run in a population by person to person transmission it would be necessary to establish a multiplying chain reaction, with each victim of a disease infecting more than one other person. Yet a typical epidemic involves no more than 25-30% of the population, even in cases where the entire population is susceptible to a new form of pathogenic attack, as in the cases of the Asian flu of 1957 and the Hong-Kong flu of 1968. There is clear evidence, obtained by four different groups to my knowledge, that husbands and wives did not infect each other significantly in these outbreaks, certainly not to more than to a 20% extent. Nor did twins living close together. If pathogens were generally transmitted by person-to-person contacts then one would expect to be more at risk in a densely-packed city population than if one lived in the remote countryside. It is a matter of simple experience that this is not true.

If this were an entirely scientific matter, there is little doubt from the evidence that the case for a fundamentally biological universe would be regarded as substantially proven. The reason why the scientific community passionately resists this conclusion is that biological systems are teleological, which is to say purposive. And if we admit the universe to be inhabited by a vast number of purposive components then the thought cannot be far away that perhaps the Universe itself might be purposive, a conclusion that not only stands astronomy immediately on its head, by making a large fraction of our efforts futile, but which offends the tenets on which modern science presents itself to the public. It is here where the biggest issue ultimately for science may lie.

To end with, let me give an example. Nowadays an immense amount of detail is known about the empirical conditions inside clouds where stars are forming, but remarkably the process of star-formation seems just as far away from resolution as it did a quarter of a century ago. The issue, as it was then seen to be, lay in an on-off coupling between the gas and a magnetic field pervading the gas. The coupling had to be on at early stages to reduce the angular momenta of spinning condensations, while it had to go off at later stages to prevent stellar magnetic fields soaring upwards to unacceptable intensities.

My last point is that the properties of bacteria with respect to electric charges and magnetic fields are such as we would never think of ascribing to inorganic particles. Bacterial membranes act as one-way systems with respect to the passage of both chemicals and electric charges. While being shielded by the cell wall from external ultraviolet light, they can serve as traps for electric charges. Second, a fraction of bacteria contain an ordered ferromagnetic domain that enables orientation to occur with respect to an external magnetic field. This leads to another interesting test, while at the same time suggesting unexpected relationships between grains and magnetic fields. A hollow bacterium containing an iron particle has a complex refractive index unlike a conducting solid particle, giving a different behaviour in the case of rod-like particles with respect to the wavelength dependence of the polarisation produced by alignment with respect to a magnetic field, the difference between an interference effect yields an undulating curve, and a smooth curve, with the iron-bacteria curve again agreeing excellently with the data.

But of course the ultimate difference between a purposive universe and a universe of "sound and fury, signifying nothing" as Macbeth said, must lie in cosmology, a subject outside the range of this contribution.

DISCUSSION

J.-C. Pecker: In the "old" days, one used to interpret the rather isotropic interstellar absorption by saying that the size distribution of dust grains is highly peaked (Pecker 1972): the "very small" grains are expelled by galaxies, the "large ones" are retained by stars, the "small ones" (5×10^{-6} - 5×10^{-5} cm) are staying in the ISM (this is related to the fact that $(L/M)_{gal} \sim 1/10(L/M)_{\odot}$). Now my question is ≪ how do you obtain the highly-peaked size distribution for your "biological grains", a distribution which may, -or may not?-, be critical for the interpretation of ISM absorption? ≫

F. Hoyle: The size distribution of particles ejected by comets could determine the size distribution in interstellar space, but subject to a truncation at about $0.5\mu m$ due to radiation pressure by starlight expelling the smaller particles out of the galactic disk. On the whole, the low value of the real part of the refractive index of hollow particles permits much more freedom for the size distribution than in the "old days". We have checked that the size distribution obtained from Comet Halley is acceptable in this sense.

G. Burbidge: You have concentrated on the composition of the general interstellar dust in our Galaxy. What about the dust that is present in large amounts when massive stars evolve? The prototype is η Carinae where we know that dust is appearing very rapidly. Is this not made by the condensation of ejected gas?

F. Hoyle: There is a question of the direction of cause and effect here. Are the particles condensing in gas flows from stars, or are the particles near stars derived from an initially-present distribution, which may have been modified chemically (as for instance by caramelisation) by the light of newly-formed stars. The former is the orthodox view but the latter may be the correct view.

The **MEGA** SETI, a major step in bioastronomy

Jean Heidmann
Observatoire de Paris
F-92195 MEUDON

ABSTRACT

After a quarter century of breakthrough developments in radio astronomy, space exploration and biology, the Search for ExtraTerrestrial Intelligence (SETI) obtained the support of the National Academies of the USA and of the USSR, of the International Astronomical Union and, more tangibly, the financial support of NASA.

If Other Beings have technical capabilities at least comparable to ours, it is possible to communicate over distances larger than a thousand light-years, radio waves being the most efficient vehicles for information propagation across interstellar space.

A million stars are then within reach but a formidable problem is raised by the tremendous number of possible channels to be investigated. We report on the most recent developments made by NASA Extraterrestrial Research Division, concentrating on the two main steps: spectral analysis and signal recognition.

Then we sketch the interest of international collaboration and present a project of SETI observations with the Nancay large radio telescope in collaboration with NASA.

> Qui sait si, bientot, nous ne communiquerons pas ensemble par un telegraphe ni plus ni moins merveilleux que celui qui nous permet actuellement de causer a voix basse et instantannement d'un bout a l'autre du globe terrestre? Camille Flammarion, "les Terres du Ciel", 1877

I - THE SETI CONTEXT

To assume that life is part of the cosmical evolution of the universe appears now as a reasonable working hypothesis. Several stages with increasing complexity begin to emerge in this evolution.

First the stellar evolution leading from the Big Bang to the formation of stars and planets with the important synthesis of chemical elements, and more especially carbon.

Then chemical evolution with the synthesis of simple organic compounds important for biology, such as the tens of organic molecules discovered by radio astronomy in interstellar space or amino acids and nitric bases of extraterrestrial origin included in some meteorites.

Third comes prebiotic evolution leading to much more complex organic compounds like those produced in laboratory experiments simulating primitive planetary environments or like those seemingly occurring in the atmosphere of the large satellite Titan of the planet Saturn.

Fourth stage, primitive biological evolution, still largely unknown but which probably led on Earth in a very short time, only a few hundred million years, to a flourishing biota.

At last, fifth stage, the appearance of evolved forms of life, but here in a very long time spanning billions of years, as those of which we are witnesses and parts.

It already appears in this simple sketch that astronomy plays a first rank role in the problems raised by life in the cosmos. So much so that at its 1982 General Assembly the International Astronomical Union set up a new specific commission to support, organize and officialize its members activities in this field: "Bioastronomy, search for extraterrestrial life".

This organization was made possible because of the extraordinary progress made these last decades by radio astronomy, solar system space exploration and molecular biology. It started a quarter century ago with the seminal paper by G.Cocconi and P.Morrison in **Nature** showing the possibility to use radio waves to propagate information over interstellar distances and by the first search made by F.D.Drake in his Ozma project.

Since then, thanks to the farsighted determination of a few pioneers and to the subsequent professed support of the national academies of the United States and of the Soviet Union, bioastronomy became a post teenage branch of science ready for a front attack of the search for life in the universe scientifically, observationally and experimentally.

It is in this general framework that dwells SETI, Search for ExtraTerrestrial Intelligence, more exactly in the fifth stage of emergence of evolved forms of life. If "Other Beings" have means at least equal to those we have with our large radio telescopes and radars it is possible to transfer signals farther than a thousand light years. This flatly means that a million stars are within our reach and that a serious technological gap has been bridged by the recent developments of radio astronomy.

II - THE QUEST FOR MEGACHANNELS

The fastest and most favorable supports for transferring information over large interstellar distances are the radio electromagnetic waves; at higher frequency a limit is set up by the corpuscular nature of waves while at lower frequencies the natural emissions of our Galaxy generate noise. The optimal range is between 1 and 10 GHz, i.e. wavelengths between 30 and 3 cm.

On the other hand, in our present state of knowledge, in order to reach far away for a given energy budget, the carrying waves should be inside the narrowest possible bandwidth. However, a natural dispersion of a few hundredths of a hertz is introduced by the interstellar medium through the Drake-Helou effect. Then, in the optimal frequency range there are a thousand billion possible channels.

There lies the major problem of SETI: when Drake made his first search his receiver had only one channel and nowadays most radioastronomical receivers have still only a thousand simultaneous channels, which is very few compared to the number of channels to be explored. In order to bypass it NASA decided to embark on a research and development study to produce a receiver in the ten millions simultaneous channel class.

When in 1982 the US Congress waived its veto for search of intelligence elsewhere, a 1.5 to 2 million dollar yearly budget line was available for this new receiver. Headed by the Life Science Division at the NASA headquarters in Washington, the Extraterrestrial Research Division managed the operation with the SETI Program Office at the Ames Research Center, in collaboration with the Jet Propulsion Laboratory and contract with the Space, Telecommunications and Radioscience Laboratory at Stanford University. The work is provided with the advice from a Science Working Group composed of leading radio scientists.

Since the beginning of 1982 a 74000 channel prototype is tested on the field using the Goldstone 64m dish of the Deep Space Network and the 300m Arecibo radio telescope. Another prototype, the Sentinel project, slightly different, is operated since 1983 behind a Harvard/Smithsonian Institution 24m dish. These two prototypes behaving satisfactorily, the 8 million channel stage is undertaken.

III - THE SPECTRAL ANALYZER

The problem of detecting artificial extraterrestrial radio signals splits into two parts: first the spectral analysis aimed at investigating in the narrowest bandwidths the waves received in the largest spectral region, and second to recognize, among all these emissions received during a search session, the existence or not of signals coded according to an artificial distribution. The first part is the spectral analysis, the second the pattern recognition. Let us start with the first.

The analyzer built by NASA in order to detect very narrow bandwidth anomalous distributions in a wide band is the MCSA, the Multi (or Mega) Channel Spectral Analyzer (figure 1). The signal received from the radio telescope is converted to an intermediate frequency centered at 30 MHz; a 10 MHz band is converted to base-band by two mixers operating in quadrature and passed through two low pass 5 MHz filters. The two signals are then sampled at 10.6 MHz and converted into digital numbers, components of a complex 4-bit signal. This one is entered in a first finite impulse response filter

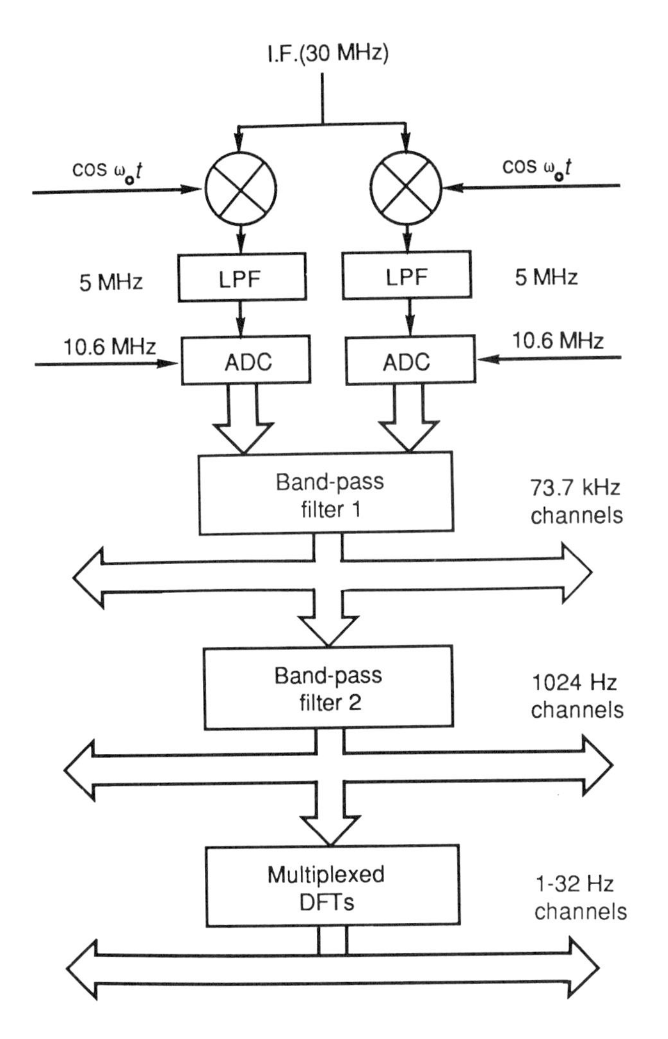

Figure 1- Block diagram of the multi-channel spectum analyzer; from top to bottom: the mixers, the low-pass filters, the analog-digital converters, the first and second filters and the multiplexed processors (doc. NASA).

and analyzed by an inverse discrete Fourier transform into 112 complex amplitudes in 112 channels each 74 kHz wide. Each of these amplitudes is then entered into a second battery of FIR filters and again transformed by DFT, leading each to 72 amplitudes in 72 channels 1024 Hz wide. All in all one gets 8064 complex amplitudes which are analyzed by multiplexed microprocessors by DFT. At this stage two choices are available: use DFT's with 32 or 1152 points giving respectively 36-32 Hz channels or 1024-1 Hz channels. At the output of these microprocessors one then gets complex amplitudes in 30000-32 Hz channels or in 8,257,536-1 Hz channels.

All along this chain outputs for complex amplitudes are available with bandwidths 1, 32, 1024 and 74000 Hz. Then these amplitudes are averaged in power for each channel and passed through a threshold dependent on the noise observed.

The building architecture chosen, with successive stages, has been preferred to a more conventional monolithic one with 8 million channels which could be obtained by a DFT processor. Thus it is easier to build a 74000 channel prototype using just one subunit of each step; this highly modular architecture with a high level of parallel processing reduces the number of different subunits, restricts to a part of the spectrum only the failure of an element and at last gives a large flexibility by just changing the minicode operating the second filter battery and the final DFT. And more important the system could be extended...

The 74000 channel MCSA prototype fits in a rack; it is shown on figure 2 with its control and acquisition VAX 11/750 computer. The complete MCSA would be much more bulky: it would incorporate 976 printed circuit cards loaded with chips filling twenty racks. As a matter of fact this shows practically the effort made by SETI to build this extraordinarily sophisticated receiver. Most fortunately there are only 13 different circuits in this complex machinery; the main microprocessor contains 680 identical units. It is thus an excellent candidate for very large scale integration. The Stanford group is developing a VLSI chip, the SETI DSP Engine, which, reproduced 680 times will considerably reduce bulk and price.

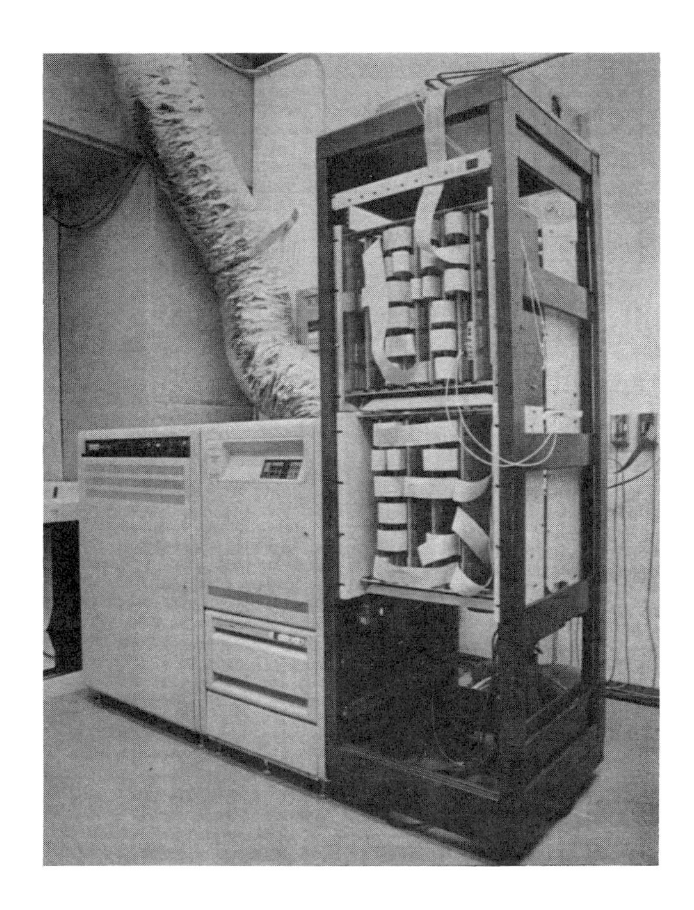

Figure 2- The prototype of the 74000 channel MCSA with its VAX 11/750 computer (doc. NASA).

IV - THE SIGNAL DETECTOR

Let us now look at the second question: to detect the existence of signals in the spectra. Here also the problem is gigantic. The data flux is enormous, of the order of ten billion bits per second, noise essentially; it is among this trash that an eventual systematic pattern has to be recognized, a weak one in addition; and as it is out of the question to preserve this noise, it has to be treated in real time.

A typical example is given by an observing run on a target, giving an 8 million point spectrum every second for 1000 seconds. A field of 8 billion pixels is produced; the screening through the threshold will reduce the number of occupied pixels, all the more since it is higher; but if it is too high signals may get lost, while if it is too low the number of false alerts will not be tolerable.

Effort is now being made to detect among the remaining pixels two types of signals: a continuous wave or a regularly pulsed wave. They should then produce vertical lines in the field. Unfortunately, with such high spectral resolutions, Doppler effects are very important. Thus a typical orbital planetary motion gives a one hertz per second shift, then a slanted line with unknown slope along which, for a pulsed train, points will be set with an also unknown interval (figure 3).

Presently algorithms are developed for detection of such alignments; to avoid memory flooding, pruning methods are investigated. A VAX 11/750 is able to treat in real time 1000 spectra with 74000 channels. The complete model will require a few hundred microprocessors, each looking at 100 kHz sections of the spectrum

Here also VLSI is a must. Stanford is developing specialized processors based on new content addressable memories. The signal detector will require ten of them, each with 50 VLSI CAM chips

Thanks to these important technological developments the ten million channel goal should be reached within two or three years; and behind doors whispers toward a hundred million channels are starting. Some will say: then why not to wait still to start SETI on a grand scale? Pointless question because if Drake with his Ozma one

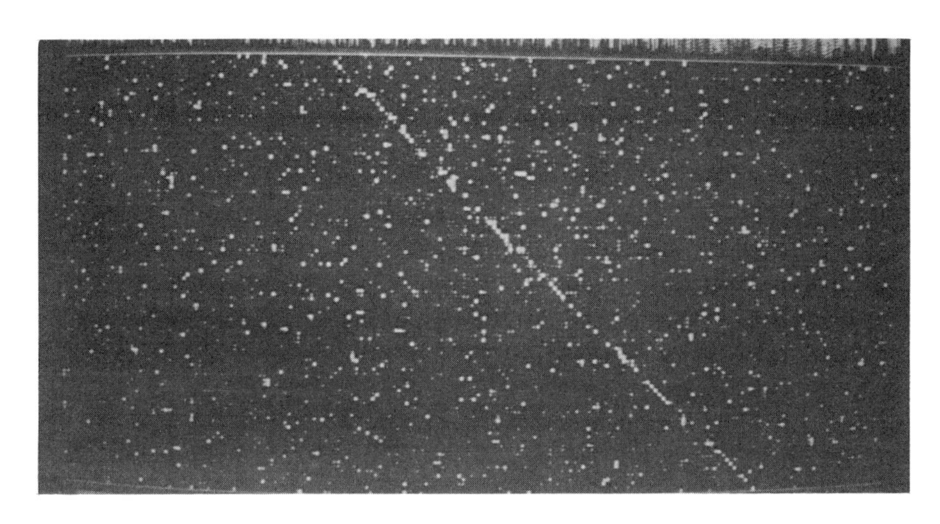

Figure 3- A field of 100 successive 206 channel spectra, a very small part of an observing run of the 74000 channel MCSA prototype aimed at the Pioneer 10 spacecraft; frequency is horizontal and time vertical. The spacecraft, with its 1 watt emitter beamed with its 33 db antenna, was more than 5 billion kilometers away and was, for the MCSA, a very powerfull "extraterrestrial" source. Because of Doppler shifts arising mainly from the Earth rotation it left a track with a 0.2 Hz/s drift (doc. NASA).

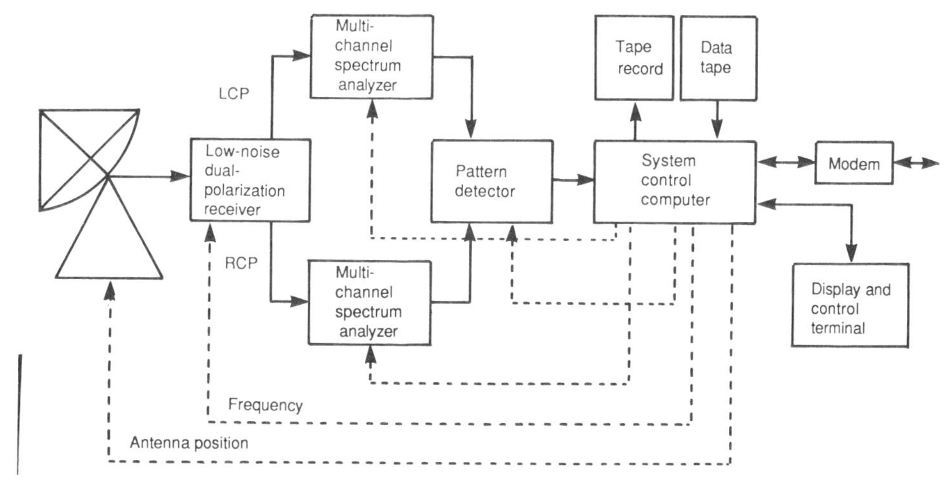

Figure 4- Block diagram of the complete SETI system; from left to right: the radiotelescope, the dual polarisation low noise receiver, the two MCSA's, the signal detector, the computer with its peripherals and pointing and frequency (doc. NASA).

and only channel 25 years ago had not been followed by those who hit the track with a hundred, then a thousand, a hundred thousand channels, technological progress would have remained at the stage of wild fantasy, while soon a one minute search of the MCSA will be equivalent of what Ozma would have taken a hundred thousand years to accomplish (yes, this is the ratio, all taken into account), and thus introducing us rightly inside the famous "cosmic haystack" to find the right signal in the right channel for the right star at the right moment.

V - THE SEARCH PROGRAM

Responsible scientific programs depending upon governmental authorities have to rest on already well established bases, as far as possible from pure speculation. Thus SETI can be supported by the only one known case of evolved life: ours on Earth.

Apparently the result achieved here in the cosmical evolution depends on the happy concurrence of three factors: a planet at the right distance from a stable star as a base, carbon for molecular structures and liquid water for chemical evolution.

The targets of the search program presently prepared are thus the nearest solar type stars. Nevertheless, in order not to skip unthought of potentialities some flexibility has been included in the program for the same price tag in order to be able to "listen" off targets.

The target program includes all the 800 type F, G and K stars up to 80 light year distance contained in the Royal Greenwich Observatory catalogue. 20 minute runs will be conducted in all of the 1-3 GHz band, with maximum resolutions 1 and 32 Hz and maximum sensitivity 10^{-28} W/m^2 for pulsed trains and 10^{-27} W/m^2 for continuous waves. In addition some globular clusters and nearby galaxies will be included, the disadvantage of their large distance being compensated by the large number of stars simultaneously seen by the radio telescope beam. A few bands between 3 and 10 GHz will also be searched. This target search will use the largest radio telescopes available: the Arecibo one and the 64m Deep Space Network one and will last 3 to 5 years.

The second part of the search, the one for unthought of emissions, will be an all sky survey between 1 and 10 GHz, with a few bands up to 25 GHz, with the medium resolutions 32 and 1024 Hz and a sensitivity 10^{-24} W/m^2. It will dwell more along the galactic plane, richer in stars, for eventual powerful interstellar radio beacons, intentional or not. This program will use moderate size radio telescopes, with larger beams, as the 34m Deep Space Network one, and will last 5 to 7 years.

VI - THE SETI SYSTEM

The full system is made up of the antenna, the receiver, the MCSA's, the signal detector and is under control of a computer (figure 4). As a matter of fact, as the interstellar medium does not perturb the radio waves left or right circular polarizations, both those could be used for coding the information at the emitting end. Thus they will be both analyzed with two MCSA's. Note that once the experimental algorithms for signal recognition are fully developed, they will be built in special, high-performance hardware in the system in order to improve power and speed. The receiver is a double polarization cooled classical one. Pointing and scanning of the antenna will be automated in order to minimize slewing.

In case a suspicious signal is received it will pass automatic preliminary tests: to compare it to an interference data bank, check it is in the main beam, it has fixed celestial coordinates. In case these tests are passed alert is immediately given to other SETI observatories in order to eliminate all local interference (or even possible hoax) and not to lose track of a message.

VII - A NASA-NANCAY COLLABORATION?

Search for extraterrestrial intelligence is a challenge for all the scientists of our planet. As said by M.D.Papagiannis, the first president of the bioastronomy committee: "We stand at a historic cross-road. We have the technical capability, the skilled scientists, and above all the strong desire to move ahead with our search for extraterrestrial life."

SETI is the outcome of the cultural development of all of mankind; in addition to the multidisciplinary bases on which it rests, it is an enterprise with a typical international aspect. One can only support collaboration in this field.

Furthermore this collaboration is technically a must. A number of observatories in the world should engage in it in order to fight against the radio interferences which are not the same for two different Earth hemispheres; also in case of detection the immediate and independent verification is of prime importance.

In February 1985 I presented to the Prospective Commission of Paris Observatory a proposal aimed at having a copy of the NASA MCSA/signal detector system be set by the US behind the Nancay radio telescope, in exchange of its use for SETI during a sizable part of the time to be defined.

One year later, and independently, the NASA SETI Program Office made a preliminary contact with the Paris Observatory through the Centre National d'Etudes Spatiales in order to investigate an eventual collaboration. Thus a possibility of SETI US/France collaboration is clearly opened at both ends.

The Nancay radio telescope (figure 5) is perfect for decimetric waves from 1 to 3 GHz; furthermore, in the decimetric class, it is by its effective area (0.70 hectare) the third largest, after the Arecibo (7.06 ha) and Effelsberg ones (0.78 ha), before Green Bank (0.66 ha) and Jodrell Bank (0.44 ha).

Since 1981, F.Biraud in our Departement de Radioastronomie, has several search runs at Nancay with cooled receivers and a 1024 channel autocorrelator filter at the H and OH frequencies, in collaboration with J.Tarter, from Berkeley University and Ames Research Center. The Departement de Radioastronomy has thus a precious know-how in the field.

Owing to its high sensitivity and high spectral resolution the SETI system could also be aimed at other celestial objects for searching for unknown astrophysical phenomena and physical processes with very narrow bandwidth radio emissions or absorptions.

A French/American collaboration for a "Mega" SETI would associate us to the effort for stepping behind the "ultimate" last frontier which has the potential to unveil civilizations more advanced than ours.

I thank with pleasure Francois Biraud for having read my manuscript. This report was prepared for the meeting of the Visiting Committee of the Nancay radio telescope in December 1986.

Figure 5- The Nancay radio telescope; in the background the 200m tiltable reflector, in front the 300m fixed spherical one and in between the movable antenna on its tracks.

REFERENCES

F.Biraud, 1983, SETI at the Nancay radiotelescope, Astronautica Acta, Academy Transactions Note, 10, 759.

B.M.Oliver, 1985, Signal processing in SETI, Communications of the Association for Computing Machinery, 28, 1151.

F.Drake, J.H.Wolfe, C.L.Seeger ed., 1983, SETI Science Working Group Report, NASA Technical Paper 2244. M.D.Papagiannis ed., 1984, The search for extraterrestrial life: recent developments, International Astronomical Union Symposium no 112, Reidel Publ.Co.

J.F.Duluk, I.R.Linscott, A.M.Peterson, J.Burr, B.Ekroot, J.Twicken, 1986, VLSI processors for signal detection in SETI, 37th Congress of the International Astronautical Federation, preprint no 489.

J.Heidmann, 1985,A la recherche de la vie dans l'univers, Revue du Palais de la Decouverte, 14, 19. D.K.Cullers, I.R.Linscott,

DISCUSSION

F. Hoyle: The starting point for exobiology of transferring the Urey-Miller experiment from the Earth to interstellar space seems at first sight to be an attractive possibility. I was turned away from it some years ago by the circumstance that neither of the strong absorption bands of water-ice near 3.1μm nor that of ammonia ice near 2.95μm are seen at any real appreciable intensity. Nor, it seems to me, does the irradiation of ices by ultraviolet in the laboratory lead to good agreement with the 3-4 μm absorption shown by grains. While it is conceivable that this was how biology began, it seems to me that by now the process has gone a long way from early beginning, to essentially a biological conclusion.

G. Chincarini: You showed that one of the fundamental steps is the formation of the COOH complex from [OH OH]. How likely is this reaction and how is it related to the environment, in space or in laboratory.

J. Heidmann: As a matter of fact I did not. I just spoke about the COOH and NH_2 linking through H_2O formation.

CAN WE DERIVE THE CHEMICAL HISTORY OF THE GALAXY FROM THE STUDY OF GLOBULAR CLUSTER STARS?

Robert P. Kraft

Lick Observatory, Board of Studies in Astronomy and Astrophysics
University of California, Santa Cruz, California

The original title of this talk, as proposed and announced by the organizers in a widely-circulated flyer, was "Can Chemical Composition Test Models of the Universe?," the grandeur of which came as a big surprise to the speaker, whose reaction was to retreat into a shell in horror and disarray. On further thought, however, considering the weight of the occasion, the sybaritic pleasures of Venice in the springtime, and the joys of Italian reds, the speaker figured he should come anyway, but to offer a more modest topic, which the organizers graciously accepted. Thus "Universe" was reduced to "Galaxy." As for history, we pick up the thread of our story essentially at the epoch of globular cluster formation, and not much earlier.

Basically, I shall be concerned with the input parameters required for the determination of an accurate history of galactic nucleosynthesis. This will mostly be limited to what can be derived from the study of abundances in the spectra of old, metal-poor stars of the galactic halo, with special emphasis on the similarities and, surprising to some of us at least, the differences between stars of the halo, sampled in the solar vicinity, and stars of globular clusters. Although unexpected, the differences are not large; nevertheless their existence may carry clues to the understanding of star formation processes during the earliest galactic epoch.

1 TESTS OF PREDICTED BIG-BANG ABUNDANCES

Backing up a little, we can first ask whether the study of stellar abundances in globular clusters provides any constraint on currently accepted scenarios of element production during the big bang. This has been reviewed recently by Boesgaard and Steigman (1985). Aside from the well known fact that globular clusters having big-bang zero metal content do not exist (the most metal-poor clusters appear at [Fe/H] ~ -2, i.e., $Z_{Fe} \approx 0.0002$), cluster stars provide information on ^{4}He only; field halo subdwarfs provide additional information on ^{7}Li from direct spectroscopy, but again provide no samples of actual zero-metal stars (extreme Pop III) despite extensive searches (Bond 1980; Beers *et al.* 1985). This is not to say that stars with [Fe/H] < -2 do not exist, but they are indeed scarce, and various scenarios have been put forward to explain why this is so (Truran and Cameron 1971; Bond 1981; Hills 1982; Carr *et al.* 1984; Jones 1985). Most recently, Cayrel (1986) has conjectured that essentially all Pop II stars were originally Pop III stars, transformed by a mechanism I will discuss briefly later.

Returning to ^{4}He, we have to admit that direct spectroscopic evidence based on old stars is intrinsically hard to come by and unreliable in any case; abundance estimates must be based on evolutionary theory coupled to morphological features of cluster HR diagrams, e.g., the luminosity differences between the horizontal branch (HB) and the turnoff point (e.g. VandenBerg 1983), the ratio of the number of HB red to blue stars (e.g., Buzzoni *et al.* 1983), or stellar pulsation theory, i.e., the location of the blue edge of the RR Lyrae instability strip (Iben 1974). The most recent treatments of these methods yield primordial helium abundances in only fair agreement with each other, Y_p ranging from 0.19 to 0.26. The most reliable methods yield $Y_p \approx 0.23$. Fortunately, this is in good agreement with the result obtained from metal poor extragalactic H II regions (Kunth & Sargent 1983), especially the famous I Zw 18, for which $Y_p \approx 0.24$ (Davidson & Kinman 1985), although there are many caveats (as discussed by the last-named authors).

^{7}Li has been detected in the hotter subdwarfs of the solar vicinity (Spite *et al.* 1984) in which one finds Li/H $\approx 10^{-10}$ with little scatter. This is to be contrasted with Li/H $\approx 10^{-9}$, ten times higher, in corresponding Pop I dwarfs of galactic clusters and the solar neighborhood (Duncan and Jones 1983; Boesgaard and Tripicco 1986). Thus either the subdwarf abundance represents the primordial ^{7}Li and the disk enriched itself in Li rather uniformly (at least in the stars of the solar vicinity), or else Li was destroyed in a rather uniform way in Pop II subdwarfs. Arguments too extensive to go into here can be given in support of both views; I return to this point later in discussion of subdwarf nitrogen and carbon abundances. One should add that detection of ^{7}Li in globular cluster upper main sequence stars defies present technology, even with CCDs (recall V $\sim$ 17 or 18 for these stars even in the nearest clusters, and high resolution is required); the element itself would be completely destroyed in the brighter evolved cluster giants. We are fond of saying in California these days, that "this is a project for the Keck Ten-Meter Telescope."

As for D and ^{3}He, old stellar populations contribute virtually no direct information either from spectroscopy or evolutionary theory.

2 ABUNDANCES OF ELEMENTS WITH $Z \geq Z_{carbon}$ IN OLD STARS OF THE GALACTIC HALO FIELD

We deal first with the results derived from samples of halo field stars, leaving cluster stars for later. However, at the outset we need to distinguish between samples of giants and dwarfs. This is because we are concerned with the "primordial" abundances as revealed by studies of stellar atmospheres; the numbers we seek are those that remain unsullied by mixing and dredgeup of nuclearly processed material from within the stars themselves. There is plenty of evidence that even in old, low-mass stars, the products of Bethe-cycle burning, which involves the abundances of such spectroscopically accessible species as ^{12}C, ^{13}C, ^{14}N and perhaps ^{16}O, may be mixed into the atmosphere of stars as they ascend the giant branch in the HR diagram. Amongst halo field stars, the samples are near enough, therefore bright enough, that we can study both giants and dwarfs. The fundamental problem is that, in clusters, we can study only ***giants***; the main sequence stars are, in almost every case, too faint for spectroscopic analysis. The question is whether studies of giants and dwarfs in the halo field give us clues from which the *ab initio* abundances of cluster subdwarfs (unmixed?) can be reconstructed from the abundances measured in cluster giants. Consequently, I summarize first what is known about the behavior of the CNO group as a function of evolutionary state amongst halo field stars.

• 1. As metal-poor halo field subdwarfs evolve into the subgiant and giant star region, the degree of carbon depletion and nitrogen enhancement greatly exceeds expectation based on classical evolutionary theory (Iben 1967); typically C $\downarrow$ 3-5X, N$\uparrow$ 10X, from the subdwarf values, as compared with predictions of C$\downarrow$ 0.5X, N$\uparrow$2X (Sneden, Pilachowski & VandenBerg 1986).

• 2. In many of the same stars, $^{12}C/^{13}C$-isotope ratios reaching the equilibrium values ~4 are achieved (Sneden, Pilachowski & VandenBerg 1986), again unexplained by conventional evolutionary models.

• 3. However, in field subdwarfs, all studies show that nitrogen-abundance is a function of T_{eff} (or M_v), in the sense that [N/Fe] is larger by a factor of 2 in stars near the turnoff (0.8 $M_\odot$, compared with stars with masses near 0.6 $M_\odot$ ($T_{eff} \sim 4700°K$) (Carbon *et al.* 1987). Could some kind of mixing (meridional circulation?, diffusion?) be responsible, and perhaps be responsible also for the lithium "depletion" seen in Pop II compared with Pop I dwarfs (unless the effect is due to a systematic error of analysis)?

• 4. Pilachowski, Sneden & Wallerstein (1983) have summarized the situation for O and conclude that O remains unchanged with evolutionary state as expected. However, systematic errors may exist (these will be discussed later in the section on clusters), and the situation is somewhat unclear.

The main conclusion we reach from the study of halo field stars is that more mixing and dredgeup occur than is predicted by classical theories of evolution, and that the ratios $^{12}C/^{13}C$ and C/N are considerably modified as a star evolves into the giant star domain. Evidence in favor of oxygen depletion in giants and diffusion (or circulation) processes in subdwarfs are much less convincing.

With appropriate allowances having been made for evolutionary effects in C and N, we turn to the other elements that have $Z \geq Z_{carbon}$, amongst which we do not expect to find evolutionary effects in low-mass stars. A recent survey by Lambert (1987) (Fig. 1) provides an excellent basis for a short summary. Illustrated in Figure 1 is the mean behavior of [el/Fe] as a function of [Fe/H]. Into Lambert's schematic diagram, I have inserted critical galactic epochs corresponding to certain [Fe/H] values; thus the lower metallicity limit for globular clusters is shown, as is the probable lower limit for the metallicity of the so-called thick disk [as defined by Zinn (1985)], and the mean metallicity associated with the old disk population. (From a metallicity point of view, the latter may be the same thing.) The dashed-line sections, in Lambert's opinion, are poorly known extrapolations; I agree.

Inspection of Figure 1 leads one to make the following points:

• 1. The solar ratios of [C/Fe], [N/Fe], r-process and s-process species seem to have been established in the halo by the time we reach the epoch of formation of the globular clusters;

• 2. Ab-initio overabundances of oxygen are well established, as expected based on the ejecta of Type II supernovae (Arnett 1978; Tinsley 1979).

• 3. s-process elements behave as if they were secondary (Luck & Bond 1985) more-or-less into the cluster-formation epoch.

• 4. α-process elements crudely track oxygen.

Figure 1 - The distribution of [eL/Fe]-ratios as a function of [Fe/H] (metallicity), after Lambert (1987), with minor modifications. Dotted sections refer to slight modifications preferred by the present author. Dashed sections are judged relatively uncertain by Lambert.

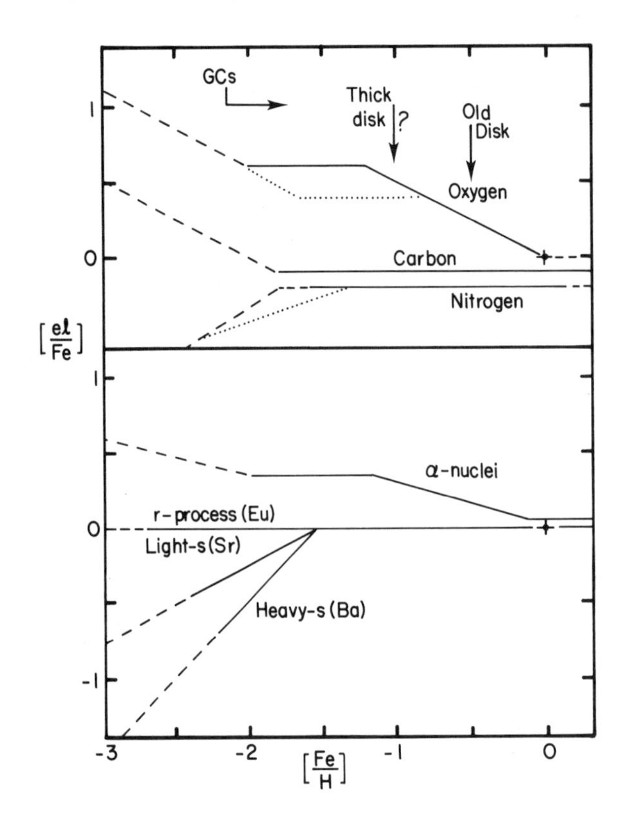

To go deeply into all the nucleosynthesis scenarios developed over the past decade to explain the leading features of Figure 1 would take me too far afield from the main point of the lecture, *viz.*, the contrast between clusters and the field. But a few surprises should be mentioned.

• 1. Nitrogen, classically thought to be a secondary element, i.e., one which is generated in the CNO cycle using ^{12}C as a seed nucleus and which should therefore satisfy [N/Fe] $\propto$ [Fe/H] (Tinsley 1979), is apparently injected in considerably quantity into the galaxy at an early epoch. Studies of extragalactic H II regions yield a similar result (cf. e.g., Pagel 1985). Where does all this nitrogen come from?

• 2. Turning to the α's, we find that Si, Mg, and Al are explosively ejected from massive stars (i.e., M $\sim$ 25 $M_\odot$) (Woosley & Weaver 1982) in quantities comparable to O, but not Ca. Yet [Ca/Fe] $\sim +0.4$ and [O/Fe] $\sim +0.3$ (or +0.6 – more on this later) for stars with [Fe/H] < -1. However, production of Ca such that

[Ca/O] ≈ 0.0 occurs in pair-instability "hypernovae," i.e., objects with masses between 100 and 300 $M_\odot$ (Woosley, Axelrod & Weaver 1983). Such objects do not produce Fe, but do produce substantial amounts of ^{12}C, and trace amounts of ^{15}N.

• 3. With transition into the disk near [Fe/H] ∼ − 1, contributors to the production of O and α-nuclei become less important relative to the producers of Fe, C, and N.

A related, but equally important point, does not show up directly in Figure 1; *that is the "scatter" in [el/Fe] at a given [Fe/H].* For α's, this is summarized in Lambert's review and plotted in the form [el/Fe] vs [Fe/H], from which one may estimate σ[el/Fe]. Similar plots for O and s-process species are given by Luck & and Bond (1985); a compilation of results from several investigators of C and N in subdwarfs given in the form [C/Fe] and [N/Fe] versus [Fe/H] is found in Carbon *et al.* (1987). All these plots lead essentially to the same result, *viz.,*:

• 4. In the halo field stars, [el/Fe] at a given [Fe/H] has no more scatter than can be accounted for by the errors of observation, usually σ ∼ 0.1 to 0.2 dex. This point has been made repeatedly by a number of investigators (Luck & Bond 1985; Nissen *et al.* 1985; Carbon *et al.* 1987).

The last is especially interesting, for if the halo abundance distribution derived from the study of local field stars is representative of a variety of nucleosynthetic sites – as almost certainly must have been the case – then nature must have conspired to "pollute" *the primordial big-bang medium in a highly uniform way over a rather large galactic volume, as early as the epoch of globular cluster formation.* Moreover, any given site (center of activity) must have mixed the elements having $Z \geq Z_{carbon}$ rather well. Thus, if the only governing factors were the relative numbers of massive stars, the slope of the luminosity function for these must have been rather similar everywhere in the halo.

There are, of course, observational caveats. Most of these results, e.g., for O, Ca, Mg, Si, are derived from high-resolution spectroscopic analysis involving only a handful of lines (for O in giants, only a single line(!), λ6300 of [O I]). Samples of stars are necessarily very small (e.g., for O, only 12 field giants and ∼20 dwarfs and subdwarfs have been studied), owing to the limitation of observable flux. Plots of [C/Fe] and [N/Fe] vs [Fe/H] seem to have a large spread (∼0.5 dex); however, for C, one has a T_{eff}-dependence driven by CO formation and for N there is a T_{eff}-dependence that may result from model deficiencies or may, in fact, be driven by unexplained mixing (Carbon *et al.* 1987). Overall, investigators agree that the real observational spread is not worse than σ ∼ 0.2 dex for most elements, a little smaller for some. There are a few anomalies worth mentioning, but they are best discussed in relation to the cluster stars.

3 ABUNDANCES ($Z > Z_{carbon}$) IN GLOBULAR CLUSTER STARS

Turning to abundance studies of globular cluster stars, we examine first the so-called "metallicity distribution" of the halo field stars compared with the clusters. Here we are concerned with the "metallicity" defined by the Fe-peak, since the results are based largely on broad-based or intermediate-based photometric studies in which UV-blocking of the emergent flux by lines principally of Fe-peak elements provides the basis for rank ordering the "metal" content. Calibration of the system obviously depends on high-resolution abundance studies of the brightest cluster giants and numerous halo field stars, including subdwarfs, giants, and RR Lyraes. It is possible also to derive accurate results from appropriately calibrated integrated cluster photometry; the most recent and most nearly complete study of this kind is by Zinn & West (1984).

The Fe-peak abundance distribution of all accessible clusters, and complete samples of halo field subdwarfs and giants have been discussed by Hartwick (1983); a comparison of halo RR Lyraes with halo clusters (and with a small part of the SMC halo) is shown in Figure 2, following Suntzeff *et al.* (1986). There are no surprises here: all distributions agree in mean and dispersion to a high degree of confidence, based on Komolgorov-Smirnov tests. They are perfectly compatible with expectations based on simple one-zone models in which their mass-loss rate is proportional to the star-formation rate (Searle 1979; Hartwick 1983); the captured blob scenario of Searle & Zinn (1978) presumably does just as well.

It is only when we examine the individual cluster stars that we find unexpected complexities. However, as mentioned in Section 2, our knowledge of stellar abundances in clusters is confined *almost* entirely to *evolved* stars – i.e., giants – since these are the only objects we can reach in spectral surveys. Consequently, we deal with stars in which the surface abundances in the CNO group may be altered from their "primordial" subdwarf values by convectively driven mixing and dredgeup of material processed or partially processed through the CNO cycle. However, in only one cluster has this been followed in detail, and then only in the case of carbon, nitrogen and oxygen being inaccessible for reasons of flux limitation. Thus, in M92 (Langer *et al.* 1986), depletion of carbon begins on or near the main sequence and continues to the red-giant tip (Fig. 3). How close to the main sequence depletion of carbon takes place in other clusters is not known, but presumably M92 is not unique. A similar effect may be seen in NGC 6752 (Bell, Hesser & Cannon 1984, Fig. 7).

We *assume*, of course, that the carbon depletion is accompanied by nitrogen enhancement; and indeed, in many clusters that have been studied, M92 among them, large overabundances of N are seen in the giants. In fact, it appears to be the case that cluster giants *on the average* have more carbon depletion and nitrogen enhancement than their field giant star counterparts (Langer & Kraft 1984), especially so in the cases of M92, M15 and M13. The behavior of M3 is more like that of field stars.

Thus we see that the study of cluster stars also drives a reexamination of the theory of convective mixing and dredgeup during the first ascent but points also to certain evolutionary differences between cluster stars and field stars. However, further examination reveals additional differences, *viz.*, the necessity for *primordial abundance fluctuations in [el/Fe] at a fixed [Fe/H] between cluster stars.* These are found with considerable certainty in the CNO group and with less certainty in Al and Na (but not Ca!). *In addition, in at least one cluster, there is an abundance spread in the Fe-peak.* [Excellent reviews of these effects are given by Freeman & Norris (1981) and by Iben & Renzini (1984).]

Figure 2 - Fe-peak abundance distributions for galactic globular clusters (with [Fe/H] < −0.8), halo RR Lyraes, and a portion of the SMC halo near NGC 121 (after Suntzeff *et al.* 1986). [Reprinted with permission of author.]

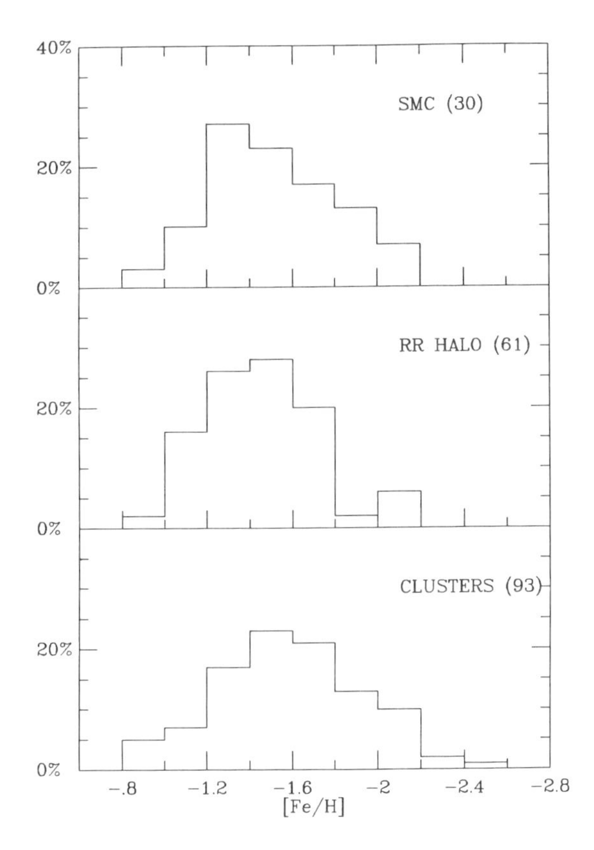

A rather brief summary of the observational "facts," which hits only the high points, is as follows:

• 1. In the most massive cluster, ω Cen, an Fe-peak abundance range is well-established (Butler *et al.* 1978; Rodgers *et al.* 1979; Cohen 1981; and many others). The Fe-peak spread is accompanied by a star-by-star correlated spread in Ca, C, and N, so that [el/Fe] for these species appears similar to that found in halo field stars (Hesser *et al.* 1985).

• 2. A less well-established range in Fe-abundance is reported for M22, but is somewhat controversial. In the words of Wallerstein, Leep & Oke (1987), the range in Fe... "is neither sufficiently large nor of adequate accuracy to confirm the reality of the difference nor is it sufficiently small to deny a difference." On the other hand, from a study of over 100 red giants, Norris & Freeman (1983) found variations and direct correlations between the strengths of Al, Ca, CN and CH features, thus suggesting that M22 shows some of the anomalous abundance patterns of ω Cen, albeit on a smaller scale.

• 3. Other clusters in which the giants have been examined spectroscopically with fairly extensive samples include M15, M13, M3, NGC 2419, NGC 4147, M5; no

Fe-peak variations have been established with certainty. Studies of "spreads" in the subgiant (e.g., Sandage & Katem 1983) and main sequence (e.g., Richer & Fahlman 1986; Bolte 1987), branches of C-M arrays also reveal no evidence for Fe-peak abundance spreads. The matter is of special interest in the case of NGC 2419, which is one of the half-dozen globular clusters with $M_v < -9.5$. (Suntzeff *et al.* 1987).

Figure 3 - Carbon depletion as a function of evolutionary state in M92 (after Langer *et al.* 1986). The mean overall decline from the main sequence value of [C/Fe] is more than an order of magnitude and the scatter is real. [Reprinted with permission of P.A.S.P.]

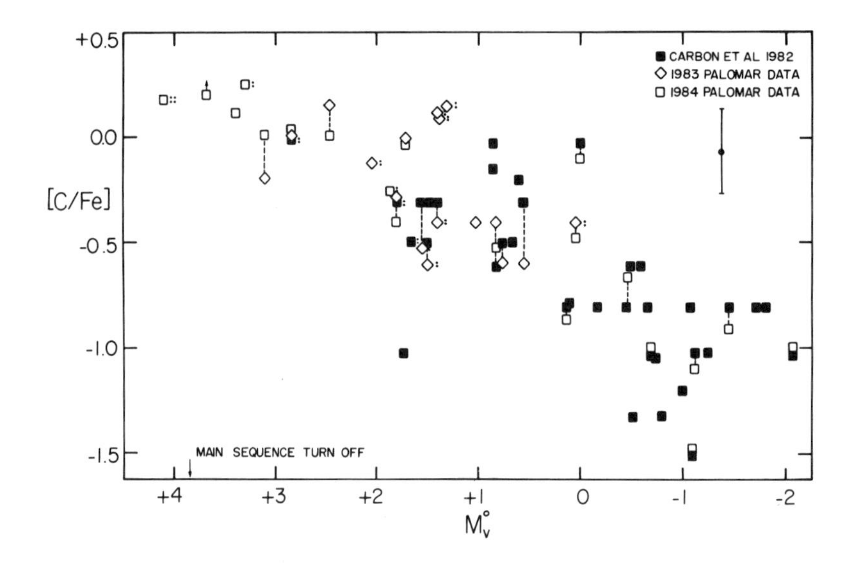

• 4. Several clusters, most notably M15 and M92, have extremely wide variations of C and N abundances and such variations are not found in the halo field giants (Kraft *et al* 1982; Langer and Kraft 1984) (Figs. 4, 5, 6). Such wide variations are not seen in M3.

• 5. In M92 and M15, the total (C+N) content is not constant from one star to another. This means there are stars that are both carbon-poor and nitrogen-poor, and stars that are both carbon-rich and nitrogen-rich. This is in addition to stars manifesting C$\rightarrow$N processing, in which C$\downarrow$ and N$\uparrow$.

• 6. Wide variations are seen in the strengths of the CN band in giants of some intermediately metal-poor clusters such as M5 (Smith & Norris 1983). Such wide variations are not seen amongst field giants (Hesser *et al.* 1977); this is probably a manifestation of the same phenomenon that gives rise to the NH and CH band variations seen in more metal-poor stars, in which the CN-bands are weak or non-existent. Wide variations in the strength of CN bands are also reported to exist in *main sequence* stars of the metal-rich cluster 47 Tuc (Bell, Hesser & Cannon 1983).

• 7. Positive correlations exist between CN-band strengths and Al I and Na I resonance line strengths in several clusters; the best evidence for this has been put forward by Norris & Pilachowski (1985), although the effect was first pointed out by Cottrell & Da Costa (1981). There are, however, no correlations between nitrogen abundances and Ca II (H and K) line strengths; Ca II H and K shows little scatter and is a good surrogate for Fe in all cluster abundance studies (Suntzeff 1980) if appropriately calibrated.

• 8. Compared with field giants, Ca, Al, Na and Si all show overabundances relative to Fe, rather similar to what one sees in the field (cf. e.g., Wallerstein, Leep & Oke 1987 and reference therein), but probably with more scatter.

Figure 4. [N/Fe] vs [C/Fe] for M92 giants. Lines of constant (C+N) are indicated. [Reprinted with permission of P.A.S.P.]

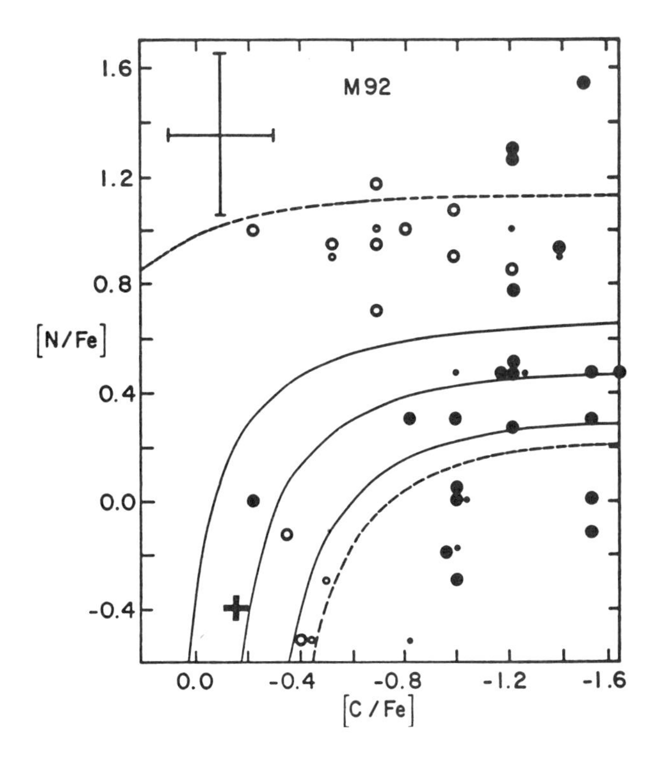

I turn at the end of this survey to oxygen, especially since there are further puzzles and also since recent investigations point rather unequivocally to primordial fluctuations in this element. What is known about oxygen is summarized in Figure 7, which is modified from the review by Pilachowski, Sneden & Wallerstein (1983). As a caveat, I note that essentially only one line can be studied in cluster giants, *viz.,* [O I] λ6300, and it typically has an equivalent width of only 15-30 mA (λ6363 is usually too weak, but see below). So we are talking high resolution, medium to low accuracy, and small samples – 1 to 4 giant stars per cluster! With this in mind, consider what we see in Figure 7:

• 1. *Clusters divided into two groups*, O-rich ([O/Fe] ~ +0.3) and O-poor ([O/Fe] ~ −0.4). This may be correlated with HB morphology.

• 2. Field halo giants have abundances [O/Fe] ~ +0.3, similar to the O-rich cluster group.

• 3. *Subdwarf [O/Fe]-values seem to be a factor of 2 larger than in giants.* Most subdwarfs are analyzed from the λ7770 O I permitted triplet which arises from 9 eV above the ground state. Pilachowski, Sneden & Wallerstein (1983) regard the discrepancy as a non-LTE effect, although this is disputed by Nissen, Edvardsson & Gustafsson (1985). If [O/Fe] = +0.3 is right, their cluster ages are actually little affected (~ 1 Gyr) by oxygen overabundance (VandenBerg 1985) and the rather older ages derived earlier are to be preferred (Sandage 1983).

Figure 5. [N/Fe] vs [C/Fe] for M3 giants. Lines of constant (C+N) are indicated. [Reprinted with permission of P.A.S.P.]

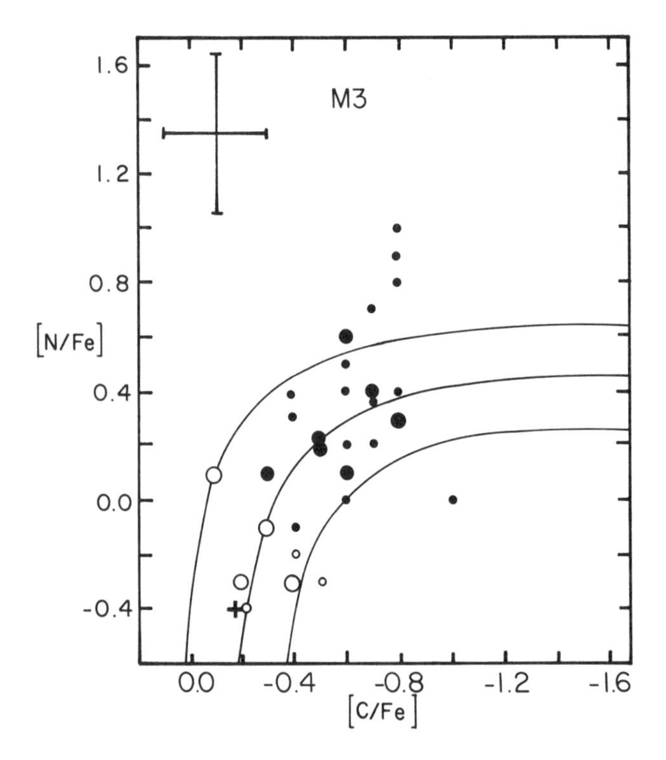

Not seen in Figure 7 is that oxygen is *variable from one star to another within a given cluster* and flops between the two poles: O-rich and O-poor! The best example seen so far is in M13, in which CCD work at [O I] λ6300 reveals three giants at the red giant tip with [O/Fe] ~ +0.3 and one star (II-67) with [O/Fe] ~ − 0.4 (Leep, Wallerstein & Oke 1986). Langer *et al.* (1987) showed that this was *not* a result of O → N processing and dredgeup, since the NH bands are unaffected.

This result is so startling, that we placed M13, II-67 on the program of objects to be looked at as a part of a program of observations to test the performance of the new Lick Hamilton high-resolution spectrograph. Results obtained by Hatzes (1987) are shown in Figures 8 and 9 – the resolution is ~50,000 and the spectra were obtained with the 3.0 m telescope in about 1/2 hour at V = 12.2. Figure 8 shows that λ6363 of [O I] was also detected in the O-rich star M13, III-56, but only very feebly in II-67.

Figure 6. [N/Fe] vs [C/Fe] for halo field giants with metallicities [Fe/H] similar to M92. Notice the field star scatter is much smaller than is the case for M92. Lines of constant (C+N) are indicated. [Reprinted with permission of P.A.S.P.]

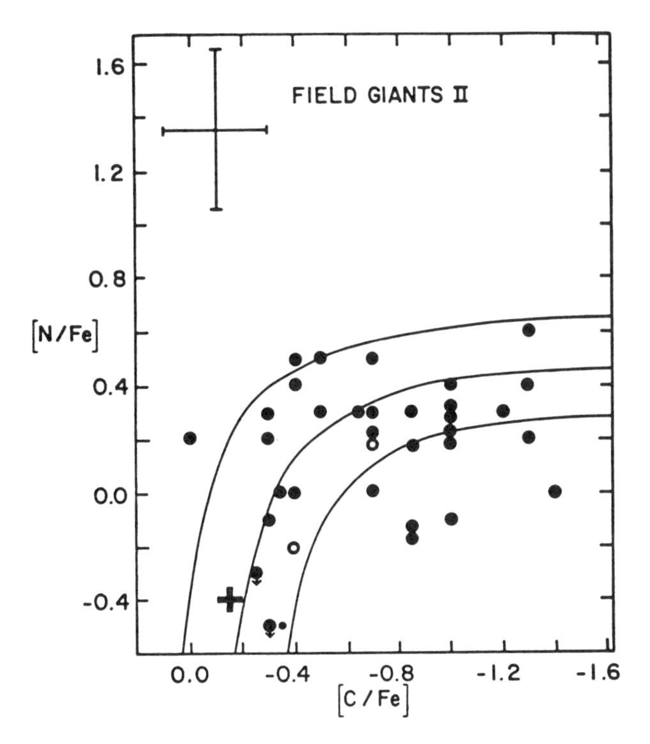

If O is underabundant by a factor of 5 in II-67, it is hard to escape the conclusion that C must also be underabundant: the star is so cool that if the C to O-ratio were overturned, C_2 bands would appear, and these are not seen. Even if O remains dominant, CO-formation would be reduced in II-67 thus freeing up carbon to produce enhanced CH-bands. This is not seen. Thus C must also be low, by a factor of at least 3. It is hard to escape the conclusion, as Wallerstein, Leep, and Oke (1987) argue, that the whole CNO group is underabundant, relative to "normal" giants in the cluster. *These results make primordial fluctuations in the CNO group amongst M13-stars practically impossible to avoid.*

4 SOME SPECULATIONS ON HOW THE ABUNDANCE ANOMALIES CAME ABOUT

To summarize the abundance situation in a somewhat oversimplified way, we need to account for certain *global* similarities between cluster and field stars while at the same time we recognize certain *local* or intracluster variations not found in the field. Thus we find the *Fe-peak abundance distribution in clusters vis-a-vis field stars is the same.* However, *within individual clusters, we need primordial abundance fluctuations in the CNO group and possibly in Na and Al.* Clusters and field stars also share in common a need for mixing processes and dredgeup that do not follow conventional evolutionary theory, but *there is some evidence that these processes go further in clusters than in the field.* Finally, in only one certain case do we find evidence of Fe-peak fluctuations within a given cluster, and *that is the most massive cluster, ω Cen.* Evidence in favor of Fe-peak fluctuations in M22 turns out to be less convincing.

Figure 7. The behavior of [O/Fe] as a function of [Fe/H] for field subdwarfs, field giants, and cluster giants. The samples are very small [after Pilachowski, Sneden & Wallerstein (1983), with minor modifications].

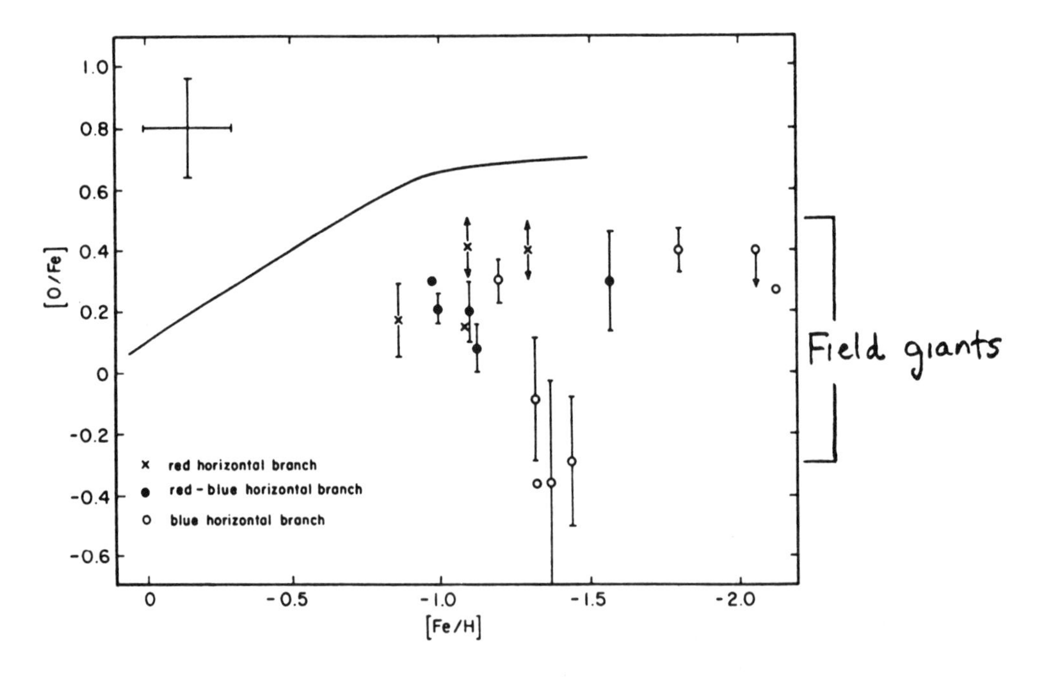

At the moment, it does not appear possible to formulate a fully satisfactory general picture that incorporates all the observational facts about cluster and field stars together. For a time (e.g, Freeman & Norris 1981; Iben & Renzini 1984; Kraft 1985), it seemed possible to explain what we see in most clusters as the result of a pollution process in which supernovae were not involved at all. That is, noting that all clusters except ω Cen (and possibly M22) have intrastellar uniformity in Fe-peak abundance, one found that the primordial abundance fluctuations were confined essentially to elements that could plausibly be produced during the evolution of intermediate mass stars. It is well known, for example, that fresh C is produced from triple-α He shell-flashes during the asymptotic giant branch evolution of intermediate mass stars (Iben 1975). It is also likely that, in interpulse phases, some of this fresh C is converted to N in hot-bottom C$\rightarrow$N burning (Renzini & Voli 1981; Iben & Renzini 1982). Cottrell & Da Costa (1981) suggested that in stellar layers having excess N, ^{22}Ne would be "overproduced" relative to the Fe-peak by the addition of two αs. If so ^{22}Ne will itself soak up the available neutrons released by the reaction ^{22}Ne (α, n) ^{25}Mg, thus in turn "overproducing" ^{27}Al and ^{23}Na.

These intermediate mass stars presumably eject this new "primary" material into the early cluster gas in the form of relatively slow moving winds; thus the ejected gas containing freshly made C, N, Na, and Al could be confined within the early cluster environment and used to pollute the still-contracting low-mass stars we presently see (D'Antona, Gratton & Chieffi 1983). On the other hand, supernova ejecta containing heavy element (Fe-peak) material would blow out the cluster gas, leaving the low-mass stars unaffected, free from Fe-peak abundance fluctuations.

Figure 8. The spectral region about [O I], λ6300A for M13, III-56 and M13, II-67, after removal of all atmospheric features [after Hatzes (1987)]. [Reprinted with permission of P.A.S.P.]

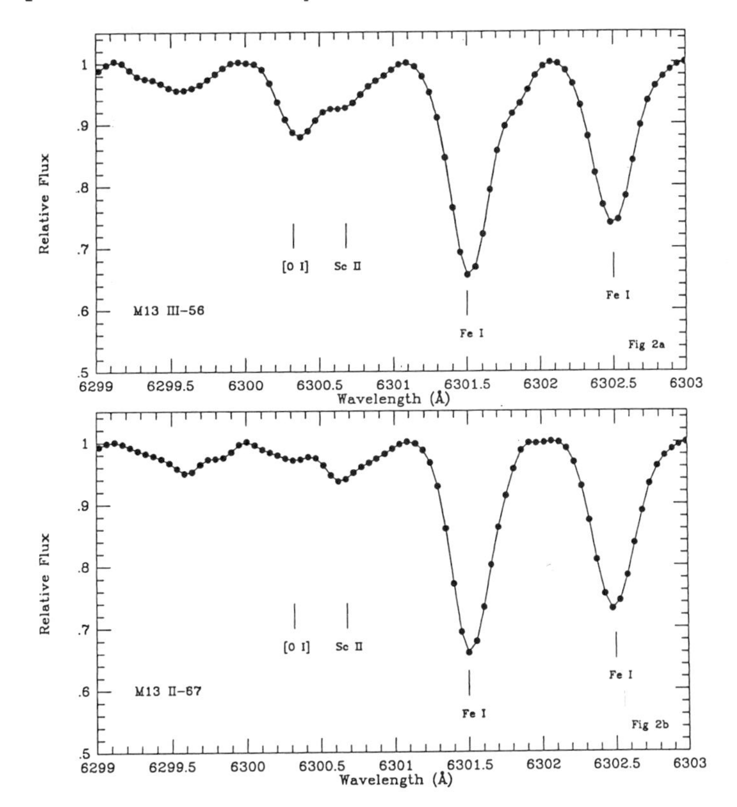

As a corollary to this picture, one can also plausibly understand why ω Cen has Fe-peak abundance fluctuations. In this case, one postulates that ω Cen is the stellar remnant of a particularly massive primordial cloud, in which significant amounts of supernova ejecta were confined by the ambient cloud gas. Dopita & Smith (1986) have shown how supernova induced shocks can be slowed down and even induce secondary star formation if the primordial mass of the cloud is sufficiently large. It seems plausible that the most massive known cluster should descend from a particularly massive primordial cloud, and these authors estimate that the primordial mass of ω Cen must have been $\sim 10^8$ $M_\odot$. Presumably the presently existing clusters descended from primordial clouds of smaller mass which were unable to contain the supernova-ejected material.

Figure 9. The spectral region around [O I], λ6363A for M13, III-56 and M13, II-67, after removal of all atmospheric features [after Hatzes (1987)]. [Reprinted with permission of P.A.S.P.]

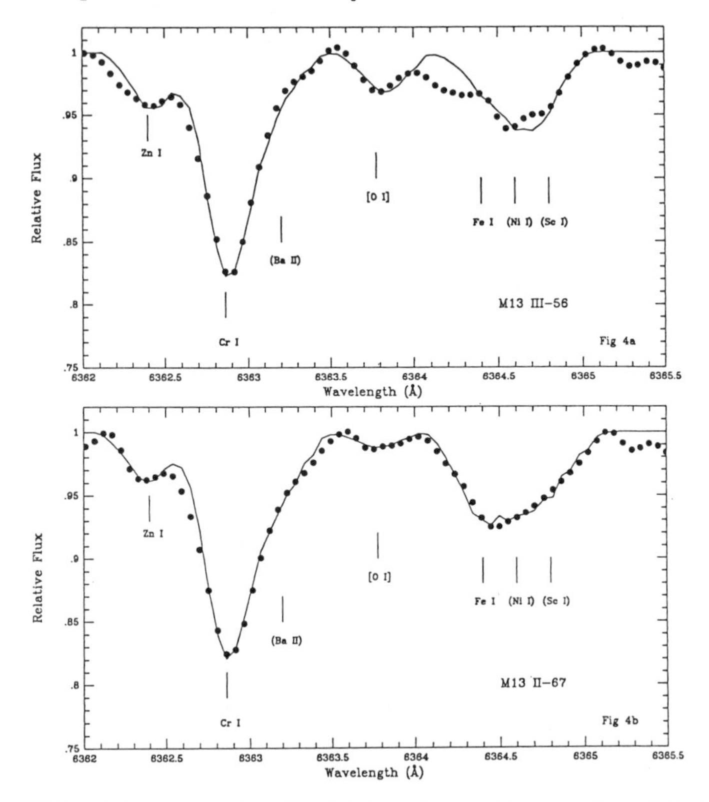

This picture may be all right as far as it goes, but it is flawed not only by its incompleteness, but also by a fundamental abundance anomaly. Essentially, it puts into the background consideration of how the cluster Fe-abundances themselves were produced: it asks us to believe these were "given" by nucleosynthetic events that took place in some pre-existing medium, presumably the same one that gave rise to the halo field stars. Now at a given cluster [Fe/H], a pollution process can produce fluctuations, but these must always basically be "add-ons," i.e., production by a pollution process must be such that $[el/Fe]_{cluster} > [el/Fe]_{field}$. But as we have seen the CNO abundances associated with M13 II-67 are much *smaller* than the typical field CNO abundances for stars having the [Fe/H]-abundance of M13. How can a pollution process produce *underabundances*?

More serious is the problem of the relationship of field stars to cluster stars in this picture. Cayrel (1986) has argued persuasively, on the basis of what we observe now in the Galaxy and in the Magellanic Clouds, that since all stars are born presently in clusters, there is no reason to believe that this was not the case in the halo formation era as well. Fall & Rees (1985) have also shown that small clouds ($M \leq 10^4\ M_\odot$) could not collapse, owing to galactic x-ray heating, until the ambient metallicity had risen to around 0.1 the solar value, which corresponds to the epoch of formation of the thick disk (cf. Fig. 1). Thus it is reasonable to suppose that gas clouds with masses in the range 10^5 to $10^8\ M_\odot$ fragmented from the early halo material. According to Fall & Rees (1977), the more massive of these clouds, after fragmenting into stars, were more easily disrupted by tidal encounters with each other or with the galactic disk than the star clusters formed from the less massive clouds. For reasonable Galactic parameters, the timescale for these interactions is in the range 1×10^8 to 5×10^9 yr for $M \sim 10^8\ M_\odot$; these massive clusters therefore provide a plausible source for the low-mass field stars we presently see. These stars would also be expected to be well-mixed in chemical composition and show the full "one-zone" model range of metallicities if we use the same arguments as earlier advanced in the case of ω Cen.

Now we can take two points of view about the metallicity content in these field stars; either the enrichment which took place in the cluster era was superimposed on a pre-existing medium more metal-rich than the big-bang material or else all the metals from $Z = Z_{carbon}$ upward were produced in the clusters themselves. If the former were true, then we put back the epoch of heavy element formation to some pre-cluster era and to processes that can at best be only dimly perceived. If the latter were true, then the enriched gaseous material left over, i.e., the stuff that did not go into field star formation, must be the "primordial" stuff as far as heavy elements are concerned, and it must have been the stuff that subsequently became the basic material from which the present globular clusters were built up. The trouble with this picture is that one does not see how such material, which must have contained a wide-range of metallicities, could have "segregated" itself so that the [Fe/H]-ratios in the present globular clusters could be so sharply defined from one cluster to another, there being essentially no spread within a given cluster!

The contrary picture is the one advanced by Cayrel (1986), in which it is argued that all the heavy elements were formed in the present globular clusters by supernova contamination of primordial gas clouds which had big-bang abundances. The trouble here is that if a given cloud were massive enough to contain the supernova ejecta, there would be produced a *range* of metal abundances, as we have already seen. Conversely, if the cloud mass is small, then the cluster gas and supernova gas would be expelled from the cloud and no enrichment would take place. Cayrel responds to this criticism by suggesting that most of the low-mass stars were in fact, formed in a single event in which the stars presently bound in the cluster were formed only where the supernova-driven outward, and still gravitationally-collapsing inward, colliding flows of gas more-or-less balanced. This is plausible enough within the total framework of Cayrel's very nice picture, but does seem to be a bit of a "balancing act."

Obviously, we need a comprehensive picture of the formation of the halo field stars as well as the globular clusters, especially when we realize that the present mass of visible stars in the halo field (Sandage 1987) exceeds that in halo globular clusters by a factor of at least 20. The clusters capture our attention because of their singularity, luminosity, and potential as "laboratories" for the study of stellar evolution, but they are a negligibly small part of the total stellar halo.

It is a great pleasure for me personally to take part in this celebration of Chip Arp's 60th birthday. Chip and I have been friends and colleagues almost from the beginning of our astronomical careers. I am especially pleased to have an opportunity to review work in an area of stellar astronomy that Chip essentially pioneered.

REFERENCES

Arnett, W. (1978). Ap. J., *219*, 1008.
Beers, T. C., Preston, G. W. & Shectman, S. (1985). A. J., *90*, 2089.
Bell, R. A., Hesser, J. E. & Cannon, R. D. (1983). Ap. J., *269*, 580.
Bell, R. A., Hesser, J. E. & Cannon, R. D. (1984). Ap. J., *283*, 615.
Boesgaard, A. & Steigman, G. (1985). Ann. Rev. Astron. Ap., *23*, 319.
Boesgaard, A. & Trippico, M. J. (1986). Ap. J. Letters, *302*, L49.
Bolte, M. (1987), in press.
Bond, H. E. (1980). Ap. J. Suppl., *44*, 517.
Bond, H. E. (1981). Ap. J., *248*, 606.
Butler, D., Dickens, R. J. & Epps, E. (1978). Ap. J., *225*, 148.
Buzzoni, A., FusiPecci, F., Buonanno, R., Corsi, C. (1983), *In* ESO Workshop & Primordial Helium, eds. P. Shaver, D. Kunth, and K. Kjär, (Garching).
Carbon, D., Barbuy, B., Kraft, R. P., Friel, E. & Suntzeff, N. (1987). Publ. A. S. P., *99*, 335.
Carr, B. J., Bond, J. R. & Arnett, W. D. (1984). Ap. J., *277*, 445.
Cayrel, R. (1986). Astron. Ap., *168*, 81.
Cohen, J. G. (1981). Ap. J., *247*, 869.
Cottrell, P. & Da Costa, G. (1981). Ap. J. Letters, *245*, L79.
D'Antona, F., Gratton, R. & Chieffi, A. (1983). Mem. Soc. Astron. Ital, *54*, 173.
Davidson, K. & Kinman, T. D. (1985). Ap. J. Suppl., *58*, 321.
Dopita, M. & Smith, G. H. (1986). Ap. J., *304*, 283.
Duncan, D. & Jones, B. F. (1983). Ap. J., *271*, 663.
Fall, S. M. & Rees, M. (1977). M.N.R.A.S., *181*, 37P.
Fall, S. M. & Rees, M. (1985). Ap. J., *298*, 18.
Freeman, K. & Norris, J. (1981). Ann. Rev. Astron. Ap., *19*, 319.
Hartwick, F. D. A. (1983). Mem. Soc. Astron. Ital., *54*, 51.
Hatzes, A. (1987). Publ. A. S. P., *99*, 369.
Hesser, J. E., Hartwick, F. D. A. & McClure, R. (1977). Ap. J. Suppl., *33*, 471.
Hesser, J. E., Bell, R. A., Cannon, R. D. & Harris, G. (1985). Ap. J., *295*, 437.
Hills, J. (1982). Ap. J. Letters, *258*, L67.
Iben, I. (1967). Ann. Rev. Astron. Ap., *5*, 571.
Iben, I. (1974). Ann. Rev. Astron. Ap., *12*, 215.
Iben, I. (1975). Ap. J., *196*, 525.
Iben, I. & Renzini, A. (1982). Ap. J. Letters, *259*, L79.

Iben, I. & Renzini, A. (1984). Phys. Reports, *105*, 329.
Jones, J. E. (1985). Publ. A.S.P., *97*, 593.
Kraft, R. P., Suntzeff, N. B., Langer, G. E., Carbon, D. F., Trefzger, C., Friel, E. & Stone, R. (1982). Publ. A.S.P., *94*, 55.
Kraft, R. P. (1985), *In* ESO Workshop on the Production and Distribution of CNO Elements, eds. F. Danziger, F. Matteucci, and K. Kjär, p. 21 (Garching).
Kunth, D. & Sargent, W. L. W. (1983). Ap. J., *273*, 81.
Lambert, D. L. 1987. J. Ap. Astron., *8*, 103.
Langer, G. E., Kraft, R. P., Carbon, D. E., Friel, E. & Oke, J. B. 1986. Publ. A.S.P., *98*, 473.
Langer, G. E. & Kraft, R. P. (1984). Publ. A.S.P. *96*, 339.
Langer, G. E., Friel, E., Kraft, R. P. & Suntzeff, N. B. 1987. Publ. A.S.P. *99* 15.
Leep, E. M., Wallerstein, G. & Oke, J. B. (1986). A. J., *91*, 1117.
Luck, R. E. & Bond, H. E. (1985). Ap. J., *292*, 559.
Nissen, P. E., Edvardsson, B., & Gustafsson, B. (1985). *In* ESO Workshop on the Production and Distribution of CNO Elements, eds. I. Danziger, F. Matteucci, and K. Kjär, p. 131 (Garching).
Norris, J. & Freeman, K. (1983). Ap. J., *266*, 130.
Norris, J. & Pilachowski, C. (1985). Ap. J., *299*, 295.
Pagel, B. J. (1985). *In* ESO Workshop on the Production and Distribution of CNO Elements, eds. I. Danziger, F. Matteucci, & K. Kjař, p. 155 (Garching).
Pilachowski, C., Sneden, C. & Wallerstein, G. (1983). Ap. J. Suppl. *52*, 241.
Renzini, A. & Voli, M. (1981). Astron. Ap., *94*, 175.
Richer, H. & Fahlman, G. G. (1986). Ap. J. *304*, 273.
Rodgers, A. W., Freeman, K., Handrey, P. & Smith, G. H. (1979). Ap. J., *232*, 169.
Sandage, A. (1983). A. J., *88*, 1159.
Sandage, A. (1987). *In* Cambridge NATO Workshop on the Galaxy, (in press).
Sandage, A. & Katem, B. (1983). A. J., *88*, 1146.
Searle, L. (1979). *In* Les Elements et Leur Isotopes dans l'Univers, Liege Colloq. No. 22, p. 437, (Institut d'Astrophysique, Liege).
Searle, L. & Zinn, R. (1978). Ap. J., *225*, 357.
Smith, G. H. & Norris, J. (1983). Ap. J., *264*, 215.
Sneden, C., Pilachowski, C. & VandenBerg, D. (1986). Ap. J., *311*, 826.
Spite, M., Maillard, J. P. & Spite, F. (1984). Astron. Ap. *141*, 56.
Suntzeff, N. B. (1980). A. J., *85*, 408.
Suntzeff, N. B., Friel, E., Klemola, A., Kraft, R. P. & Graham, J. A. (1986). A. J., *91*, 275.
Suntzeff, N. B., Kinman, T. D. & Kraft, R. P. (1987). A. J. (submitted).
Tinsley, B. (1979). Ap. J., *229*, 1046.
Truran, J. W. & Cameron, A. G. W. (1971). Ap. Space Sci., *14*, 179.
VandenBerg, D. (1983). Ap. J. Suppl. *51*, 29.
VandenBerg, D. (1985). *In* ESO Conference on the Production of Distribution of CNO Elements, eds. I. Danziger, F. Matteucci & K. Kjär, p. 73 (Garching).
Wallerstein, G., Leep, E. M. & Oke, J. B. (1987). A. J., *93*, 1137.
Woosley, S. E. & Weaver, T. A. (1982). *In* Essays on Nuclear Astrophysics, eds. C. Barnes, D. Clayton & D. Schramm, p. 377 (Cambridge Univ. Press).
Woosley, S. E., Axelrod, T. S. & Weaver, T. A. (1983). *In* Proceedings of Erice Conference on Nucleosynthesis.
Zinn, R. (1985). Ap. J., *293*, 424.
Zinn, R. & West, M. J. (1984). Ap. J. Suppl., *55*, 45.

DISCUSSION

J.-C. Pecker: Two arguments favor a longer history for clusters (or stars in clusters) than previously thought of: (a) The mixing in central parts of stars is noticeable, according to Schatzman & Maeder, and increases the age of the clusters as determined from the "turnover" of their H-R diagram; the "physical" explanation of the low solar neutrino flux is less likely now that the SN 1987 in LMC has shown the neutrino mass not to be higher than the photon mass: hence the Schatzman-Maeder theory appears as necessary in this respect. (b) The case of ωCen and M22 (Smith 1987) implies that "primordial matter" is already quite inhomogeneous. Would not that lead us to accept a much longer history of the matter from which clusters are formed, than previously considered?

R. Kraft: (1) I agree. We have evidence from the dependence of [N/Fe] vs T_{eff} along the subdwarf m.s. that some kind of mixing of material on a "low-exposure" basis occurs even on the main sequence itself. This may also be related to the ^{7}Li problem. (2) You may well be right. The real problem is how the clusters succeed in "segregating" themselves in [Fe/H]-abundance domain from the (presumably) well-mixed background.

H. Arp: In a generally metal-poor galaxy like the SMC can you naturally explain the *current* metal abundance by stellar evolution in a low density environment?

R. Kraft: All halos of which I'm aware (SMC, NGC 147, NGC 185, Galaxy, etc.) have more-or-less log normal metallicity distributions with the same σ– sometimes the *mean* metallicity is different but σ is always the same. This is consistent with simple one-zone enrichment models in which:

$$\alpha = rate\ y\ mass\ loss/rate\ y\ star\ formation$$

takes on different values (following Searle).

M. Burbidge: Isn't it true that theoretical calculations on mixing and dredge-up of products of nuclear reactions in central regions have never satisfactorily accounted for the amount of mixing that seems to be needed by the observations?

R. Kraft: It seems that real stars do more mixing than has normally been predicted by models involving thermally-drive convection.

CONSIDERATIONS ON THE SN 1987-A AND HIS CLASSIFICATION AS A TYPE II SUPERNOVA

L.ROSINO

Department of Astronomy of the University of Padova

ABSTRACT : The spectral and photometric characteristics of the SN 1987-A in the LMC and the comparison with other supernovae indicate that the object very likely is a peculiar type II SN with "plateau". Some peculiarities of the light curve hitherto unexplained are remarked.

1.INTRODUCTION

One of the most intriguing problems after the recent discovery of the SN 1987-A in the LMC has been that of his classification among the other SNe hitherto known. Restricting our considerations to the optical band (380 - -800 nm), a simple look to the light curve and spectrum of this supernova is sufficient to show that it cannot be classified as a type I (or type Ib as it was claimed shortly after its discovery),so that only two other alternatives are left:

a) The SN 1987-A is a type II supernova with some peculiarities. Or:

b) It represents a unique case, as it was,for instance, the SN 1961-v in NGC 1058, classified of type V by Zwicky.

We shall now examine the first alternative.

A supernova is in general attributed to type II when it shows at least one of the following characteristics:

a) Broad P Cyg features in the spectrum, due to H , HeI , FeII , NII and other atoms or ions. Their blueshift indicates extremely high expansion velocities of the order of at least 10^4 km s^{-1} .

b) Typical light curves with "plateau" or , more rarely, with "linear" decline .

c) Strong ultraviolet continuum at maximum or shortly after maximum. Color index : $(U-B)_0 \leqslant -0.5$; $(B-V)_0 \leqslant 0$.

Rapid decline of the ultraviolet emission.

The position in a spiral arm or in a HII region is considered a favourable element for classification in the type II.

2. LIGHT CURVES AND COLOURS OF TYPE II SNe

The total number of SNe discovered in external galaxies from 1885 to the end of 1986 is 620 of which nearly one third have been classified. Those attributed to type II on the base of the precedent criteria are 73, but acceptable light curves are available only for a part of them. Although there is a certain variety in the shape of the light curves of type II SNe, it was possible to select 15 SNe having similar light curves, with a "plateau" of constant brightness after the first decline (SNe II-P), and 6 SNe with "linear" decline or with just a trace of "plateau" (SNe II-L) (Barbon et al. 1979). Their average light curves are reproduced in Fig.1.

Fig.1 . Average B light curves of SNe-II

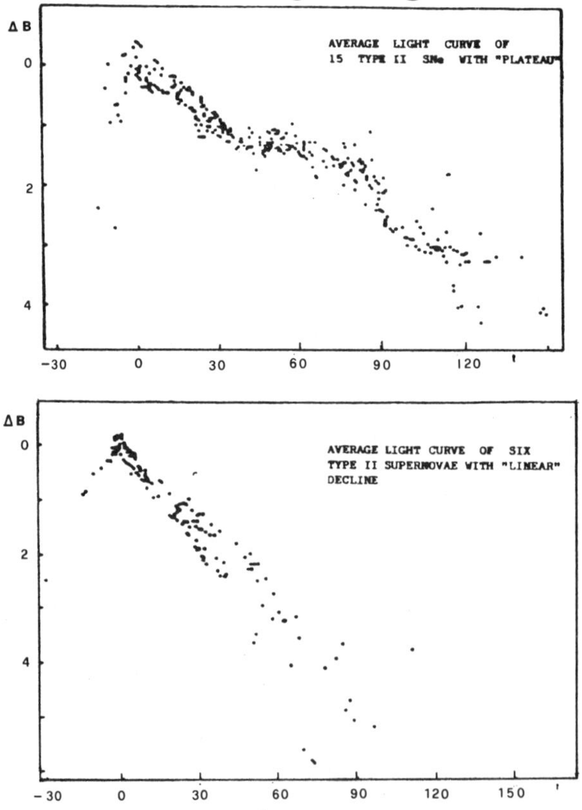

As shown in Fig.1 the brightening to maximum of a type II-P supernova is in general very fast. Porter (1987) has recently remarked that there is no evidence of a premaximum plateau in the light curve of SNe-II. The maximum is attained rapidly, then , after a week or so, the brightness begins to weaken at a rate of about 0.034 mag.d-1 , reaching in 30-35 days a "plateau". After 50-60 days of nearly constant brightness the supernova fades again ,loosing 2.5 magns. in one-two weeks. Finally,120-150 days from maximum it initiates his last decline at a constant rate of about 0.006 $mag.d^{-1}$.

The majority of the SNe-II follow,more or less,this pattern, except a few (SNe II-L) which decline from maximum at a rate of 0.05 mag d^{-1} without passing for a "plateau" or just showing a short trace of plateau. After 100-120 days the rate of decline slows down and the brightness starts to decrease as in the SNe with plateau.

The average $(B-V)_0$ color index of the type II SNe, corrected for absorption, is nearly zero or slightly negative at maximum, increasing to +1.2 in about 100 days, while the brightness decreases. Then, during the final decline it remains nearly constant. The $(U-B)_0$ is strongly negative (-0.9) at maximum, but it becomes soon positive, reaching +1 or more in a few weeks (Fig.2).

Fig.2 : Variations of color index in type II SNe.

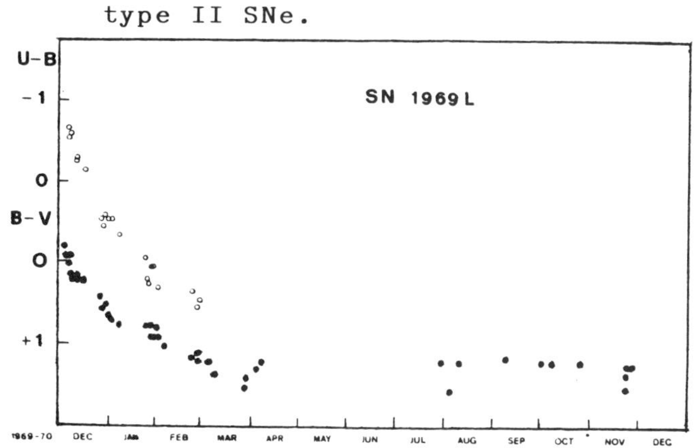

Typical representative of the SNeII with "plateau" is SN 1969 L in NGC 1058 (Ciatti et al.,1971).Its light curves in U,B and V are reproduced in Fig.3 .What is note--worthy is the fact that while in B and V the maximum is fairly well defined, in the V colour there is not a clear evidence of maximum. The V light curve begins with a "plateau" of nearly constant magnitude lasting for about 80 days before the final decline.It may also be interesting to notice that in the U light curve there is no trace of "plateau. .

Fig.3 : U,V,B light curves of the type II SN 1969-L

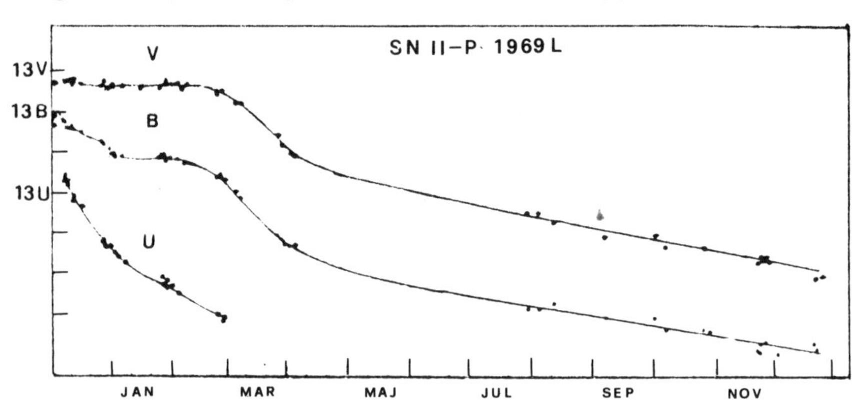

Other type II Sne have shown visual light curves characterized by a long permanence in a "plateau" just slightly weaker than maximum. The following Fig.4 represents the V and B light curves of the SN 1973-r (Ciatti-Rosino , 1977) which has also had a long permanence near maximum.

Fig.4 : V ,B light curves of the type II SN 1973-r

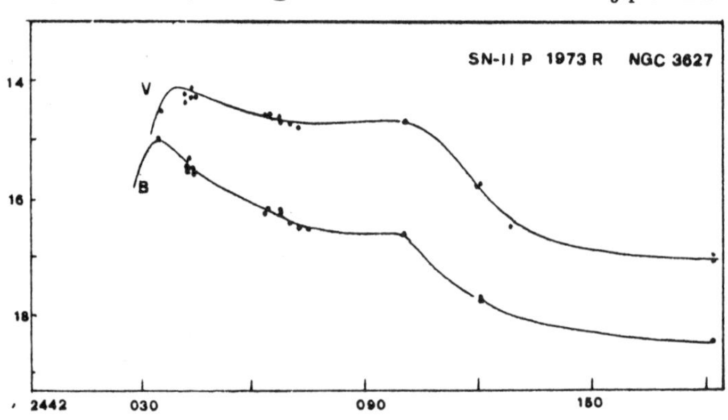

At this point we can try to answer the question:which kind of light curve have been followed up to now by the SN 1987-A ?

Fig.5 : V and B light curves of SN II 1969L AND SN 1987-A (same scale).

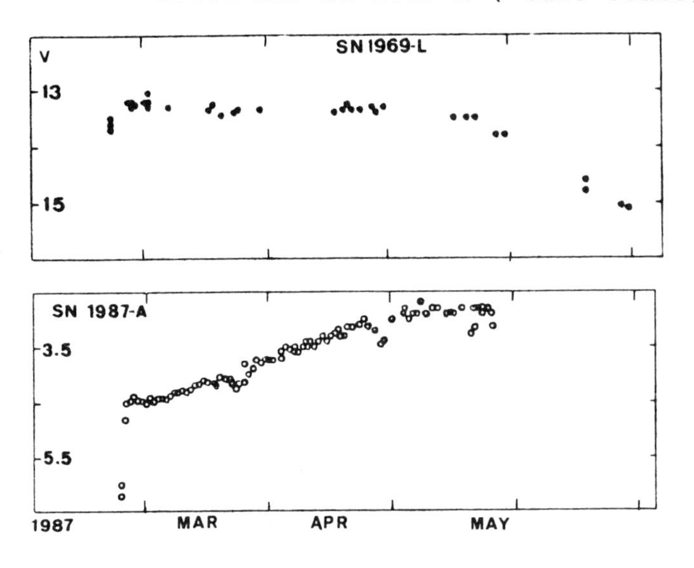

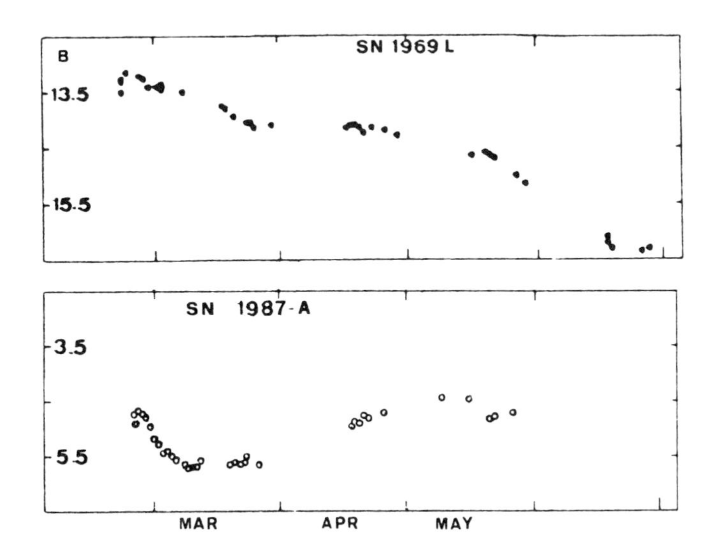

Fig.5 reproduces in the same scale the V and B light curves of 1969-L, a classical type II SN with plateau, and those of the SN 1987-A. It is quite apparent that the light curves of the two supernovae have some far resemblance,the main difference being that while the brightness of SN 1987-A has been slowly increasing in both colors during the "plateau" phase,the other type II SNe with "plateau" remain more or less constant in this phase. Evidently, after the explosion there has been in SN 1987-A an additional source of energy, responsible of the post-maximum brightening. Very likely a decline will follow by the next weeks.

A comparison of the U light curves(Fig.6) is also significant. The drop of brightness of SN 1987-A, in about 15 days, from U =4 to 7.8 has been much faster than in other type II SNe with "plateau". However, after having reached magn.U =7.8 the supernova has maintained a constant o slightly increasing brightness, forming a sort of plateau,which is never observed,in the U,in other SNeII.

Fig.6 : U light curves of SN 1969-L and 1987-A.

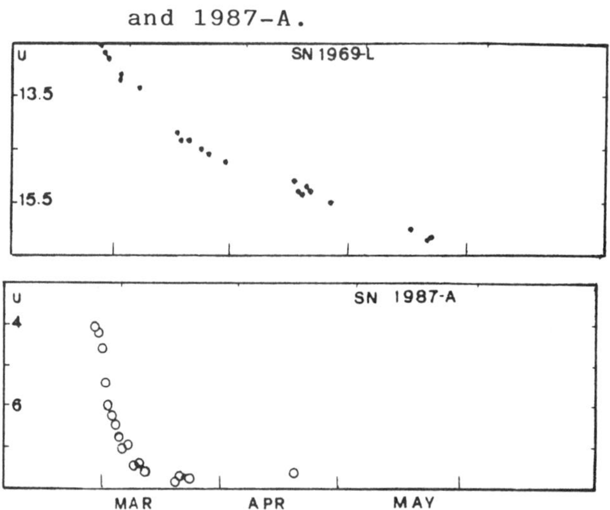

3. SPECTRA OF TYPE II SNe AND SN 1987-A

The optical spectra of type II SNe of both subtypes L and P are characterized in the first days after maximum by a featureless continuum well extended in the ultraviolet with traces of P Cyg features very broad and shallow, blueward displaced ($V \sim -12$-13 000 km s^{-1}). In the next two weeks the P Cyg absorption bands attain a major contrast. Balmer lines and the HeI-NaI feature (λ 589 nm) are easily identified.

From 15 to 30 days after maximum while the blue continuum weakens and all of the P Cyg features are slowly shifting towards the red, the bands due to H (particularly H_{α}), FeII (mult.: 37,38, 42,73,74) , NII, NIII, HeI, NaI, CaII, partly blended together to form a pseudo continuum, can be detected in the spectra of type II SNe(Fig.7).

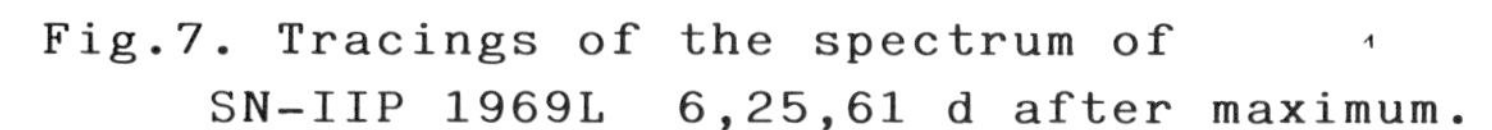
Fig.7. Tracings of the spectrum of SN-IIP 1969L 6,25,61 d after maximum.

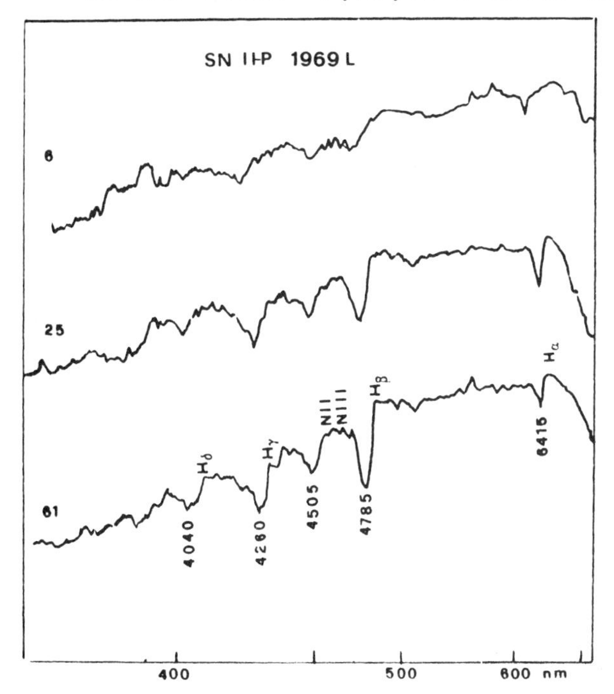

The structure of the spectrum one-two months from maximum looks somewhat alike to that of a bright nova near maximum.

The following points characterize the spectral evolution of the type II Sne:

a) the gradual drift of all the emission bands from their initial blueward position to their laboratory wavelengths.

b) The decreasing velocity of expansion of the ejected shells with time, from 12-13 000 km s^{-1} (H_{α}) in the first days to 5-6000 km s^{-1} after some weeks.

It can also be remarked that the blueshift of the Balmer absorptions decreases in each spectrum from H_α towards the higher members of the series.

Coming back to SN 1987-A, there is no doubt that his optical spectrum is alike to that of other type II supernovae. The presence, over a continuum well strong in the blue-violet region, of broad shallow P Cyg absorptions of H_α , H_β , H_γ ; the extremely high expansion velocity (more than 20 000 km s^{-1} at maximum); the successive reddening of the object and the emerging of broad emission bands of CaII, HeI, FeII, NII, NIII, all accompanied by P Cyg absorptions; their gradual drift towards the red (Danziger et al. 1987; see also IAU Circ. 4317-4374); the further spectral evolution; all these are facts extremely significant in order to classify SN 1987-A among the type II SNe, with only some minor peculiarities.

Fig.8 - Spectra of SN-II 1959-D and SN 1987-A 58 and 39d from maximum.

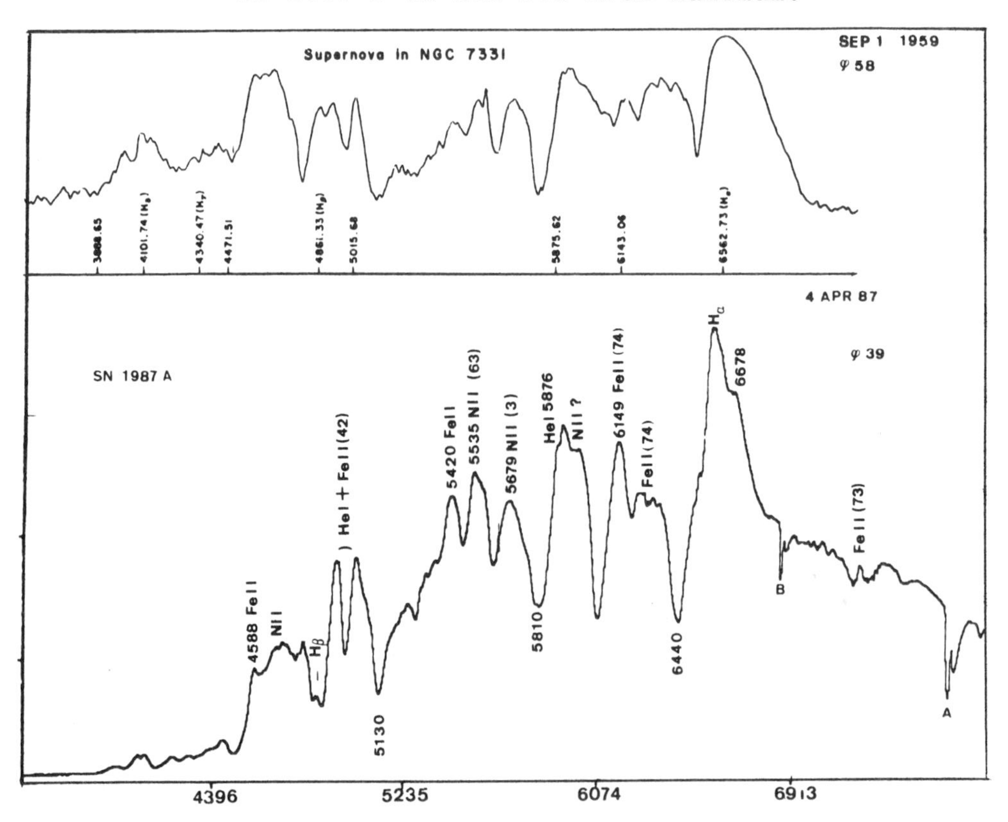

In Fig.8 are represented, in the same scale, the spectrum of the SN-II 1959-D obtained on Sep 1,1959, 58 days after maximum, by Greenstein (1965) and that of SN 1987-A, kindly taken on my behalf by Rafanelli with the CCD applied to the 1.52 m telescope of ESO at La Silla on Apr 4 , 35 d from maximum. A preliminary identification of the main emission features of SN 1987-A is shown in Fig.8. A peculiarity in this spectrum is the weakness of all the hydrogen lines, except H_α . Apart from this anomaly, the spectra of the two supernovae match so closely that it is very difficult to escape the conclusion that SN 1987-A belongs to the class of type II Sne. On the other hand the classification of SNe on the basis of their UV spectra cannot be conclusive, since we have not sufficient data on the behaviour of SNe in the far ultraviolet (see: Wamsteker et al.,1987).

4. ABSOLUTE MAGNITUDES.

The determination of the absolute magnitude of supernovae at maximum is always a difficult problem for the following reasons:

a) Unaccountable effects of the absorption within the parent galaxy. They can be significant for SN-II which are located in late type spirals.

b) Uncertainties in the distances of the parent galaxies, even when their are derived by mean of the Hubble law.

c) Errors in the determination of the magnitudes,particularly when the SNe are embedded in bright knots or located in the bright central region of the parent galaxy.

d) Uncertainties in the backward extrapolations for SNe discovered days or weeks after maximum. The corrections are somewhat arbitrary for type II SNe which not always follow a standard light curve.

Notwithstanding these and other sources of error, most of the Authors (Minkowsky ,Kowal ,Tamman, Barbon et al.) agree in estimating the average absolute magnitude of type II SNe at maximum : $M \sim -16.5$ ($\sigma=0.8$) (Barbon et al.,1979).

Now, let us consider SN 1987-A as a type II supernova with "plateau". If we assume that the "true" maximum (B=+ 4.6) was attained by the SN on Feb 26-27, two days after its first appearance, when the color indices $(B-V)_o$ and $(U-B)_o$ were both negative and the spectrum displayed an extended ultraviolet continuum, then , considering that the distance modulus of the LMC is 18.6 , after a correction of 0.8 to the B magnitude of the SN (its color excess E (B-V) being +0.2) we obtain the following value of his absolute magnitude : $M = -14.8$, decidedly less than the average luminosity of type II SNe (-16.5). However, we must consider that, in the systematic search of SNe, those with luminosity less than $M_B \sim -15$ or -14 can be very easily overlooked , particularly when they appear in distant galaxies. SN 1987-A represents a fortunate case, because the vicinity of the LMC has made possible his discovery. But we cannot exclude that SNe with absolute magnitudes comparable to that of SN 1987-A may be present in other galaxies and yet undiscovered. If this is the case, the relatively low luminosity of SN 1987-A at his early maximum by no means becomes exceptional, the absence of supernovae of so low luminosity being only an effect of selection in the discovery.

An alternative is that the "true" maximum, without any reference to spectrum or colour, may have been reached later, when the SN brightened during the "plateau" phase attaining visual magnitude ~ 2.9 . In this case the absolute magnitude of the SN would be $M_V = -16.3$ not so different from the average value found for type II SNe.

We believe ,however, that the first alternative may have a much better foundation than this one .

5. CONCLUDING REMARKS .

In conclusion, while there seems not to be any doubt that SN 1987-A must be classified as a type II supernova, some minor differences remain when it is compared with the other type II SNe with plateau. The most significant are:

a) The slow increasing brightness from V~4.5 on March 2-3 to V~2.9 on May 15-20.

b) The too rapid decrease of the UV emission and of the expansion velocity,faster than in other SNeII (Fig9).

c) The extremely high P Cyg velocity at the early maximum (from-30 to-20 thousand km s^{-1}).

d) The exceptional weakness of the Balmer emissions (H_α excluded)

e) The relatively low luminosity at maximum.

Finally a mystery is still that of the progenitor of the SN 1987-A. If it is confirmed that the star SK -69202-1 of spectral type B3-I has disappeared, as substained by Sonneborn and Kirshner (1987), then some of the theories on the origin of the type II SNe should be reviewed or modified.

Fig.9 . Decrement of P Cyg velocities in SN 1969-L and SN 1987-A.

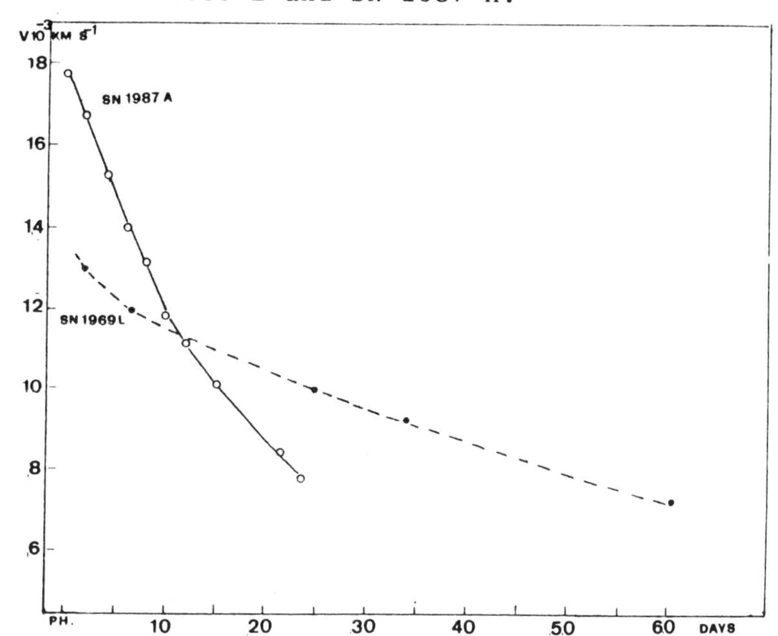

Observations and theoretical interpretations are still being carried out and very likely will continue for years. They will certainly solve most of the problems concerning this extraordinary supernova and his collocation among the other type II SNe.

REFERENCES

Barbon,R., Ciatti,F.,Rosino,L. 1979,Astron.Astrophys. 72,287.

Ciatti,F.,Rosino,L.,Bertola,F. 1971,Mem.Soc.Astron. Ital. 42,163.

Ciatti,F.,Rosino,L. 1977, Astron.Astrophys. 56,59.

Danziger,I.J. *et al*.1987, ESO Preprint No.500.

Greenstein,J. 1965, Stellar Structure, p.413. Univ. of Chicago Press.

Porter,A.C. 1987, IAU Circ.No.4359.

Sonneborn,G.,Kirshner,R.1987, IAU Circ.No.4366.

Wamsteker,W. *et al*.,1987 ,Vilspa IUE Reprint No.21.

DISCUSSION

G. Chincarini: Comment: You mention possible uncertainties in the validity of the Hubble law since M is computed using it. It seems that the small dispersion you give σ=0.8 shows that things work out fairly well given also the uncertainties in internal absorption correction etc... Question: You said the Hα is very strong with respect to the other H lines. How anomalous is such a decrement once ABS and physical conditions as we know them are taken into account?

L. Rosino: When I was speaking of uncertainties of the Hubble law, in deriving the absolute magnitude of SN's by the redshift of the parent galaxies, I was referring to the fact that the assumed distance depends on the value of the Hubble constant, rather than an eventual failure of the Hubble law. For instance, the absolute magnitude of a type II supernova is, on the average, -16.5 if H= 100 km/s/Mpc and -18 if H= 50 km/s/Mpc. On the contrary, the absolute magnitude of SN 1987-A has been derived, not from the redshift but from the direct knowledge of the distance of the LMC.
It is difficult to give an answer to the second question. The weakness of Hβ and the invisibility of Hγ (in emission) seems to be rather peculiar when we compare the spectrum of this SN with that of other type II SN's at about the same phase. But it can be accounted for by considering the effect of the P Cyg absorptions of nearby emission lines.

L. Woltjer: Of the six galactic supernovae, two (SN1181 and Cas A) seem to have had luminosities as low as that of the LMC supernova. This may indicate the existence of a rather numerous class of such supernovae with M$\sim$ -15.

G. Burbidge: My comment is similar to that just made by L. Woltjer. Are we not seeing in SN 1987A a supernova fainter at maximum light than normal type II which is only seen because it is not very far away. Several galactic SN remnants were not so bright. Were they not similar SN to 1987A. Does this not change our views about the frequency of type I and type II and this "new" type. Is this new type not much more common than type I and type II and energetically less powerful?

L. Rosino: Perhaps it is still too early to give an answer to these questions. At the present moment (April 30, 1987) the brightness of SN 1987-A is still slowly but steadily increasing in V as in B. Considering that its color excess E(B-V) = 0.2 and that its present apparent V magnitude is +3.2, the absolute M_V is equal to -16, not so far from that of "normal" type II supernovae with plateau ($\sim$-16.5). Moreover, the brightness is still increasing. So that we may affirm that the principal anomaly of this SN is not that of having an absolute magnitude weaker than other SN-II P, but that of displaying a light curve that has never been observed in other type II SN. On the other hand the spectrum is undoubtedly similar to that of SN-II.
Now, all will depend on the behaviour of this SN by the next weeks. If its brightness will start to decline, then we could consider the SN as a peculiar type II. But if this does not happen, then we must conclude that it represents a new class of supernovae.

ON THE STELLAR FLARES OF VERY SHORT DURATION

V. A. Ambartsumian
Byurakan Observatory of the Academy of Sciences
Armenian SSR

During the last decades, astronomers have paid considerable attention to the photo-electric observations of stellar flares in K-M dwarfs. As a result, we now have the photo-electric light curves of at least several thousand flares of a number of such stars. Often the flares were monitored in different colors simultaneously.

However, the enormous quantity of information contained in this large number of curves has not yet been analyzed sufficiently. It is true that some statistical conclusions have been obtained about the distribution of amplitudes, on the duration of flares, etc.; however, the conclusions about the regularities of light variations during the individual flares, about the shapes of light curves, as well as on the connection of these regularities with phases of early stages of stellar evolution, are relatively scarce. This makes difficult the understanding of the physics of flares and their role in the stellar evolution. But we can expect the discovery of new interesting facts just about the early phases of stellar evolution.

It seems, therefore, that the study of light curves of individual flares must be a very broad field of investigation. However, in the present note we would like to discuss only two circumstances.

I. As is well known, each flare as a rule begins with a sharp increase in the brightness of the star. This property is more strongly expressed in U-color. But according to the further behaviour of the light curve we can divide all flares in three classes:

A) The flares showing only one maximum.

B) Those which show two or more maxima, of which the first is the highest (main maximum).

C) Those in which the first maximum is not the highest.

In cases A and B, the increase of brightness from the minimum till the main maximum proceeds monotonically.

It is an important statistical fact that the majority of light curves belong to the classes A or B. This means that in more than half the flares the brightness increases monotonically from minimum to the highest point of the light curve.

Of course, the inclusion of a given flare in one or another of classes A, B, and C will, to some degree, depend on the time resolution of the monitoring instrument used. Therefore, the obtained percentages of flares belonging to different classes are more or less conditional. Our conclusion that the majority of flares belong to the classes A or B was obtained on the basis of examination of curves which have the time resolution of about 1 or 2 seconds.

Better time resolution might reveal some small secondary minima between the minimum and main maximum, obliging us to assign such a flare to class C. But it seems to us that the possibility of some corrections of this kind does not disprove the significance of the fact that at resolutions on the order of 1 or 2 seconds the majority of curves show monotonous increase from the minimum to the main maximum.

The classification described above may serve as a basis for the assumption that each flare is some kind of superposition of several more or less distinct elementary processes (subflares or explosions). We can imagine that in case A the flare consists of only one such process (elementary explosion), in case B it consists of several explosions of which the first is the strongest, and in case C we have again several explosions of which the first in time may not be the first in strength.

This, of course, does not mean that the superposition of explosions takes place linearly (as regards the quantities of radiated energy) or that they happen independently of each other.

We suppose that νdt is the probability of occurence of elementary explosion during the time dt in the quiet star and that $\nu_1 dt$ is the probability of occurence of a secondary explosion during the time interval between some t and t + dt after the first explosion. Then it is possible to conclude from observational data that $\nu_1 > \nu$. This means that the occurence of the first explosion increases the probability of the next one. And, of course, $\nu_1(t)$ is the function of time that elapsed since the first explosion (which is the beginning of the flare). Thus, it is impossible to consider the secondary explosions as spontaneous events.

Therefore, the true theory of the formation of a complex flare must take into account the interaction of elementary processes. The construction of such a theory will be possible either on the basis of understanding of the physical nature of these processes or on the grounds of careful analysis of observational data, as the solution of some very complicated and non-linear inverse problem.

II. Since each A flare consists of only one elementary explosion, it is interesting to concentrate attention first on the study of their properties.

According to the observational data, the duration of increase of brightness from minimum to maximum of these flares is, as a rule, shorter than one minute. In most cases, it takes only several seconds.

It is natural to ask about the possible lower time limit of observed flares. Evidently the flares of smallest amplitude can have exceedingly low value of such limit. But we can restrict our question asking about the lower limit for "big flares" in which the luminosity of the star in U color is more than doubled at maximum.

In this connection, it is interesting that at the observatories in Rogen (Zwetkow) and in Byurakan (Tovmassian and Zalinian) big flares of EV Lacertae have been observed in which the time of increase in the colour U was only of the order of several tenths of a second. Particularly at the Byurakan Observatory special observations have been made with very high time resolution, during which cases have been registered when the doubling of brightness has occured in the time interval of the order of 0.1 second. At least in one case the observers have succeeded in monitoring a flare of EV Lacertae in which the doubling of the brightness has been achieved in less than 0.1 second.

Since the velocity of propagation of the flare in the atmosphere of a star or above its atmosphere (and of course the component of that velocity perpendicular to the line of sight) cannot exceed the velocity of light, we can conclude from these observations that the big flares of class A, at least in some cases, occupy only a small part of the surface of the star (of the order of one hundreth of the disk or less).

This means that in such cases the surface brightness of the region of the flare in U must exceed the surface brightness of the normal region of the stars photosphere in that colour by at least several hundred times.

The question remains: can the velocity of propagation of the flare really reach values of the order of light-velocity?

Of course, this question requires deep study. But in any case, we can imagine two possibilities:

a) The flares take place in the coronal or even in more diluted layers of the surroundings of the star. In such layers, there are more possibilities to acquire very high velocities. But one must take into account that it is hardly possible to imagine in such regions the very concentrated sources of energy unless we suppose that the energy is brought there in some non-radiative form and then released high above the photosphere in an explosion or decay.

b) The flares are taking place in the photosphere or in the reversing layers and the transfer of energy occur in accordance with laws of gas-dynamics and hydrodynamics. In this case, it is difficult to imagine that velocities can be attained which are many times larger than the velocity of escape.

If the velocity of propagation reaches even 4000 km/sec, we should assume that the linear size of the flare region will hardly exceed 100 km. In this case, the surface brightness of the region in U colour should exceed 10^6 times. or more the normal surface brightness of the star.

In both cases considered, we are forced to agree that we deal here with extremely interesting processes. From the outside, they remind us of the phenomenon of lightning. If we take this similarity more seriously and assume that the initial increase of brightness proceeds along a line, (i.e. the process is one-dimensional), we reach the conclusion that in a narrow strip of stellar surface the surface brightness increases by factors of hundreds of millions.

* * * I am very sorry that my duties in the Academy of Sciences of Armenia have prevented me from presenting this personally to the symposium in honour of Dr. Halton Arp who created during the last decades many new ideas in Astronomy, as a real innovator. I thank deeply Dr. E. Khachikian who has agreed to read the initial version of this paper at the Symposium.

RR LYRAE LUMINOSITIES, AND GLOBULAR CLUSTER AGES, FROM SUBDWARF FITTING

R G Noble
Department of Physics,The University, Leeds LS2 9JT, England
R J Dickens
Astrophysics and Geophysics Division, Rutherford Appleton Laboratory, Chilton, Didcot, Oxfordshire OX11 0QX, England

Abstract. Distances to ten globular clusters with modern CCD colour-magnitude diagrams have been obtained by fitting their main sequences to nearby subdwarfs. This results in an empirical calibration of $M_V^{RR} = 0.591 + 0.126$[Fe/H] for the cluster RR Lyraes. When combined with a new derivation of the Oosterhoff period-shift versus metallicity relation of $\Delta \log P = -0.079\Delta$[Fe/H], this yields $\Delta M_V^{RR} \approx -1.6\Delta \log P$. Pulsation theory indicates such an effect to be due to a higher mass for the more metal-rich RR Lyraes. Ages for the ten globular clusters, based on the subdwarf moduli, range between 10 and 16 billion years, the lower ages being found for the more metal-rich clusters. These ages are on average about two billion years younger than those found from the $\Delta M_{TO}^{RR}(V)$ method.

Colour-magnitude (CM) diagrams of a number of globular clusters extending several magnitudes down the main sequence are now available. Based on CCD photometry, they are of much higher precision than earlier photographic photometry, especially around the turnoff region, and go sufficiently faint that good fits of the lower main sequence to nearby subdwarfs are now possible. It seems fitting at this meeting, which commemorates Chip Arp's 60th birthday, to discuss the current status of these observations since he was one of the first astronomers to observe a globular cluster main sequence.

One advantage of the subdwarf-fitting method over others is that it provides an essentially empirical means of deriving the cluster distance, the only appeal to theory being the use of models to correct the subdwarf colours to the appropriate cluster metallicity. This then provides an empirical calibration of the luminosity of any cluster star; in particular the cluster horizontal branch (HB) stars, thus yielding M_V^{RR} for the RR Lyraes in that cluster, and stars at the main-sequence turnoff. Given the cluster distance, its age may be determined using stellar models either to interpret the turnoff absolute magnitude (M_V^{TO}), or to fit the CM diagram with theoretical isochrones. A third method, independent of cluster distance, uses the magnitude difference, $\Delta M_{TO}^{RR}(V)$, but this requires further (HB) models to provide the theoretical M_V^{RR}. In this note we draw attention to the implications of the results of using the subdwarf-fitting method to calibrate the RR Lyraes and turnoff stars, and make a brief comparison with the results of the $\Delta M_{TO}^{RR}(V)$ method. These methods have been discussed extensively in the literature (*e.g.* see Iben & Renzini (1984) and references therein).

Table 1 summarises the results from our survey of published or in press CCD CM diagrams. For all clusters, we have re-assessed the relevant parameters of the clusters and their CM diagrams, correcting the subdwarf colours to the cluster metallicity using the VandenBerg & Bell (1985) isochrones and allowing for the nonlinear behaviour of $(B - V)$ with [Fe/H] on the main sequence. [Fe/H] is taken from Frogel *et al.* (1983), except for NGC7492 where the value from Buonnano *et al.* (1987) is used. The third column gives the apparent distance moduli deduced from the subdwarf fits, which lead to the absolute magnitudes M_V^{RR} and M_V^{TO} given in columns 4 and 5. $\Delta M_{TO}^{RR}(bol)$ is obtained using bolometric corrections appropriate to turnoff and HB stars, and the last two columns give the derived ages in Gyr. t_{TO} comes from the dependence of M_V^{TO} with age and composition determined by interpolation in the isochrones of VandenBerg & Bell, and t_Δ from equation (9) of Iben & Renzini (1984). Note that the latter was derived from the Yale isochrones and the Sweigart & Gross

Table 1: Colour-magnitude data and cluster ages from subdwarf fits

Cluster	[Fe/H]	$(m-M)_V$	M_V^{RR}	M_V^{TO}	$\Delta M_{TO}^{RR}(bol)$	t_{TO}	t_Δ
7078 (M15)	-2.21	15.53	0.38	3.67	3.22	13.5	14.2
6341 (M92)	-2.01	14.60	0.40	3.90	3.43	15.5	16.5
4590 (M68)	-1.96	15.35	0.27	3.65	3.30	13.0	14.5
6205 (M13)	-1.47	14.50	0.29	3.90	3.51	14.0	16.0
362	-1.39	14.92	0.51	3.88	3.34	13.0	13.4
6752	-1.35	13.29	0.39	4.01	3.52	15.0	15.8
7492	-1.34	17.15	0.48	4.09	3.51	16.0	15.6
6121 (M4)	-0.72	12.90	0.42	3.90	3.36	9.5	11.9
104 (47 TUC)	-0.59	13.51	0.58	4.06	3.44	10.5	12.5

Figure 1.(Left). Correlation between cluster age, t_{TO}, obtained using M_V^{TO} from subdwarf fitting and the models of VandenBerg & Bell (1985), and metal abundance, [Fe/H], from Frogel *et al.*(1983).

Figure 2.(Right). Correlation between cluster age, t_Δ, obtained from the magnitude difference between turnoff and horizontal branch, and equation (9) of Iben & Renzini (1984), and [Fe/H].

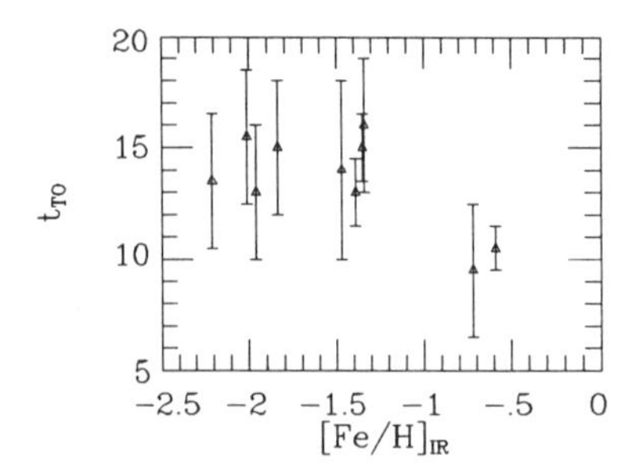

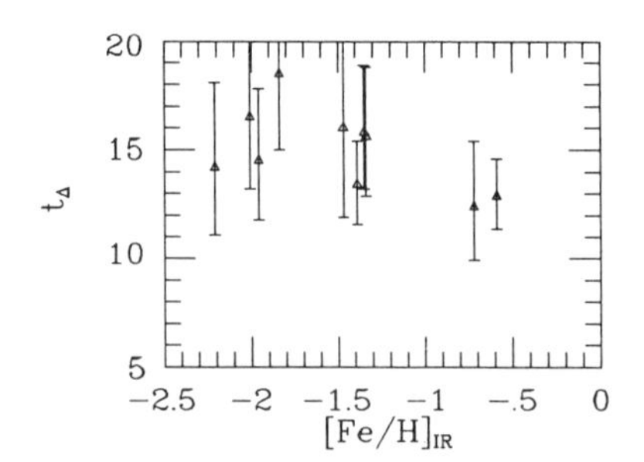

(1976) HB models; this should provide a useful point of comparison with earlier work based on these models, although it will clearly be important to apply the same method using the VandenBerg & Bell isochrones with new HB models when the latter become available.

Figure 1 shows the correlation of t_{TO} with [Fe/H]; estimated one-sigma error bars (essentially "internal" errors, estimated from the various uncertainties in fitting procedures) in the ages are shown. Errors in [Fe/H] will be around ± 0.15 (one sigma). It can be seen that the clusters with $[Fe/H] < -1.2$ group around a mean age of about 14 billion years, whereas the two metal-richest clusters (47Tuc and M4) appear to be younger. In comparison, Figure 2 shows the ages from the ΔM_V method to be on average 1 to 2 billion years older, and showing less of a trend with metallicity. A full discussion of these results is outside the scope of this paper, and will be given elsewhere. However the new subdwarf results do suggest a downward trend in globular cluster ages, *e.g.* as compared to those derived by Sandage (1982) and this stems mainly from the systematically brighter RR Lyrae luminosities derived from the subdwarf fits (see below). The reality of the trend of age with metallicity depends crucially on the result for 47 Tuc. For further discussion of this see Hesser *et al.* (1987).

Figure 3 shows the correlation of M_V^{RR} with [Fe/H], again including estimated "internal" one-sigma error bars. The mean M_V^{RR} is around $0^m\!.4$, with a suggestion of a trend with metallicity. The line shows the formal least-squares solution, $M_V^{RR} = 0.591(\pm 0.083) + 0.126(\pm 0.053)[Fe/H]$. Such a relation can be combined with the Oosterhoff period-shift versus metallicity relation (Sandage 1982) to find the direct dependence of period shift on M_V^{RR}. This is of physical significance, and of interest to stellar evolution theory because the period shift arises essentially through a variation between

Figure 3.(Left). Relationship between the absolute magnitude of the RR Lyrae region on the horizontal branch, M_V^{RR}, as derived from the subdwarf fitting, and [Fe/H].
Figure 4.(Right). The dependence of the Oosterhoff mean period shift, $\Delta \log P$ on metal abundance, using only clusters with more than ten ab-type RR Lyraes.

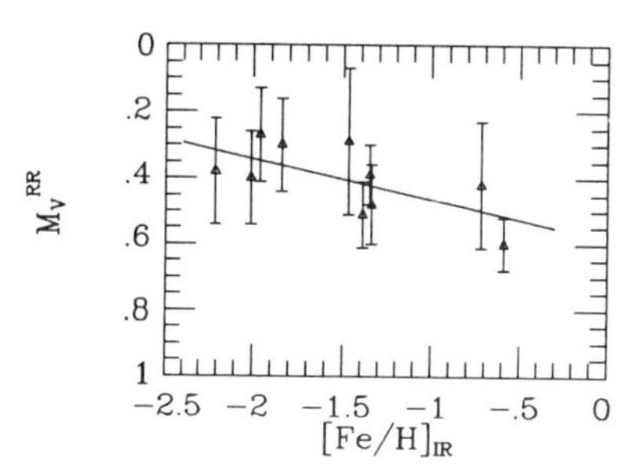

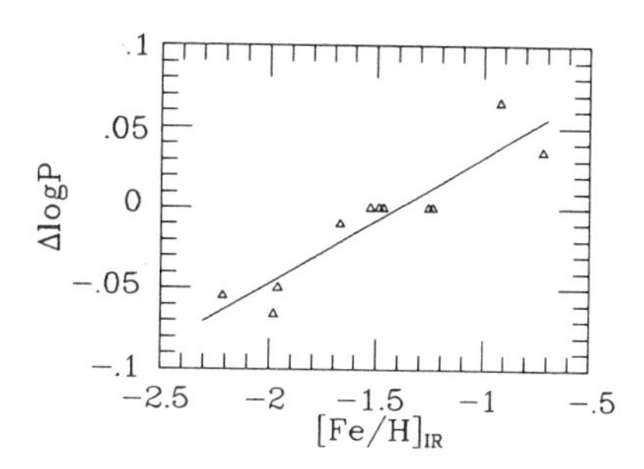

clusters in the mean mass-to-light ratio, $\mathcal{M}/L$, of their RR Lyraes (*i.e.* HB stars in the instability strip). Thus, given the luminosities via the above subdwarf calibration, an estimate of the RR Lyrae mass can be made. Furthermore, the RR luminosities can themselves be used to calibrate cluster distances, once the details of their behaviour from cluster to cluster are understood physically.

Figure 4 shows a new determination of the correlation of period shift with abundance, where only clusters with more than ten ab-type RR Lyraes have been included, and again using [Fe/H] from Frogel *et al.* The line shows the least squares fit, $\Delta \log P = -0.079(\pm 0.011)$[Fe/H]. This is in excellent agreement with the completely independent determination for field RR Lyraes by Lub (1987) of $\Delta \log P = -0.084$[Fe/H]. Combining our derivation with the correlation of Figure 3 yields $\Delta M_V^{RR} \approx -1.6(\pm 0.7)\Delta \log P$. From the pulsation relation of Van Albada & Baker (1971), $\Delta \log P = -0.68\Delta \log \mathcal{M} - 0.336\Delta M_V^{RR}$ at constant T_{eff} (*i.e.* comparing $\mathcal{M}/L$ from one cluster to another), from which we find $\Delta \log \mathcal{M} \approx 0.4\Delta M_V^{RR} \approx 0.07\Delta$[Fe/H]. This implies that RR Lyraes in the more metal-rich clusters have a higher mass than those in more metal-poor ones. These coefficients are still rather poorly determined, but the differences implied are up to about $0.1\mathcal{M}_\odot$ over the range of globular cluster abundances. This is not unreasonable—similar differences in turnoff masses are indicated over this abundance range, even for constant cluster age (see VandenBerg & Bell 1985).

REFERENCES

Buonanno,R.,Corsi,C.E.,Ferraro,I.,& Fusi Pecci,F. (1987). *Astron. Astrophys. Suppl. Ser.*, **67**, 327.
Frogel,J.A.,Cohen,J.G.,& Persson,S.E., (1983). *Astrophys. J.*, **275**, 773
Hesser,J.E.,Harris,W.E.,VandenBerg,D.A.,Allwright,J.W.B.,Shott,P., & Stetson,P. (1987). *Pub. A.S.P.*, in press.
Iben,I.,& Renzini,A., (1984). *Physics Reports*, **105**, 329
Lub,J.(1987). Field RR Lyrae Stars. *In* Stellar Pulsation, a Memorial to John P Cox, Springer-Verlag, in press.
Sandage,A.R., (1982). *Astrophys. J.*, **252**, 553
Sweigart,A.V.,& Gross,P.G., (1976). *Astrophys. J. Suppl.*, **32**, 367
Van Albada,T.S.,& Baker,N., (1971). *Astrophys. J.*, **169**, 311
VandenBerg,D.A.,& Bell,R.A., (1985). *Astrophys. J. Suppl.*, **58**, 561

DISCUSSION

R. Kraft: What O-abundance did you adopt in your fitting procedure?

R. Dickens: The oxygen abundances were standard, that is solar oxygen to iron as used in the Vandenberg and Bell isochrones.

G. Burbidge: Surely the question for cosmology is not the *average* ages of the globular clusters, but the age (with its uncertainties) of the *oldest* one you can find.

R. Dickens: Yes indeed, but the uncertainties in each age determination are such that I felt at this stage, the mean age would give the best estimate. No attempt has been made so far to carefully assess the relative weights of each age determination, but I agree this should be done when looking for an upper limit to globular cluster ages for cosmology.

EXTRAGALACTIC OBSERVATION

INTERGALACTIC HYDROGEN CLOUDS

Morton S. Roberts
Institute of Astronomy, University of Cambridge, Cambridge, England
National Radio Astronomy Observatory,* Charlottesville Virginia, U.S.A.

PROLOGUE

The concept, the possibility of intergalactic matter is not a new idea in astronomy. And, as we shall see, there is indeed matter between the galaxies, in special places and under special circumstances. What would be new and exciting would be the finding of an isolated cloud of such matter. One well separated from galaxies, having no stellar component, a cloud of gas alone. Such clouds appear to be frequent at high redshifts. They have yet to be found locally.

1. INTRODUCTION

The search for intergalactic matter parallels the study of galaxies themselves. Such matter may take many forms and distributions: uniform or clumped; hot or cold; it may be material as familiar as hydrogen atoms and dust or as exotic as magnetic monopoles and axions. With such a range of possibilities I might appear somewhat restrictive in addressing intergalactic hydrogen clouds but here there are available extensive observational results. The picture is complex and its interpretation far from unanimous.

Briefly, intergalactic hydrogen clouds (IHC's) are unambiguously seen in galaxy groups, e.g. M81, Leo, and the Local Group. The general geometrical form of such features has been successfully modelled using tidal theory and there is little doubt that such intergalactic hydrogen has its origin in tidal interactions. The extensive HI surrounding some galaxies, e.g. NGC 628 and NGC 2146 may also be tidal in origin or, alternatively, the remains of galaxy formation or the explosive ejection from galaxies. There are also the many hydrogen clouds revealed by their Lyman absorption signature in the spectra of high-redshift ($z \gtrsim 2$) quasars. These clouds are distinguished by their low column density, typically $\sigma \sim 10^{13} - 10^{14}$ but reaching 10^{17} atoms cm^{-2}. There is also an apparent redshift - dependent evolution, d(number of clouds)/dz $\propto (1 + z)^{\gamma}$ where γ is of order 2 (e.g. Murdoch, _et al_. 1986).

* The National Radio Astronomy Observatory is operated by Associated Universities, Inv., under contract with the National Science Foundation.

This z-dependence is mild enough that such clouds could be present locally if the extrapolation from $z \sim 2$ to $z \sim 0$ is applicable. A much stronger evolution with redshift of the neutral hydrogen column density appears to hold. Observationally such clouds are difficult to find at low z. The ultraviolet high altitude/space instrumentation employed to-date does not have adequate spectral resolution and ground based 21-cm observations, where $\sigma \sim 10^{18}$ atoms cm^{-2} represents a typical lower limit, are not sensitive enough. Finally, and for completeness, there is a hot intergalactic medium known to pervade galaxy clusters and is visible from its x-radiation.

Intergalactic hydrogen clouds relatively distant from galaxies, i.e. in the general field, are frequently postulated but never seen. Many searches for such clouds have been made, all with negative results and only upper limits to their number are available. Examples of IHC's which are in proximity to normal galaxies but appear not to be tidal debris have been proposed, e.g. clouds near M31, M33 and in the Sculptor group. This class of IHC's is distinguished by having velocities similar to those of the high velocity clouds and the Magellanic Stream, objects which cover about a sixth of the sky. Similar clouds have not been found in any of eight other groups searched and these local examples are most likely instances of confusion with foreground material.

2. TIDAL INTERGALACTIC HYDROGEN

Intergalactic matter originating from the tidal interaction of two or more galaxies is the most common form of cold IHC; it may be the only form. It is seen as bridges of 21-cm line radiation spanning two or more galaxies in a group or as broad, jet-like features extending from one of the tidally interacting galaxies. High angular resolution shows significant spatial structure within these broader features. Such tidally-induced intergalactic clouds are relatively common with over twenty instances known, although there has been only one systematic 21-cm search for such tidal effects (Haynes 1981).

With newer, more sensitive data one can look with hindsight at the results of the very first extragalactic hydrogen study, that of the Magellanic Clouds (Kerr and Hindman, 1953; Kerr, Hindman and Robinson, 1954, Fig. 1) and see the suggestion of tidal effects at work in the region between the two clouds. The newer results (Hindman, Kerr and McGee 1963, Figs. 1 and 7) clearly outline an HI bridge between the two, a bridge which at the low resolution (HPBW $2^{\circ}.2 \simeq 2$kpc) of the survey shows a fairly smooth transition in radial velocity from one Magellanic Cloud to the other. Other prominent galaxies also show similar intergalactic features, e.g. M51, (Haynes, et al. 1978) M81 (Cottrell 1976; van der Hulst 1977) and NGC4631 (Roberts 1968). In each case the companions are clearly visible, NGC 5195 for M51, M82 and NGC 3077 for M81 and NGC 4656 for NGC 4631. Modelling of these features has been quite successful (e.g. Toomre and Toomre 1972, Rots 1978) though details are not always fit (Haynes, et al. 1978).

The only systematic search for neutral hydrogen streams in groups is by Haynes (1981). She studied 15 loose systems each of two or more galaxies with a projected separation less than $190h^{-1}$ Mpc, of radial velocity less than 3000 km s^{-1} and within the Arecibo declination range ($-1° \leq \delta \leq 38°$). Eight of these groups show no evidence of tidal disruption to within the 3'.3 HPBW used for this study. One system, NGC 672/IC1727 has the HI centroid in each system shifted towards one another, a feature suggestive of tidal action and previously noted by Combes, *et al.* (1980). The six remaining groups show hydrogen features consistent with tidal disruption. Thus, at least 40% of the groups studied by Haynes show evidence of intergalactic hydrogen with a morphology whose gross spatial and velocity features can be simulated by modelling of tidal effects.
The best known and nearest example is the Magellanic Stream, a band of HI extending from the Magellanic Clouds over more than a quadrant of sky. The first clue was found by Nan Dieter (1965) who mapped the south galactic pole region in HI and found the velocity anomaly that separates the Stream from foreground galactic hydrogen. The study of this region was extended by Wannier and Wrixon (1973) still using telescopes in the northern hemisphere. It was the necessary southern hemisphere survey (Mathewson *et al*. 1974) that found a "long filament of HI which extends from the region of the Magellanic Clouds to the south galactic polar cap and beyond" to the region of the earlier northern hemisphere studies. This recognition of a link to the Magellanic Clouds was vital and the entire feature was appropriately named "the Magellanic Stream" (Mathewson *et al*. 1974).

Observations made with higher resolution (angular and velocity) and with greater sensitivity show a wealth of detail in various regions of the Stream (e.g. Mathewson, *et al*. 1977, Mirabel, *et al*. 1979, Haynes 1979, Mirabel 1981, Cohen 1982 a, b). The initial picture of a smoothly varying velocity field and spatial density holds only in the broadest sense. These newer data show significant structure in many regions of the stream; some are best described as "shreds". Seen as a continuation of the northern end of the Magellanic Stream, these shreds could very well account for at least some of the very high velocity clouds found near $l \simeq 90° \pm 40°$ and $b \simeq -30 \pm 20°$. These shreds as well as those seen through the Magellanic Stream in the direction of the Sculptor group figure importantly throughout the discussion of intergalactic hydrogen clouds.

The prominent HI clouds in the Leo group(s) of galaxies show the varied features to be found in more distant examples of tidal encounters. These are in the M66 (NGC 3627) group and the M96 (NGC 3368) group. (Sandage and Tammann (1975) place both in the "Leo Group", de Vaucouleurs (1975) suggests calling it the "Leo Cloud" containing subgroups as noted above. To avoid any ambiguity the Messier notation will be used.) NGC 3628 in the M66 group has an HI plume extending eastward over 100 kpc from its centre (for an adopted distance of 10 Mpc). There is both spatial and velocity structure along this plume though little velocity gradient and remarkably narrow velocity profiles, 17 km s^{-1} full width at 20% intensity (Rots 1978, Haynes, et al. 1979).

This plume is one of the few IHC's with an optical counterpart. In this instance, the optical feature (Zwicky, 1956) was known before the 21-cm feature was discovered. Various explanations have been proposed for the rarity of optical counterparts to HI tidal features. What is needed are HI observations of the many optical features seen in interacting systems (e.g. Arp 1966). Another example of optical/HI coincidence is found in the HI stream between the two Magellanic Clouds where a blue stellar population has been found by Irwin, et al. (1985). Either a young population was removed together with the HI in this tidal encounter or star formation does occur in tidal debris.

Dynamical studies, stellar population studies (as in the Magellanic bridge) and radiative effects (Sunyaev 1969) on the neutral component of tidal features all can serve as independent clocks. Their evaluation will increase our understanding of each mechanism and of the tidal material itself.

Perhaps the most remarkable of intracluster hydrogen clouds found thus far is that in the M96 group (Schneider, et al. 1983; Schneider 1985): here six clouds are found distributed along an elliptical ring of semi-major axes $\sim$ 100 kpc and centered on two bright galaxies NGC 3379 (M105) and NGC 3384. A seventh cloud reaches off of the ring towards yet another bright galaxy, NGC 3368 (M96). Schneider (1985) offers two possible explanations for these clouds: primordial material remaining from the group's formation or repeated tidal disruption of a galaxy in orbit within the group, the resultant debris being distributed along the elliptical path. Rood and Williams (1984) attribute the gas to material stripped during collision between two prominent members of the M96 group.

3. INDIVIDUAL GALAXIES WITH EXTENSIVE HI ENVELOPES

High sensitivity 21-cm measurements of galaxies are usually able to trace HI to an extent greater than the conventional optical boundaries. This has an immediate practical application in the determination of rotation curves where the last detectable HII region fixes the largest radius for an optical velocity measurement. In many instances 21-cm measurements have been able to significantly extend the radial range of such measurements. It is by such techniques that flat rotation curves were first recognized (Roberts 1976).

A comparison in HI and optical of well-defined measured quantities for a large sample of galaxies is that of Hewitt, et al. (1983). They compare an effective diameter, D_{70}, a diameter within which 70% of the HI mass is contained, to UGC blue (isophotal) diameters, a_{UGC}, measured to 25.9 ± 0.7 blue magnitude $(arcsec)^{-2}$ (van der Kruit 1987). They find that, with significant scatter, $D_{70} = 1.2\ a_{UGC}$. There appears to be no dependence with type for systems later than Sa. The

UGC diameter (or the Holmberg diameter, measured to 26.5 blue magnitude (arcsec)$^{-2}$) is thus a good comparative basis as to whether the HI in a galaxy is extensive.

Many of the first described examples of significant outlying HI: e.g. M81, M101, NGC 4631 were clearly tidal in origin. Others like NGC 2146 (Fisher and Tully 1976) with HI at six Holmberg radii and NGC 628 (Briggs 1982) at 3 Holmberg radii are far less obvious. Tide-inducing candidates can be found because truly isolated galaxies, if they exist at all, are rare. Such candidates often strain possibilities because of a lack of specific information, e.g. mass, true separation. Tully and Fisher (1976) summarize the case for NGC 2146. They can only conclude that the extensive HI envelope associated with this galaxy may be attributed to (i) tides, (ii) explosive events from within NGC 2146, or (iii) primordial hydrogen near a young galaxy. Choosing from among these requires further information.

The above two examples of relatively isolated galaxies having extensive HI are rare. NGC 628 is particularly interesting because it was uncovered in a survey looking for just such examples (Briggs, *et al.*, 1980). The initiating motive for their survey is related to the origin of quasar absorption lines. Only NGC 628 was found and they concluded that HI envelopes surrounding spiral galaxies showing column densities $\gtrsim 3 \times 10^{18}$ cm^{-2} at 2 to 3 Holmberg radii are not common. In fact they are uncommon enough that these outer regions (as sampled locally, i.e. $z \simeq 0$) cannot form quasar-like Mg II absorption lines.

4. HYDROGEN CLOUDS NEAR GALAXIES

There are a number of instances in which HI clouds are found near galaxies. These are neighbors as measured by 3 of the coordinates of phase space: right ascension, declination and radial velocity, the remaining coordinates, especially distance, are unknown. The examples are M31 (Davies 1974), M33 (Wright 1974), NGC 55 and NGC 300 (Mathewson, *et al.* 1975) and a faint dwarf irregular galaxy in Sculptor (Cesarsky, *et al.* 1977). The last 3 of these systems are in the southern part of the Sculptor group. These instances of HI companions are distinguished by their low radial velocity, most are negative, and by their association with nearby galaxies, members of the local group or of the next nearest group, Sculptor. These HI clouds, if at the distance of their neighboring galaxy, have dimensions typical of the largest of galaxies 30-70 kpc. M31 is an exception, its HI companion is $\sim$ 5 kpc in size while the M33 companion, $3^\circ \times 5^\circ$ on the sky, dwarfs its galaxy. With the M31 cloud an exception again, the typical HI mass, at their neighboring galaxy distance, is in the range $1 - 8 \times 10^8$ $M_\odot$. Such HI masses, with properties of the above HI clouds, and contained within an antenna beam would easily be detected out to a distance at least four times that of the Sculptor group (generally taken to be 3 kpc). Three HI surveys (Haynes and Roberts 1979; Lo and Sargent 1979; Materne, *et al.* 1979) examining a total of 8 different groups failed to uncover a single new HI companion. The dilemma is startling, almost Ptolemaic in appearance. Our Local Group and the next nearest group show intergalactic hydrogen clouds in relative abundance but eight other nearby groups when examined with adequate sensitivity do not show any such clouds of similar nature (i.e. in size or HI mass). Is some subtle additional factor operating or have we misread the initial observations?

Only with additional data and with hindsight have we been able to resolve the dilemma. More sensitive and more complete surveys for high velocity clouds (Giovanelli 1980; Mirabel and Morras 1984; Hulsbosch 1985) have shown a much larger number of such clouds; Giovanelli (1980) estimates that 10% of the sky is covered by high-velocity clouds. Velocities reaching $\sim$ -500 km s^{-1} have been found by Hulsbosch (1985). Many of these very high velocity clouds fan out from the northern end of the Magellanic Stream where velocities $\sim$ -400 km s^{-1} are already seen. And central to this region are the two prominent Local Group galaxies M31 and M33. Their "companion" HI cloud velocities are $\sim$ -400, and are thus typical of the very high velocity cloud velocities throughout this general area.

In like vein, a deeper and more complete survey of the Sculptor region (Haynes and Roberts 1979), shows HI clouds not only near NGC 55, NGC 300 and the dwarf irregular but throughout the southern region of the Sculptor group. A wide range of positive and negative radial velocities separate these clouds from the conventional values of galactic and Magellanic Stream velocities. The mean value of the positive velocity clouds does not agree with the mean value of the Sculptor group or of any reasonable subset in this group, down to the 3 galaxies noted above. If the anomalous negative velocity clouds are also included in forming the HI mean the comparison is even worse.

Various selection biases have conspired to make it appear that HI clouds are to be found as neighbors of at least some of the nearest of galaxies. It is ironic that much of the confusion here is due to intergalactic hydrogen, but in this instance material so relatively close, $\lesssim$ 50 kpc, that it is to be found over large areas of the sky. If we omit all spatial coincidences of clouds having a heliocentric radial velocity $\leq$ 300km s^{-1} we are left with only one instance of "isolated" HI companions, those in the M96 group and here a tidal origin is possible.

5. THE SEARCH FOR HI CLOUDS IN THE NEAR FIELD

At radio wavelengths, neutral hydrogen is detectable in both emission and absorption. In the latter instance the detectability of a cloud is independent of its distance but does require an adequately strong background source. There have been many experiments with each technique looking for both distributed and clumped material, only the latter is discussed here.

Both types of experiments give null results. Upper limits to absorption detections yield an upper limit to the product ρ A where ρ is the volume density of HI clouds of cross sectional area A. For an optical depth $\tau < 0.03$ and H = 50 km s^{-1} Mpc^{-1}, $\rho A < 2.2 \times 10^{-3}$ Mpc^{-1} (Roberts and Steigerwald 1977). Shostak (1977) evaluates this product graphically for a range of volume densities and masses. These upper limits are not as restrictive as those obtained from emission searches.

The results of five such searches by various groups are summarized in Fisher and Tully (1981). They derive a combined result, summarized in the first two columns of Table 1. For comparison, absolute magnitudes corresponding to M(HI)/L = 0.5, a value typical of late-type galaxies are listed in column 3 and the corresponding luminosity function (number of galaxies / Mpc^3 / luminosity decade) from Kirshner, et al. (1983; also see Felton 1985) in column 4.

Table 1

Upper Limits to the Number of Isolated Hydrogen Clouds

(1)	(2)	(3)	(4)
M (HI)	$\phi[M(\text{HI})]$	M_{pg}	$\phi(M_{pg})$
$10^7 h^{-2}$	$< 5.9\ h^3$	-12.9	0.043
$10^8 h^{-2}$	$< 0.18\ h^3$	-15.4	0.026
$10^9 h^{-2}$	$< 0.006\ h^3$	-17.9	0.016
$10^{10} h^{-2}$	$< 0.002\ h^3$	-20.4	0.0025

(1) Mass of hydrogen clouds, $h = H_o/100$
(2) Number of clouds / Mpc^3 / Mass decade
(3) Absolute magnitude of galaxy with $M(\text{HI})/L_{pb} = 0.5$
(4) Number of galaxies / Mpc^3 / luminosity decade

From Table 1 we see that massive hydrogen clouds, $\sim 10^{10}\ h^{-2}\ M_o$, have an upper limit to their spatial density of 2×10^{-3} Mpc^{-3}, similar to the frequency of galaxies of absolute magnitude -20. For less massive HI clouds the upper limits are well above the actual number density of galaxies. Since the comparison is made between optically luminous matter, galaxies, and by definition non-luminous material, the use of 0.5 for M(HI)/L is rather arbitrary.

In the various searches for isolated hydrogen clouds, positive signals have been detected in a number of instances. In every instance of adequate signal-to-noise ratio a galaxy previously catalogued or one visible on the Palomar Sky Survey atlas has been found at the location of the 21-cm signal.

6. SUMMARY AND CONCLUSIONS

a. Intergalactic hydrogen clouds in the form of tidal bridges and tails are common in galaxy groups. There are more than 20 examples and the frequency is $\sim$ 40% within groups containing at least one late-type system. The morphology of this material is consistent with a tidal origin.

b. All known examples of hydrogen clouds as separate, non-tidal companions disappear if the sample is restricted to clouds with heliocentric radial velocities > 300 km s^{-1}. This limit is chosen to avoid confusion with the Magellanic Stream and with high-velocity clouds. This is not a particularly restrictive condition since hydrogen clouds with properties similar to those of the purported companion clouds would easily be visible to redshifts of order $\sim$ 1200 km s^{-1}. None have been found in this latter, much larger search region.

c. The elliptical ring of clouds in M96 is a critical and unique case of separate hydrogen clouds within a galaxy group. As well as primordial, two possible galaxy-interaction mechanisms have been proposed for their origin. Chemical abundance measurements via a background source could clarify this situation.

d. The extensive hydrogen envelope found surrounding some galaxies is most likely tidal in origin. However, in some instances the perturbing galaxy is not at all obvious and a subset of such envelopes could be primordial hydrogen or due to explosive ejection.

e. Isolated hydrogen clouds have been sought for in many experiments. None have been found and upper limits to the spatial density, as a function of hydrogen mass, are derived. Positive signals are always identifiable with galaxies. Though more stringent upper limits are needed it is possible that the fully gaseous cloud does not exist, that they have either formed stars or dissipated. The latter instance is disappointing and the former puzzling for where are they? If these are the dwarf irregulars why are they only forming stars now and why are they underabundant in the heavier elements? Where are those which are about to turn on? If they are rare, then the upper limits in Table 1 are uninteresting. Or perhaps there are indeed HI clouds near to galaxies, primordial or as explosive ejecta but of a lower mass than considered above. We may have been thrown off the scent by the confusion with foreground clouds.

As can be seen there is still much to be done in this area of research. There are obvious clues but we may not have been asking the right question.

REFERENCES

Arp, H. 1966, Atlas of Peculiar Galaxies (Pasadena: California Institute of Technology).
Briggs, F.H. 1982 Ap.J. **259**, 544.
Briggs, F.H., Wolfe, A.M. Krumm, N., and Salpeter, E.E. 1980, Ap. J., **238**, 510.
Cesarsky, D.A., Falgarone, E.G., and Lequeux, J. 1977 Astron. Astrophys., **59**, L5.
Cohen, R.J. 1982a, Mon. Not. R. Astr. Soc., **199**, 281.
Cohen, R.J. 1982b, Mon. Not. R. Astr. Soc., **200**, 391.
Combes, F., Foy, F.C., Gottesman, S.T., and Weliachew, L. 1980, Astron. Astrophys. **84**, 85.
Cottrell, G.A. 1976, Mon. N. R. Astr. Soc., **174**, 455.
Davies, R.D. 1974, Mon. Not. R. Astr. Soc., **170**, 45.
de Vacouleurs, G. 1975, in Galaxies and the Universe, ed. A. Sandage, M. Sandage and J. Kristian (Chicago: Univ. of Chicago Press), p.557.
Dieter, N.H. 1965, Astron. J., **70**, 552.
Felton, J.E. 1985, Comments Astrophys., **11**, 53.
Fisher, J.R., and Tully, R.B. 1976, Astron. Astrophys., **53**, 397.
Fisher, J.R., and Tully, R.B. 1981, Ap. J. (Letts.), **243**, L23.
Giovanelli, R. 1980, Astron. J., **85**, 1155.
Haynes, M.P. 1979, Astron. J., **84**, 1173.
Haynes, M.P. 1981, Astron. J. **86**, 1126.
Haynes, M.P., Giovanelli, R., and Burkhead, M.S. 1978, Astron. J. **83**, 938.
Haynes, M.P., Giovanelli, R., and Roberts, M.S. 1979, Ap. J., **229**, 83.
Haynes, M.P., and Roberts, M.S. 1979, Ap. J. **227**, 767.
Hewitt, J.N., Haynes, M.P. and Giovanelli 1983, Astron. J., **88**, 272.
Hindman, J.V., Kerr, F.J., and McGee, R.X. 1963, Australian J. Phys. **16**, 570.
Hulsbosch, A.N. 1985 in The Milky Way Galaxy, IAU Symp. No. 106, Ed. H. van Woerden (Dordrecht: D. Reidel), p.409.
Irwin, M.J., Kunkel, W.E. and Demers, S. 1985, Nature, **318**, 160.
Kerr, F.J., and Hindman, J.V. 1953, Astron. J. **58**, 218.
Kerr, F.J., Hindman, J.V., and Robinson, B.J. 1954, Australian J. Phys., **7**, 297.
Kirshner, R.P., Oemler, Jr. A., Schechter, P.L., and Schectman, S.A. 1983, Astron. J., **88**, 1285.
Lo, K.Y., and Sargent, W.L., 1979, Ap. J., **227**, 756.
Materne, J., Huchtmeier, W.K., and Hulsbosch, A.N. 1979, Mon. Not. R. Ast. Soc., **186**, 563.
Mathewson, D.S., Cleary, M.N., and Murray, J.D. 1974, Ap. J., **190**, 291.
Mathewson, D.S., Cleary, M.N., and Murray, J.D. 1975, Ap. J. (Letts.), **195**, L97.
Mathewson, D.S., Schwarz, MP., and Murray, J.D., 1977, Ap. J. (Letts.), **217**, L5.
Mirabel, I.F. 1981, Ap. J., **250**, 528.
Mirabel, I.F., Cohen, R.J., and Davies, R.D. 1979, Mon. Not. R. Ast. Soc., **186**, 433.
Mirabel, I.F., and Morras, R. 1984, Ap. J., **279**, 86.

Murdoch, H.S., Hunstead, R.W., Pettini, M., and Blades, J.C. 1986, Ap. J., **309**, 19.
Roberts, M.S. 1968, Ap. J., **151**, 117.
Roberts, M.S. 1976, Comments on Astrophys., **6**, 105.
Roberts, M.S., and Steigerwald, D.G. 1977, Ap. J.,**217** 883.
Rood, H.J., and Williams, B.A. 1985, Ap. J., **288**, 535.
Rots., A.H. 1978, Astron. J. **83**, 219.
Sandage, A., and Tammann, G.A. 1975, Ap. J., **196**, 313.
Schneider, S.E., Helou, G., Salpeter, E.E., and Terzian, Y. 1983, Ap. J. (Letts.),**273**, L1.
Schneider, S.E., 1985, Ap. J. (Letts.), **288**, L33.
Shostak, G.S. 1977, Astron. Astrophys., **54**, 919.
Sunyaev, R.A. 1969, Astrophys. Letts., **3**, 33.
Toomre, A., and Toomre, J. 1972, Ap. J., **178**, 628.
van der Hulst, J.M. 1978, Bull. Amer. Astron. Soc., **10**, 428.
van der Kruit, P.C. 1987, Astron. Astrophys., **173**, 59.
Wannier, P., and Wrixon, G.T. 1972, Ap. J. (Letts.), **173**, L119.
Wright, M.C. 1974, Astron. Astrophys., **31**, 317.
Zwicky, F. 1956, Erg. d. exakt. Naturwiss, **29**, 344.

DISCUSSION

G. Chincarini: You emphasize the lack (locally) of HI isolated clouds. How is this related to the presence of the Lyα forest observed at high redshift?

M. Roberts: A number density-redshift evolution does appear to be present. It is mild enough that if applicable over the range $z \simeq 3$ to $z = 0$, HI clouds should be present locally. However these clouds would be of very low surface density, orders of magnitude below the detection limits attainable by current radio techniques.

H. Arp: I would emphasize that the optical appearance of NGC 2146 is very disturbed-I would say explosive in appearance. The observed extremities of HI would not have had time to arrive so far from the center if the presently observed optical disturbance was caused by a currently occuring collision.

M. Roberts: I agree that the optical appearance of NGC 2146 is indeed very peculiar. As to the time scale for "spreading" the hydrogen, clearly it cannot be a recent event by any of the mechanisms proposed by Fisher & Tully.

HIGH VELOCITY CLOUDS IN M101

Renzo Sancisi
Kapteyn Astronomical Institute

Thijs van der Hulst
Netherlands Foundation for Radio Astronomy: Dwingeloo

With the exception of the HVCs of our galaxy no other evidence has been found in the past for z motions of the gas in spiral galaxies. The HI line studies of face-on galaxies (van der Kruit and Shostak 1984) have served to determine the z-velocity dispersion, but have not revealed any peculiar vertical motions.

21 cm line observations of the ScI galaxy M 101 with the Westerbork Synthesis Radio Telescope show now for the first time the presence of gas moving at high velocity in the direction perpendicular to the disk (van der Hulst and Sancisi 1987). These observations reveal two large gas complexes at high velocity, one 10 arcmin (=20 kpc, assuming a distance of 7.2 Mpc to M 101) to the north-east and the other 8 arcmin (=16 kpc) to the south-east of the centre of M 101. The two complexes are clearly visible in Figure 1, which displays the distribution of HI brightness temperature as a function of position and radial velocity. Feature A is a huge, loop-like complex about 8 arcmin in size (16 kpc) and extending in velocity up to about 130 km/s from the rotating, "local" HI layer to which it is connected. In the direction of this high velocity complex, the HI disk shows a hole in its density distribution, a bifurcation and a kink in the spiral arms. The mass of the whole HI complex taking part in this peculiar motion is 1.6 x $10^8 M_\odot$. The other complex (B) has a smaller angular size ($\sim$2.5 arcmin = 5 kpc), but reaches higher velocities, up to about 160 km/sec from the rotating HI. It is also connected with the gas in the disk. Its total HI mass is about 2 x $10^7 M_\odot$. These two are the only features with such large velocity deviations from rotation.

The origin of these two gas complexes is not clear. Unfortunately it is not known whether the observed motions are infall of gas towards the plane on the near side of the galaxy or gas outflow on the far side. Any explanation, however, must account for: i) the enormous energies involved which are about 10^{55} ergs, ii) the observed space-velocity structure, in particular the connection with the rotating HI, and iii) the corresponding density and kinematical discontinuities in the HI disk and the holes.

A **tidal** origin, as in the case of the bridges and tails often seen in interacting systems, seems unlikely in the present case mainly because it cannot easily explain the connection of the high velocity gas with the "local"HI disk.

Supernova explosions as advocated in the past to explain the high velocity clouds and the HI supershells in our galaxy (cf. e.g. Heiles 1984), are a possibility, but 10^4 such events would be needed for each HI complex. Furthermore, there is no evidence of violent star formation or other unusual activity taking place in the disk at the regions close to the observed high velocity features.

Cloud-galaxy collisions, as discussed by Tenorio-Tagle et al. (1987) for the HVCs and the Milky Way in relation to Heiles' supershells, could explain the present observations. Infalling gas clouds could create holes in the gaseous layer and also build up large HI structures with high velocities. Such intergalactic clouds could be tidal debris from past interactions of M 101 with one or more companions, for example with NGC 5474. It is not clear whether these large HI complexes in M 101 are indeed a population similar to the HVCs of our galaxy. They are probably much more massive than the largest HVC complexes, unless these are further than 10 kpc from the Sun. There is, however, the possibility that the analogues to the HVCs of our galaxy are not the huge gas complexes just discovered, but the same gas which will undoubtedly spread out in smaller clouds and eventually rain down towards the disk of M 101.

REFERENCES

Heiles, C. (1984), *Astrophys.J.Suppl.*, **55**, 585.
van der Hulst, J.M. and Sancisi, R. (1987), preprint.
van der Kruit, P.C. and Shostak, G.S. (1984), *Astron. Astrophys.*, **134**, 258.
Tenorio-Tagle, C., Franco, J., Bodenheimer, P. and Rozyczka, M. (1987), *Astron.Astrophys.*, **179**, 219.

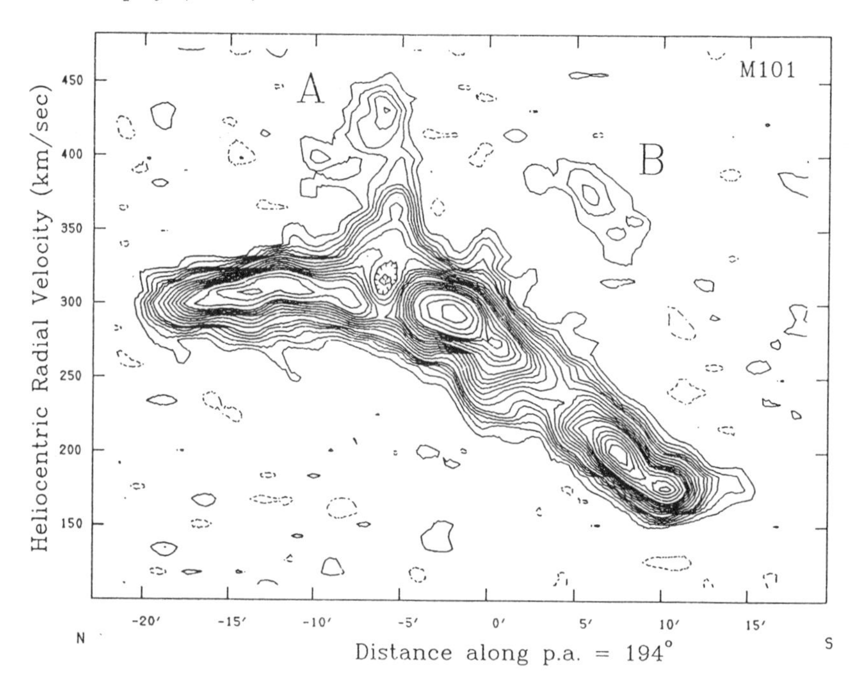

Figure 1. Map showing the distribution of HI emission in M 101 as a function of position and radial velocity at position angle 194° and centered at α= 14^h 02^m 11.56^s, δ=54° 36' 04.3"(1950). The contour levels are -5, -2.5 (dashed), 2.5, 5, 7.5, 10, 12.5, 15, 20, 25, 30, 35, 40, 45, 50, 60, 70, 80, 90, 100, 125, 150 and 175 mJy/beam area (HPBW = 80" x 80", 1 mJy/beam area = 0.1 K). The bulk of the HI emission originates from the differentially rotating disk of M 101. The high-velocity gas complexes labelled A and B are redshifted with respect to that disk.

DISCUSSION

W. Tifft: Is the HI mass in the "clouds" consistent with the HI mass missing in the "holes" in the disk?

R. Sancisi: A crude estimate of the missing gas in the holes shows that the high-velocity gas would indeed be sufficient to fill the observed holes in the disk.

H. van Woerden: Could you indicate how your observations might clarify the nature and origin of high-velocity clouds in our galaxy? Do you get a clue on the distance?

R. Sancisi: One of the striking properties of the high-velocity gas complexes in M101 is their clear connection with the "local" rotating disk. If one believes that the HVC's in our galaxy are indeed their analogues then the obvious thing to do would be to look for density-velocity connections between the HVC complexes and the HI in the rotating disk of our galaxy. This would also lead naturally, to a distance determination.

J. C. Pecker: Would you say that the existence of large velocity HI clouds (existing, I presume, in every "normal" galaxy) does affect at least the dispersion of the Tully-Fisher relation?

R. Sancisi: If such high-velocity gas clouds, as discovered in M101, were indeed present in every spiral galaxy, they would represent a negligible addition to the HI global profiles (a low-level wing as in M101) and would not affect, therefore, the determination of the profile half-intensity (or even 20%) width.

H. Arp: That is a very exciting observation! The most significant aspect to me is that HI discontinuities occur at the ends of each major hydrogen spiral arm. The most crucial question I can ask you is why the redshift excursion of the HI is predominantly positive in both A and B.

R. Sancisi: The possible origins I have discussed do not explain why both A and B are redshifted with respect to the "local" rotating HI disk. I am not particularly worried about it, since there are only two of them, but maybe I do not fully appreciate the significance of the facts you are pointing out.

P. Shaver: Do you see any high-velocity gas associated with the known holes in the HI disk.

R. Sancisi: The complexes A and B are the only high-velocity gas detected so far in M101. We do see, however, some HI at intermediate velocities ($\lesssim$50 km/s) at various positions. It is not clear yet whether such HI is associated with the holes seen in the Allen and Goss map and it is not even sure that those intensity minima are real holes in the HI disk. New 21cm observations of higher sensitivity and angular resolution, which are now being carried out at Westerbork, should provide an answer to such questions.

INFRARED MEASUREMENTS OF INTERACTING GALAXIES

Ramon D. Wolstencroft
Royal Observatory, Edinburgh, EH9 3HJ, Scotland

ABSTRACT

Transfer of matter within and between galaxies can occur during close encounters of galaxies. Collisions between infalling and ambient molecular clouds precipitates star formation, especially in the central regions of galaxies, and these effects can be observed in the infrared. Infrared mapping of galaxies is particularly useful in studies of the distribution and development of star formation in individual galaxies: we discuss two examples, NGC 4038/9, a classical interacting system, and M83, a starburst galaxy. The central theme of this paper is the statistics of infrared enhancement of interacting galaxies. A sample of 203 interacting galaxy pairs with IRAS counterparts taken from the Arp-Madore Catalogue has been examined by Telesco, Wolstencroft and Done (1987). For interacting pairs in which the two galaxies are of comparable size the most intense star formation, as measured by R, the 60 to 100 μm colour temperature, is greatest for pairs with the smallest separation and the highest interaction strength Q. There are about ten times as many pairs containing one IRAS galaxy as there are pairs for which both pairs were detected by IRAS: this supports earlier conclusions that the conditions for maximum tidal disruption are much less likely to hold for both galaxies because of the strong dependence on interaction geometry. For pairs with a large and a small galaxy it is always the larger galaxy that is detected by IRAS.

1. INTRODUCTION

During close encounters galaxies undergo tidal disturbances which can lead to matter being transferred both between galaxies and from the outer to the inner regions of a galaxy on plunging orbits. The visible consequences of these interactions are well documented in Arp's (1966) 'Atlas of Peculiar Galaxies': they consist of bridges, tails and other distortions. These encounters can have a spectacular effect on the level of star formation which in many cases leads to powerful infrared emission from so-called starburst galaxies (Lonsdale, Persson and Matthews, 1984; Cutri and McAlary, 1985). Collisions between molecular clouds are suspected to be the trigger that precipitates the formation of massive stars in these galaxies (Scoville, Sanders and Clemens, 1987; Scoville and Good, 1987), and thermal emission from dust heated by these young stars is believed to produce the majority of the far infrared emission. Interactions also lead to enhancements of Seyfert activity (Dahari, 1984; Keel et al., 1985) and may play a role in triggering quasars (Hutchings and Campbell, 1983; Stockton and Mackenty 1983; Mackenty and Stockton, 1984; Shara, Moffet and Albrecht, 1985). Sanders et al (1988) have suggested that the ultraluminous IRAS galaxies, which are believed to be primarily the products of interactions and ultimately of mergers, may represent the initial dust embedded stages of quasars; a similar idea was also considered earlier by Roos (1984). The intensity of the interaction between galaxies depends sensitively on the geometry and orbital parameters of the encounter and on the mass ratio of the two galaxies. Toomre and Toomre (1972), in a classical paper, showed, for example that a companion galaxy moving on a parabolic orbit in the plane of the victim disk galaxy and in the direction of the victim's rotation (direct passage) is much more disruptive than is the case of retrograde passage, because the victim feels the tidal force for a much longer period.

Although the models of Toomre and Toomre (1972), and of later authors such as Miller and Smith (1980) and Negroponte and White (1983), provide convincing reconstructions of the optical appearance of some well known interacting galaxies, these simulations describe the stellar component only and do not predict the spatial distribution of regions of newly formed stars. Recently progress has been made in this direction by Noguchi and Ishibashi (1986) who have carried out numerical calculations of galaxy encounters: they use cloud particles in the victim disk galaxy, which become stars when the cloud particles collide. These models show that high star formation rates can occur in the outer parts of the victim disk. When the self gravity of the disk is allowed for (Byrd et al. 1986; Byrd, Sundelius and Valtonen, 1987) it is also evident that substantial amounts of disk gas can be transferred into the central regions of the galaxies, where enhanced star formation is likely to occur. It is clear that the determination of the spatial distribution of the regions of star formation in interacting galaxies by infrared mapping can help to provide a more complete understanding of how gas is both transferred and compressed to form extended regions of massive young stars.

The application of infrared mapping to galactic star formation is still in its infancy and very few galaxies, either interacting or isolated, have yet been mapped. In section III of this paper we shall discuss two examples: firstly NGC 4038/9 (the 'Antennae'), a classical interacting pair, which has been mapped in the near infrared by Wright and McLean (1987) and by Joy (1986); and secondly M83, a barred spiral galaxy, with vigorous star formation at its centre, which has been mapped at $10\mu m$ and $20\mu m$ by Telesco et al (1987). The main emphasis in this paper (section II) will be on the IRAS measurements of a large sample of interacting galaxies selected from the recently published 'Arp–Madore (1987) Catalogue of Southern Peculiar Galaxies and Associations' (hereafter the AMC). This is a southern and more comprehensive version (6446 entries) of Arp's (1966) northern 'Atlas of Peculiar Galaxies' (333 entries).

2. INTERACTING GALAXIES FROM THE ARP–MADORE CATALOGUE

Arp and Madore (AM) inspected all extended objects with diameters $\gtrsim 10$ arc sec and $\delta \leqslant -22.5^{\circ}$ on the IIIa J plates of the SERC Southern Sky Survey. Out of 77,838 objects inspected, 6446 (8%) were included in the AMC. The objects were assigned to one or more of 25 categories (see Table 1) of which 22 describe galaxies (3 categories contain objects other than galaxies namely categories 21, 22 and 25).

Wolstencroft and Done (1987) identified the IRAS counterparts of the AM objects by searching for IRAS sources lying within 90 arc sec of each AM object position. They found that 1513 out of 6446 (23%) have IRAS counterparts. 4398 of the 6446 objects are classified into only one category and thus represent the most unambiguous classifications. Some useful statistics on the infrared properties of these single category objects are listed in Table 1. The quantities tabulated are the average values of the fraction, f, of objects detected by IRAS; the mean temperature, as measured by R, the ratio of 60 to $100\mu m$ flux densities; and g, the fraction of 'hot' galaxies. It is of interest to compare the two categories of interacting galaxy to be discussed later in this section, namely 1 (galaxy with a small interacting companion) and 2 (interacting double – diameter of companion at least half that of the main galaxy), with category 8 (galaxy with an

TABLE 1

STATISTICS OF ARP-MADORE GALAXIES WITH IRAS COUNTERPARTS

(Single Category Objects Only)

	Category	N_1	N_2	$f = \frac{N_2}{N_1}$	N_3	$g = \frac{N_3}{N_2}$	$< \frac{F_{60}}{F_{100}} >$	Comments
1	Int'g Pairs(one small)	224	70	.313	59	.843	.449	
2	" " (~same size)	636	133	.209	102	.767	.521	Hot
3	Int'g Triples	120	19	.158	16	.842	.512	Hot
4	Int'g Quartets	19	5	.263	4	.800	.491	
5	Int'g Quintets	6	2	.333	1	.500	-	
6	Ring Galaxies	77	11	.143	9	.818	.441	
7	Gals. with Jets	91	19	.209	14	.737	.542	Hot
8	Apparent Companions	373	137	.367	107	.781	.371	Cold,High *f*
9	M51 Companions	88	18	.205	15	.833	.515	Hot
10	Peculiar Sp. Arms	116	45	.388	39	.867	.390	Cold,High *f*,*g*
11	3 armed Spirals	18	7	.389	7	1.000	.406	Cold,High *f*
12	Peculiar Disks	61	24	.393	19	.792	.444	High *f*
13	Compact Gals.	255	38	.149	27	.711	.491	
14	Prominent Absorption	38	18	.474	16	.889	.422	Highest*f*,high *g*
15	Tails or Loops	86	21	.244	20	.952	.644	v.hot,highest *g*
16	Disturbed but isolated	143	46	.322	39	.848	.494	
17	Chains (>4 aligned)	174	17	.098	11	.647	.452	
18	Groups (>4 unaligned)	198	21	.106	14	.667	.414	Low *f*
19	Rich Clusters	110	1	.009	1	1.000	-	Low *f*
20	Dwarf Galaxies	456	15	.033	11	.733	.427	Low *f*
*21	Stars with Nebulosity	57	15	.263	10	.667	.446	Low *f*
*22	Miscellaneous	63	11	.175	7	.636	.370	
23	Close Pairs (not int'g)	618	105	.170	76	.724	.387	Cold
24	Close Triples (not int'g)	301	46	.153	34	.739	.429	
*25	Plan. Nebulae	70	44	.629	26	.591	.980	
	Total	4398	888	.202	684	.770	.468 (.449)←galaxies only)	

N_1 = Number of single category AM objects in that category (with or without IRAS counterparts)
N_2 = Number of N_1 objects detected by IRAS
N_3 = Number of N_2 objects detected by IRAS at <u>both</u> 60μm and 100μm
f = N_2/N_1 = Far infrared red detectability of a given category
g = N_3/N_2 = Fraction of "hot galaxies"
* → not galaxies

apparent companion). Category 8, for which there is no obvious visible evidence of interaction contains the coldest galaxies (R=0.37) but shows a high detection rate (f=0.37). Category 2, which like category 1, shows clear visible evidence of a physical encounter, is much hotter (R=0.52) but has only an average detection rate (f=0.21). Category 1 is an intermediate case with R=0.45 and f=0.31. It is interesting to note that this trend of decreasing f with increasing R holds for those 13 categories with above average values of f; the hottest category is 15 (galaxy with tails and loops) for which R=0.64 and f=0.24. However, selection effects are clearly at work here.

The AM Catalogue provides valuable samples of well defined classes of interacting galaxies, which allow the relations between pair separation, S, temperature (measured by R), interaction strength, Q, interaction geometry and morphology to be quantified. In a first approach Telesco, Wolstencroft and Done (1987) examined the relations between R, S and Q for samples taken from AM categories 1 and 2: their results are described in this section. These two categories contain only pairs of galaxies in which both members of the pair are distinguishable. Since redshifts were not available for many of these galaxies the ratio S/D_p was used as a measure of physical pair separation, where S and D_p are respectively the angular separation of the pair and the angular diameter (major axis) of the largest of the two galaxies. The strength, Q, of the interaction between the galaxies is, following Dahari (1984), taken to be the tidal force acting on unit mass of the primary which is produced by the companion of mass M_c at distance L from the primary, viz $Q \propto M_c/L^3$. Using the relation between galaxy mass and diameter found by Rubin et al. (1982), Dahari showed that $Q=k\ (D_cD_p)^{1.5}/S^3$, where D_c is the angular diameter of the companion and k is the constant of proportionality, taken by Dahari to be unity. In numerical simulations Byrd et al. (1986) and Byrd, Sundelius and Valtonen (1987) have shown that significant inflow of disk gas into the central regions of the perturbed galaxy can occur provided Q exceeds the critical value $Q_c \approx 0.05$ (this applies strictly for the case of direct passage of the companion past a halo free galaxy): the inflow rate for Q=0.05 is $\sim 0.2 M_\odot\ yr^{-1}$ averaged over about 6×10^8yr. If the perturbed galaxy has a significant halo, which stabilizes the disk, the critical value at which inflow begins is higher. Although this average inflow rate is lower than the star formation rate of a few $M_\odot\ yr^{-1}$ and greater deduced for many interacting/starburst galaxies, these simulations imply periods $\sim 10^8$yr for which the inflow rate can be one to two orders of magnitude higher than average.

For pairs of comparably sized galaxies (AM category 2) the sample was divided into a hot set ($R \geqslant 0.50$) and a cold set ($R<0.50$), which contained 49 and 44 pairs respectively. By dividing the sample into close or distant pairs according to whether S/D_p <1 or >1, it was clear that pairs with small separations are more common in the hot sample than in the cold sample: 23 of 49 (47%) of the hot pairs are close and 11 of 44 (25%) of the cold pairs are close. A chi-squared test indicated that this difference between the hot and cold distributions is significant at the 97.2% probability level. If $R>0.60$ is used to define the hot pairs the difference is significant at the 99.8% probability level. The true difference between the hot and cold pairs will be larger when projection effects are taken into account. Unfavourable geometry of the interaction, as well as asymmetry of the infrared emission relative to the epoch of closest approach, may also mask or dilute the effect. It therefore can be concluded that interacting galaxies of comparable diameters exhibit the most intense star formation, as manifested by the colour temperature R, when their separations are comparable to or closer than their diameters.

Telesco, Wolstencroft and Done (1987) also examined the dependence of colour temperature, R, on interaction strength, Q, for AM Categories 1 and 2. This approach should in principle be more 'physical' than using separation alone. They found that 21 out of 26 (81%) of the hottest pairs ($R \geqslant 0.60$) have $Q \geqslant 0.05$ compared to 27 out of 44 (61%) of the cold pairs ($R < 0.50$), a result significant only at the 90.8% level (χ^2=2.9). However for higher interaction strengths, there is a marked contrast: 17 out of 26 (65%) of the hottest pairs have $Q \geqslant 0.30$ compared to only 14 out of 44 (32%) of the cold pairs. Thus the hottest galaxy pairs are twice as likely as cold pairs to exceed $Q>0.30$, a result significant at the 99.3% level.

When galaxies with small companions (AM category 1) were considered the results were different. No statistically significant difference between the hot and cold pairs was found either in terms of S/D_p or Q. However it was found that the histogram of S/D_p for the whole category 1 sample was very similar to that of the category 2 hot pairs

(R>0.50). This could indicate that the larger category 2 companions are more vulnerable to obvious tidal disruption than are the smaller category 1 companions. If so category 1 pairs which have larger values of S/D_p, and hence are probably cold pairs, may show little evidence of tidal effects and be preferentially classified by AM as category 8, galaxies with apparent companions; we note from Table 1 that these galaxies are exceptionally cool. Since there is a continuity between classes 1 and 2, the dividing line being when $D_c=0.5\ D_p$, it would be of interest to examine the dependence of the (R, S/D_p) and (R,Q) correlations on D_c/D_p for the combined sample (203 IRAS sources).

In a study of the near-infrared colour excess (K–L) of 22 interacting galaxies selected from Arp's (1966) 'Atlas of Peculiar Galaxies', Joseph et al. (1984) found that of the 18 pairs showing a measurable excess, none showed an excess in both galaxies: their interpretation was that the conditions for maximum tidal disruption are much less likely to hold for both rather than just one of the galaxies. Telesco, Wolstencroft and Done (1987) examined their sample of 93 category 2 pairs to see whether this result holds for the IRAS emission. For 42 of the 93 pairs, either both galaxies were contained within the IRAS positional error ellipse (38) or the ellipse was displaced from both galaxies (4). Of the remaining 51 usable pairs, just one galaxy was the IRAS source in 47 pairs (92%) and both galaxies were IRAS sources in 4 pairs (8%). This result is influenced by the fact that the 42 excluded pairs have smaller separations ($S/D_p = 1.0 \pm 0.2$) than the 51 usable pairs (2.6 ± 0.4). Since the interaction is stronger for smaller separations the chance of both galaxies being IRAS sources should be either greater or no worse for the excluded pairs than for the usable pairs. This provisionally suggests that both galaxies are IRAS sources in at least 8% of all pairs. A similar result was found for category 1 pairs: 43 out of 57 pairs were usable and of these one galaxy was an IRAS source in 41 cases (95%) and both galaxies were IRAS sources in 2 cases (5%). In all 41 cases the larger galaxy was the IRAS source. Considering the small sample size (18) of Joseph et al. (1984) their results and those of Telesco, Wolstencroft and Done (1987) are consistent.

3. TWO CASE STUDIES

A The Antennae (NGC 4038/9)

This classical example of an interacting galaxy system has very long tails (see figure 1 of Schweizer, 1978): the projected distance between the ends of the two tails is 19 arcmin or 83Kpc at the assumed distance of 15Mpc. Toomre and Toomre (1972) satisfactorily modelled this interaction as two disk galaxies of equal mass which approached on a direct elliptical orbit (e = 0.5): the galaxies were at their closest (25 Kpc) about 7 x 10^8 yr ago. Matter was pulled from the outer parts of the disks to form the tails which are observed to contain both stars (Schweizer, 1978) and neutral hydrogen (van der Hulst, 1979). The system is viewed nearly edge on to the orbital plane hence giving the unusual appearance of crossed tails.

The recent history of star formation in this system is undoubtedly complex but we can begin to understand what has happened by studying maps obtained in the radio continuum by Hummel and van der Hulst (1986) and in the near infrared by Joy (1986) and Wright and McLean (1987). Before discussing these maps it is worth noting that the global IRAS colours ($\log (F_{100}/F_{60}) = 0.28$, $\log (F_{60}/F_{25}) = 0.99$) of NGC 4038/9 imply that the star formation in this system is less than in any of the 29 out of our sample of 93 AM category 2 interacting galaxies which have good quality fluxes at 25 μm, 60 μm and 100μm. This is based on the analysis by Rowan-Robinson (1987) of IRAS galaxy colours in terms of three pure components – cold disk, starburst and Seyfert. This conclusion is corroborated by the CO measurements of Young et al (1986) who find that NGC 4038/9 has the smallest value of $L_{IR}/M(H_2)$ = 8.6 $L_\odot/M_\odot$ of their sample of 13 interacting/merging galaxies (mean ratio = 78 $L_\odot/M_\odot$) and in fact the value is more typical of isolated galaxies: Young et al regard $L_{IR}/M(H_2)$ as a measure of star formation efficiency. In the simulations of close galaxy encounters by Noguchi and Ishibashi (1986) the maximum rate of star formation occurs typically 3 x 10^8 yr after the epoch of closest approach, so that at 7 x 10^8 yr after this epoch a relatively low star formation rate is not unreasonable.

The 6cm and 20cm radio maps obtained at the VLA by Hummel and van der Hulst (1986) show 13 radio knots together with a smooth underlying component. The peak of the latter component, which has a steep radio spectrum, is centred close to a region of strong dust absorption between the two galaxies. The position of the radio knots and the optical knots are superposed on a deep photograph of the two galaxies in figure 1. The majority of the radio knots coincide with optical knots seen in Hα and have spectral indices between −0.3 and −0.6; assuming the non-thermal component has a spectral index of −0.7 a thermal contribution can be deduced which ranges between 10 and 60% (average = 20%). Two of the knots can be identified as the nuclei of the northern (NGC 4038, knot 7) and southern (NGC 4039, knot 1) galaxies. Hummel and van der Hulst (1986) argue that the half life of the relativistic electrons in the two nuclei is ~ 5 x 10^6 yr, implying that stars must have formed there recently. However, the near infrared colours of the nuclei measured by Joseph et al (1984) and by Joy (1986) are consistent with emission from evolved stars. For example for the nucleus of NGC 4038 the observed colours are

(6 arcsec)	J−H = 0.87, H−K = 0.31, K−L = 0.20±0.12	(Joy)
(12 arcsec)	J−H = 0.79, H−K = 0.34, K−L = 0.30±0.05	(Joseph et al)

Assuming that the observed K−L is the sum of a stellar contribution with K−L = 0.30 and a thermal component of temperature 300K (see figure 1 of Joseph et al, 1984) we conclude that the maximum thermal contribution at L (3.5μm) is about 5%, and that the current level of star formation is minimal. The K−L colours of the nucleus of NGC 4039 are discordant (−0.04±0.12, Joy; and +0.34±0.06, Joseph et al) but in either case the thermal contribution is small. Joy has also measured K−L for 4 other knots: however, because of the large errors in K−L (±0.36) these yield mostly upper limits to f, the thermal contribution, viz

Knot	4(C/D)	11(S)	12(G)	E	4038N	4039N
f(%)	48±19	<42	<21	<19	<5	<9

The letter designation of the knots is from Rubin et al (1970). With the possible exception of knot 4 it appears that the star formation in these knots is relatively weak.

The complex morphology of the two galaxies can be judged from figure 1. Several of the radio and optical knots in NGC 4038 lie along an arc about 3Kpc in diameter which appears to originate in the nucleus and to spiral counterclockwise along the western boundary of the galaxy at least to knot 13. This feature may even continue as far as knot 6. The near infrared emission for the most part delineates the emission from evolved stars: this conclusion is based on JHKL colours of the knots and of parts of the extended emission obtained and interpreted by Joy (1986). Wright and McLean (1987) have published an excellent H(1.65μm) image of the Antennae (their figure 1) which they

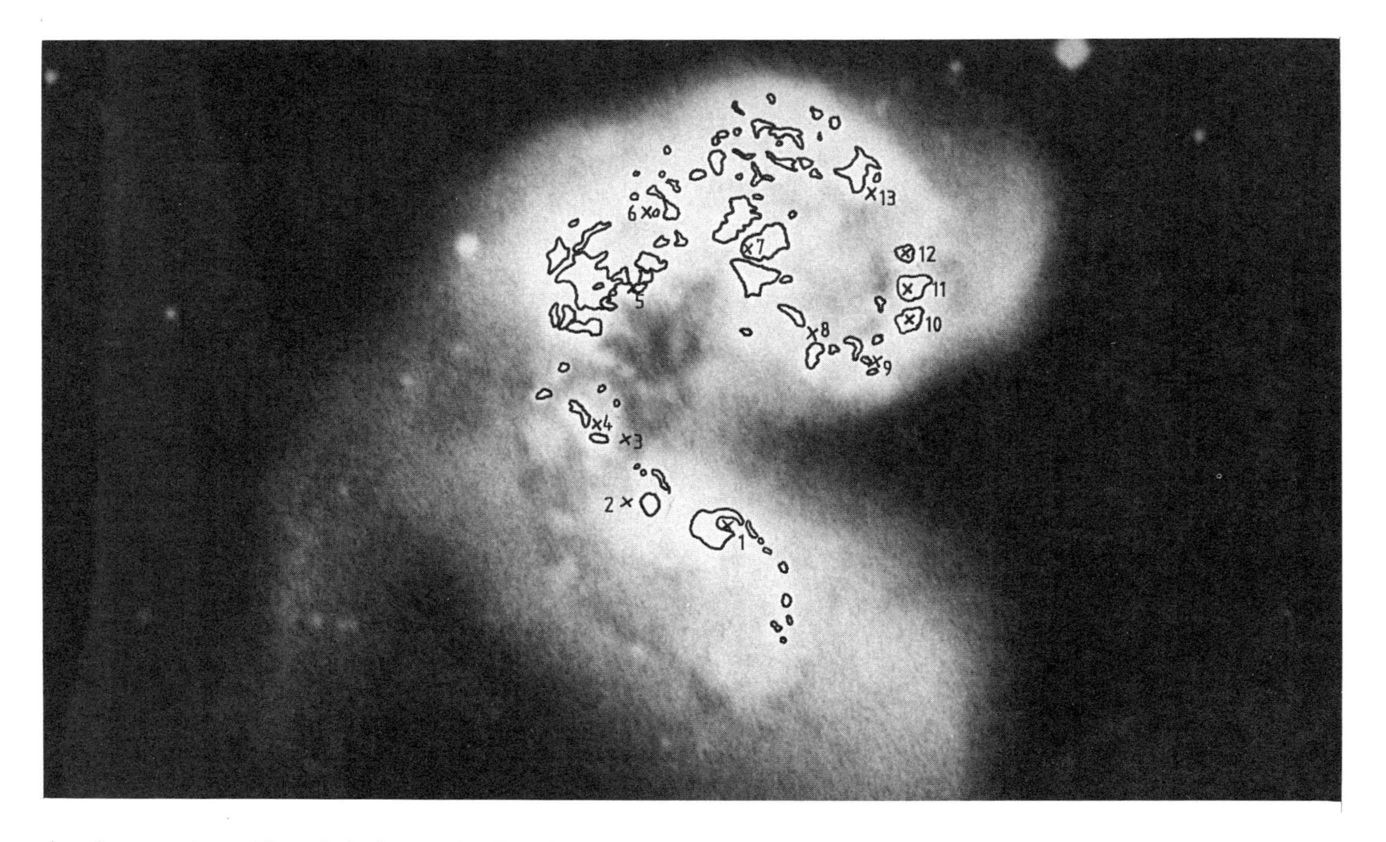

Figure 1   Superposed on this optical photograph of NGC 4038/9, taken on IIIaJ emulsion with the UK Schmidt Telescope, are the outlines of the brightest optical knots and the positions of the 13 radio knots (marked with an X) taken from figure 3 of Hummel and van der Hulst (1986). North is at the top and east is to the left. The nuclei of the two galaxies coincide with knot 7 (NGC 4038) and knot 1 (NGC 4039).

obtained with the two dimensional infrared camera, IRCAM, on the UK Infrared Telescope. This shows that the brightest region in NGC 4038 covers a circular area centred on the nucleus together with a tail that coincides in position with the line of optical knots to the SE, including knot 5: while the interaction $\sim 7 \times 10^8$ yr ago is believed to have stripped stars from the outer parts of the disk to form the tail, it would also have disturbed the inner parts of the galaxy and it is tempting to think of this tail of old stars, which is directed towards the tidal tail, as a relic of that interaction. Wright and McLean (1987) however describe this feature as bar-like and suggest that the non-axisymmetric gravitational potential associated with it might be responsible for driving gaseous material towards the nucleus. The knots in NGC 4039 are confined close to the northern boundary of the galaxy, as judged from the optical photograph, in an essentially linear arrangement but with a spiral tail on the SW side which can also be seen in the optical photograph. The brightest emission at $1.65\mu m$ in the image of Wright and McLean is elongated along the line of knots and contains knot 2 at its NE corner. Rubin et al (1970) suggested, on the basis of the progressive velocity gradient of 235 kms^{-1} between knots 1 and 4, that these knots define the major axis of the galaxy with the NE side approaching the observer. If so it is not clear what produces the extensive optical emission to the south. Judging from the optical image there is variable extinction across both galaxies which complicates the interpretation; and Wright and McLean (1987) note that the extinction is particularly strong in NGC 4039. Further progress in understanding the history and current pattern of low level star formation in this system will benefit greatly from imaging in all near infrared passbands especially in L and K-L (such work is planned by Wright and McLean); and from detailed numerical simulations of this system which should incorporate star formation as described by Noguchi and Ishibashi (1986).

B M83

A variety of observations indicate that the central region (~ 500pc diameter) of this barred spiral galaxy is a site of active star formation. These observations include strong and extended emission in the optical and ultraviolet continuum (Bohlin et al, 1983), at 10 μm (Rieke 1976) and in the Brackett lines (Turner, Ho and Beck, 1987). From the $158\mu m$ [CII] and 2.6mm CO emission (Crawford et al, 1985; Sofue et al, 1987) it is clear that the central molecular density is high. Enhanced star formation in the central regions of barred spiral galaxies may be very common (Hawarden et al, 1986; Puxley et al, 1987) and numerical models (Tubbs, 1982; Schwarz, 1984; Combes and Gerin, 1985) suggest that inflow of gas and dust along the bar towards the nucleus may occur. Specifically orbit crowding in the non-axisymmetric bar potential may lead to collisions between molecular clouds and hence to shocks that coincide with the offset dust lanes in the bar; the loss of angular momentum at each shock allows the gradual infall of matter to the inner Lindblad resonance, where it may feed a circumnuclear torus or disk. This infall may be considerably enhanced by the presence of a bound companion or during an encounter.

Is this scenario consistent with the observations of M83? Offset dust lanes are visible in the bar and they coincide with non-thermal emission seen at 6cm and 20cm by Ondrechen (1985). The 6cm emission defines two narrow linear features which terminate close to the nuclear region and are offset by 32 arcsec (590pc at the assumed distance of 3.7 Mpc). The optical morphology of the inner kiloparsec is complex. Turner, Ho and Beck (1987) show V and Hα images in which a dark lane of absorption crosses this region at a position angle of about 15°: based on Brackett line ratios they conclude that A_V lies in the range 15 to 30 magnitudes. The position of this dark lane is marked on the $10.8\mu m$ map (figure 2(a)) obtained by Telesco et al (1987) using the 20-pixel bolometer

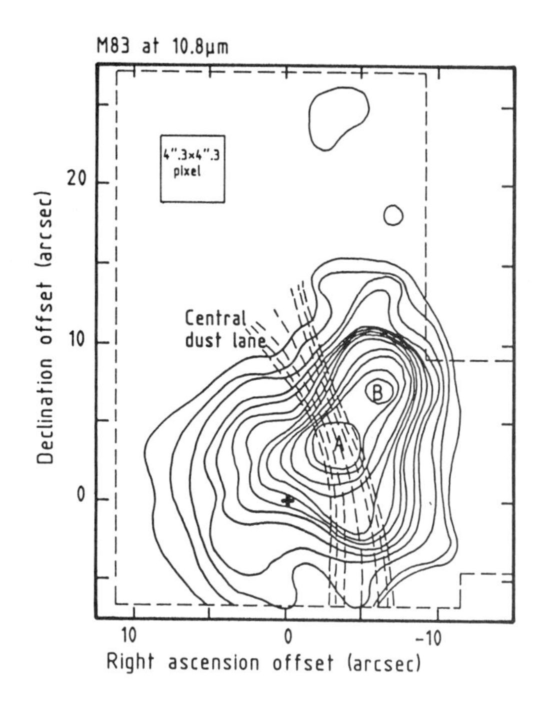

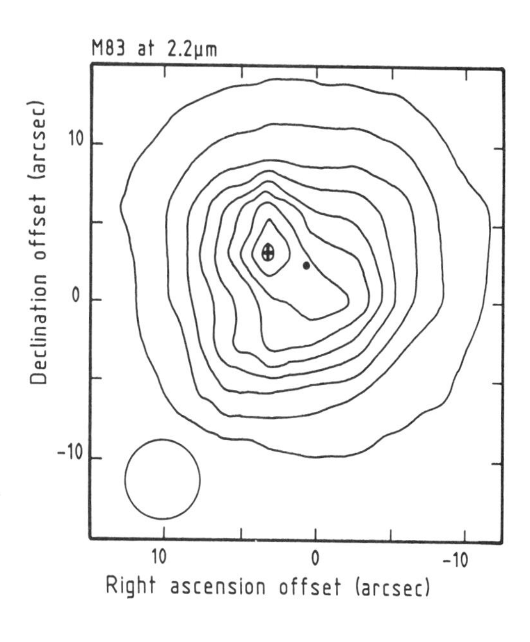

Figure 2(a) (above). Fully sampled map of M83 obtained with the 20-pixel bolometer (developed at the NASA Marshall Space Flight Center) at 10.8μm ($\Delta\lambda$ = 5.2μm) by Telesco et al (1987): this map (4.3 x 4.3 arcsec pixels) incorporates additional data not shown in their figure 1. The area mapped is defined by the dashed lines and extends to $\Delta\alpha$ = −30 arcsec. The infrared bar containing sources A and B is about 10 arcsec (185pc) long. The highest contour shown is 425 mJy per pixel. Some contours are not shown because of the steep brightness gradient north of the bar. The plus sign indicates the position of the peak of optical brightness.

Figure 2(b) (below). Map of M83 at 2.2μm obtained by raster scanning a single element detector (5 arcsec aperture) (Wolstencroft and Davies, 1987). As in figure 2(a) the plus sign indicates the position of the optical peak: the optical and 2.2μm peaks coincide to within an uncertainty of less than 2 arcsec. The circle (with dot) marks the centre of the approximately circular outer contours indicating what is probably the centre of the distribution of evolved stars.

array developed at the NASA Marshall Space Flight Center: the pixel size is 4.3x4.3 arcsec (square) and the map was fully sampled at one half pixel intervals in both coordinates. The dark lane passes through the south eastern (source A) of two sources located in a bar shaped feature ~ 10 arcsec (185pc) in length. The infrared bar with a position angle of 140° is close to being orthogonal to the optical bar (position angle of 60°). If matter from the optical bar is feeding the nuclear region it is unclear whether inflow occurs along the dust lane or along the infrared bar. The total 10.8μm flux density over this inner 500pc is 3.3±0.2 Jy, which is 70% of the global 12μm IRAS flux, with the corresponding total luminosity being ~ 1 x 10^{10} $L_{\odot}$ of which A and B each contribute ~ 1 x 10^{9} $L_{\odot}$.

The 2.2μm map shown in figure 2b was obtained by Wolstencroft and Davies (1987) using a single element detector on the UK Infrared Telescope (5 arcsec aperture). The peak in this map coincides with the peak of optical brightness and the tail is parallel to the close 10μm contours to the SW; there is also the suspicion of a component that follows these contours to the north. Assuming that the outermost contours of the 2.2μm emission are principally measuring the spatial distribution of late type stars the centre of this distribution, and presumably the centre of mass, is 3 arcsec from the 2.2μm peak (at position angle 255°). The distribution of 20cm radio continuum emission obtained by Cowan and Branch (1982) at 1x2 arc resolution is quite similar to the 10.8μm map, showing both the bar and the two sources A and B in the bar, but with the extension to the east of source A being more extended and showing additional structure. However, while the two sources are of comparable flux density at 10μm, source B is very much stronger at 20cm. This may indicate (Telesco et al, 1988, in preparation) that the star formation history of A and B is quite different: for example source A may contain very young massive stars, few of which have yet evolved to the supernova stage, whereas source B may be an older region of star formation where many supernova remnants are contributing to the total radio emission. Some support for this interpretation of source A comes from the analysis of Turner, Ho and Beck (1987) who conclude that the anomalously high ratio of Brackett line to radio flux emission in their region M83-3 may be because it is a very young (10^{4} yr) ultra-compact HII region which is optically thick in the radio. However, Thompson (1987) has pointed out that young stellar objects often show unusually strong Brackett emission for the observed source luminosity which can be explained in terms of high mass outflow near the object: this would weaken the age constraint for M83-3 to a more acceptable estimate of 10^{5} to 10^{6} yr.

If there is a progressive change in the epoch of star formation along the infrared bar we would expect a bar of uniform surface brightness. To produce two relatively discrete regions of strong emission may require an infall rate that is variable with time. This might be achieved by a close encounter which can enhance the infall rate. The closest galaxy, NGC 5253, is 100 Kpc distant and could have had an encounter 5 x 10^{8} yr ago assuming a velocity relative to M83 of 200 kms^{-1}. Such an encounter was suggested by Rogstad, Lockhart and Wright (1974) to explain the warp which they found in the outer part of the HI distribution surrounding M83 (see also van den Bergh, 1980). They argued that the warp may persist ~10^{9} yr. However, it seems unlikely that the current high level of star formation activity in the centre of M83 could be a relic of such an ancient encounter unless the period between perigalacticum and the onset of star formation was delayed appreciably. Nevertheless it is of interest to note that the core of NGC 5253 is itself a site of vigorous star formation (Moorwood and Glass, 1982) with the brightest starburst having an age $\lesssim$ 3x10^{6} yr (Gonzalez-Riestra, Rego and Zamorano, 1987). The reason for this activity in a non-barred galaxy also requires an explanation.

ACKNOWLEDGEMENTS

I am grateful to Charles Telesco for several fruitful discussions and to Marjorie Fretwell for valuable assistance with the illustrations.

REFERENCES

Arp, H. (1966). Atlas of Peculiar Galaxies. Pasadena: California Institute of Technology.

Arp, H.C., & Madore, B.F. (1987). A catalogue of southern peculiar galaxies and associations. Cambridge: Cambridge University Press.

Bohlin, R.C., Cornett, R.H., Hill, J.K., Smith, A.M., & Stecher, T.P. (1983). Images in the rocket ultraviolet: the starburst in the nucleus of M83. Astrophys. J. 274, L53-56.

Byrd, G.G., Valtonen, M.J., Sundelius, B., & Valtaoja, L. (1986). Tidal triggering of Seyfert galaxies and quasars: perturbed galaxy disk models versus observations. Astron. Astrophys. 166, 75-82.

Byrd, G.G., Sundelius, B., & Valtonen, M. (1987). Tidal triggering of Seyfert galaxies and quasars: occurrence in multiple systems. Astron. Astrophys. 171, 16-24.

Combes, F., & Gerin, M. (1985). Spiral structure of molecular clouds in response to bar forcing: a particle simulation. Astron. Astrophys. 150, 327-338.

Cowan, J.J., & Branch, D. (1982). A search for radio emission from six historical supernovae in the galaxies NGC 5236 and NGC 5253. Astrophys. J. 258, 31-34.

Crawford, M.K., Genzel, R., Townes, C.H., & Watson, D.M. (1985). Far-infrared spectroscopy of galaxies: the 158 micron C^+ line and the energy balance of molecular clouds. Astrophys. J. 291, 755-771.

Cutri, R.M., & McAlary, C.W. (1985). A statistical study of the relationship between galaxy interactions and nuclear activity. Astrophys. J. 296, 90-105.

Dahari, O. (1984). Companions of Seyfert galaxies - a statistical survey. Astron. J. 89, 966-974.

Gonzalez-Riestra, R., Rego, M., & Zamorano, J. (1987). Star formation in the nucleus of NGC 5253. Astron. Astrophys. (in press).

Hawarden, T.G., Mountain, C.M., Leggett, S.K., & Puxley, P.J. (1986). Enhanced star formation - the importance of bars in spiral galaxies. Mon. Not. R. astr. Soc. 221, 41p-45p.

Hummel, E., & Van der Hulst, J.M. (1986). NGC 4038/9: interacting spiral galaxies with enhanced radio emission. Astron. Astrophys. 155, 151-160.

Hutchings, J.B., & Campbell, B. (1983). Are QSO's activated by interactions between galaxies? Nature 303, 584-588.

Joy, M. (1986). A multicolour infrared study of interacting galaxies. Ph.D. thesis, University of Texas at Austin.

Joseph, R.D., Meikle, W.P.S., Robertson, N.A., & Wright, G.S. (1984). Recent star formation in interacting galaxies - I.Evidence from JHKL photometry. Mon. Not. R. astr. Soc. 209, 111-122.

Keel, W.C., Kennicutt, R.C., Hummel, E., & van der Hulst, J.M. (1985). The effects of interactions on spiral galaxies. I. Nuclear activity and star formation. Astron. J. 90, 708-730.

Lonsdale, C.J., Persson, S.E., & Matthews, K. (1984). Infrared observations of interacting/merging galaxies. Astrophys. J. 287, 95-107.

Mackenty, J.W., & Stockton, A. (1984). Images and spectra of the host galaxy of the QSO Markarian 1014. Astrophys. J. 283, 64-69.

Miller, R.H., & Smith, B.F. (1980). Galaxy collisions: a preliminary study. Astrophys. J. 235, 421–436.

Moorwood, A.F.M., & Glass, I.S. (1982). Infrared emission and star formation in NGC 5253. Astron. Astrophys. 115, 84–89.

Negroponte, J., & White, S.D.M. (1983). Simulations of mergers between disc–halo galaxies. Mon. Not. R. astr. Soc. 205, 1009–1029.

Noguchi, M., & Ishibashi, S. (1986). Simulations of close encounters between galaxies: behaviour of interstellar gas clouds and enhancement of star formation rate. Mon. Not. R. astr. Soc. 219, 305–331.

Ondrechen, M.P. (1985). Radio continuum observations of the bar and disk of M83. Astron. J. 90, 1474–1480.

Puxley, P.J., Hawarden, T.G., Mountain, C.M., & Leggett, S.K. (1987). Enhanced star formation – the importance of bars in spiral galaxies. In Star Formation in Galaxies, ed. C.J. Lonsdale, pp 619–622.

Rieke, G.H. (1976). The sizes of galactic nuclei at 10 microns. Astrophys. J. 206, L15–17.

Rogstad, D.H., Lockhart, I.A., & Wright, M.C.H. (1974). Aperture–synthesis observations of HI in the galaxy M83. Astrophys. J. 193, 309–319.

Roos, N. (1985). Galaxy mergers and active galactic nuclei. I. The luminosity function. Astrophys. J. 294, 479–485.

Rowan–Robinson, M. (1987). Models for infrared emission from IRAS galaxies. In Star Formation in Galaxies, ed. C.J. Lonsdale, pp133–152. NASA Conference Publication 2466.

Rubin, V.C., Kent Ford, W., D'Odorico, S. (1970). Emission–line intensities and radial velocities in the interacting galaxies NGC 4038–4039. Astrophys. J. 160, 801–809.

Rubin, V.C., Kent Ford, W., Thonnard, N., & Burstein, D. (1982). Rotational star formation in interacting galaxies – I. Evidence from JHKL photometry. Mon. Not. R. astr. Soc. 209, 111–122.

Sanders, D.B., Soifer, B.T., Elias, J.H., Madore, B.F., Matthews, K., Neugebauer, G., & Scoville, N.Z. (1988). Ultraluminous infrared galaxies and the origin of quasars. Astrophys. J. (in press).

Schwarz, M.P. (1984). How bar strength and pattern speed affect galactic spiral structure. Mon. Not. R. astr. Soc. 209, 93–109.

Schweizer, F. (1978). Galaxies with long tails. In Structure and Properties of Nearby Galaxies (IAU Symposium No. 77), ed. E.M. Berkhuijsen and R. Wielebinski, pp 279–285. Dordrecht: Reidel.

Shara, M.M., Moffat, A.F.J., & Albrecht, R. (1985). Narrow band imaging of the QSO 4C 18.68: a tidal tail revealed? Astrophys. J. 296, 399–401.

Stockton, A., & Mackenty, J.W. (1983). 3CR249.1 and Ton 202 – luminous QSO's in interacting systems. Nature 305, 678–682.

Scoville, N.Z., Sanders, D.B., and Clemens, D.P. (1987). High–mass star formation due to cloud–cloud collisions. Astrophys. J. 310, L77–81.

Scoville, N.Z., & Good, J.C. (1987). High mass star formation in the Galaxy. In Star Formation in Galaxies, ed. C.J. Lonsdale, pp3–20. NASA Conference Publications 2466.

Sofue, Y., Handa, T., Hayashi, M., & Nakai, N. (1987). CO observations of galaxies with the Nobeyama 45m telescope. In Star Formation in Galaxies, ed. C.J. Lonsdale, pp 179–196.

Telesco, C.M., Decher, R., Ramsey, B.D., Wolstencroft, R.D., & Leggett, S.K. (1987). The infrared morphology of galactic centres. In Star Formation in Galaxies, ed. C.J. Lonsdale, pp 497–500. NASA Conference Publication 2466.

Telesco, C.M., Wolstencroft, R.D., & Done, C.J. (1987). The enhancement of infrared emission in interacting galaxies. Astrophys. J. (in press).

Thompson , R.I. (1987). A new ionization model for galaxies with dominant star formation regions in the nucleus of M83. Astrophys. J. 321. 153–155.

Toomre, A., & Toomre, J. (1972). Galactic bridges and tails. Astrophys. J. 178, 623–666.

Turner, J.L., Ho, P.T.P., & Beck, S.C. (1987). Recombination spectroscopy of star formation regions in the nucleus of M83. Astrophys. J. 313, 644–650.

Van den Bergh, S. (1980). The post-eruptive galaxy NGC 5252. Publ. astr. Soc. Pacific 92, 122–124.

Van der Hulst, J.M. (1979). The Kinematics and distribution of neutral hydrogen in the interacting galaxy pair NGC 4038/39. Astron. Astrophys. 71, 131–140.

Wolstencroft, R.D., & Davies, R.D. (1987). Near infrared mapping and multi-aperture photometry of Sbc galaxies. In preparation.

Wolstencroft, R.D. & Done, C.J. (1987). The IRAS counterparts of the Arp-Madore galaxies. In preparation.

Wright, G.S., & McLean, I.S. (1987). Near Infrared imaging of interacting galaxies. Proceedings of the Workshop on Ground-Based Astronomical Observation with Infrared Array Detectors, ed. G. Wynn-Williams and E.E. Becklin pp 355–359.

Young, J.S. et al (1986). The molecular content of interacting and isolated galaxies: the effect of environment on the efficiency of star formation. Astrophys. J. 311, L17–21.

SPIRALING INTO THE $R^{1/4}$ LUMINOSITY LAW OF ELLIPTICALS

G. Bertin, R.P. Saglia & M. Stiavelli
Scuola Normale Superiore, Pisa 56100, Italy

Abstract. Global self-consistent equilibrium sequences of slightly oblate collisionless stellar systems are constructed under a physically plausible selection criterion that requires the center of the galaxy to be isotropic and the periphery to be characterized by a quasi-radial pressure tensor. As one considers increasingly deep central potential wells, these equilibrium sequences "spiral into" a small region of parameter space where the corresponding luminosity profiles (for constant Mass-to-Light ratio) have the empirical $R^{1/4}$ law as a built-in property.

Out of all the possible equilibria $f=f(E,J^2)$ for spherical stellar systems, a large class of stellar dynamical models is identified by a selection criterion [1,2] that requires the center of the galaxy to be isotropic and the periphery to be characterized by a quasi-radial pressure tensor; in addition, at large radii the volume density $\rho(r)$ is constrained to decrease as r^{-4}, consistent with N-body simulations of collisionless collapse. This latter condition was actually suggested by continuity arguments from the oblate case [1].

A simple distribution function of this constrained class is

$$f_\infty = A(-E)^{3/2} \exp(-aE-cJ^2/2) \qquad \text{for} \quad E \leq 0$$

$$= 0 \qquad \text{for} \quad E > 0$$

where E and J^2 are the specific star energy and angular momentum (squared). Of the three dimensional parameters A, a, and c, two can be chosen to fix the relevant scales. Thus, a one-parameter equilibrium sequence is generated. The potential $\Phi(r)$ appearing in E is determined self-consistently by integrating the non-linear Poisson equation under the appropriate boundary conditions. The sequence is parametrized by the dimensionless central potential $\Psi = -a\,\Phi(0)$. For low values of Ψ, these models exhibit a flat density distribution inside the half-mass radius and may be unstable [2]. For $\Psi \gtrsim 6$, the overall structure of the models quickly converges in the direction of a singular model, characterized by an inner nucleus with $\rho \sim r^{-2}$. Some of the relevant properties are illustrated by the left Figures below. It should be noted that these are global self-consistent models, so that the full phase space properties are fixed (apart from two scales) once the value of Ψ is given. In other words, for a constant mass to light ratio, all observable profiles are fixed by the value of Ψ and they show minor changes when Ψ exceeds the value of 6.

The surprising feature is that, under the assumption of constant mass to light ratio, concentrated models ($\Psi \gtrsim 6$) have luminosity profiles that resemble observed photometries, in the sense that they are well fitted by the $R^{1/4}$ law [3].

A survey of various analytical ways of implementing the above-described selection criterion, i.e. an exploration of the class of constrained distribution functions, shows that the interesting properties of the f_∞ -models are "structurally" stable. The limiting singular models have a fairly simple density distribution resembling that originally proposed by Jaffe; they also give a relatively simple energy distribution N(E). For Ψ as large as 7.7, the f_∞ -models have been studied by N-body simulations using van Albada's code with up to 60,000 particles. For $\Psi > 2$, they are found to be dynamically stable. In addition, based on the original expression of the distribution function in terms of three integrals of the motion, N-body equilibria have been constructed with finite flattening (ellipticity up to 0.25). Details of this work are given in Ref. [2].

The right Figures presented below give an example from a study of eight bright ellipticals [4], with available photometric and kinematical data and zero or small apparent flattening. The photometric fits based on the anisotropic f_∞ -models are found to be not only very satisfactory, but more accurate than fits based on King models or a strict application of the $R^{1/4}$ law. In addition, the fit to the kinematical data has been used to determine the relevant mass to light ratios.

Statistical interpretations of these dynamical studies have been attempted [5] within the context of incomplete violent relaxation processes. Under the physical constraint of the conservation during collapse of the angular momentum distribution N_J for large values of J^2, an equation has been derived for the maximum (classical) entropy distribution function within the (E,J) partition of phase space. In the limit of small values of the binding energy the solution of this equation has been shown to reduce to f_∞.

References

[1] Bertin, G. & Stiavelli, M. (1984). Astron. Astrophys. 137, 26.
[2] Bertin, G. & Stiavelli, M. (1986). In Proc. IAU Symp. 127, ed. T. de Zeeuw, Dordrecht: Reidel, and Report SNS 86/1003 (Pisa).
[3] de Vaucouleurs, G. (1948). Ann. Astrophys. 11, 247.
[4] Bertin, G., Saglia, R.P. & Stiavelli, M. (1987). Report SNS 87/51 (Pisa).
[5] Stiavelli, M. & Bertin, G. (1986). In Proc. IAU Symp. 127, ed. T. de Zeeuw, Dordrecht: Reidel, and Report SNS 86/1004 (Pisa).

Upper left: For large values of the dimensionless central potential Ψ ($\Psi > 6$), the f_∞ equilibrium sequence "spirals into" a small region of the (r_α/r_M, q) parameter space. Here r_α/r_M is the ratio of the anisotropy scale r_α to the half-mass radius r_M. At $r = r_\alpha$, the anisotropy parameter $\alpha = 2 - \langle v_T^2\rangle/\langle v_r^2\rangle$ has unit value. The form factor q is the ratio between the gravitational binding energy $|W|$ and the natural energy scale GM^2/r_M. Lower left: Behavior of the volume density profile ρ (r) for increasing values of Ψ. These models tend asymptotically to a singular model characterized by $\rho \sim r^{-2}$ for $r \to 0$. Right: For $\Psi > 6$, models become "realistic"; here we show the photometric fit for the galaxy NGC 7626 by the $\Psi = 12$ model (Crosses: data from Kent 1984. Arrow: 2FWHM seeing radius. Solid line: seeing-unconvolved model). The lower frame shows the residuals $\Delta\mu = \mu_r - \mu_\infty$ for the same fit ($\Psi = 12$).

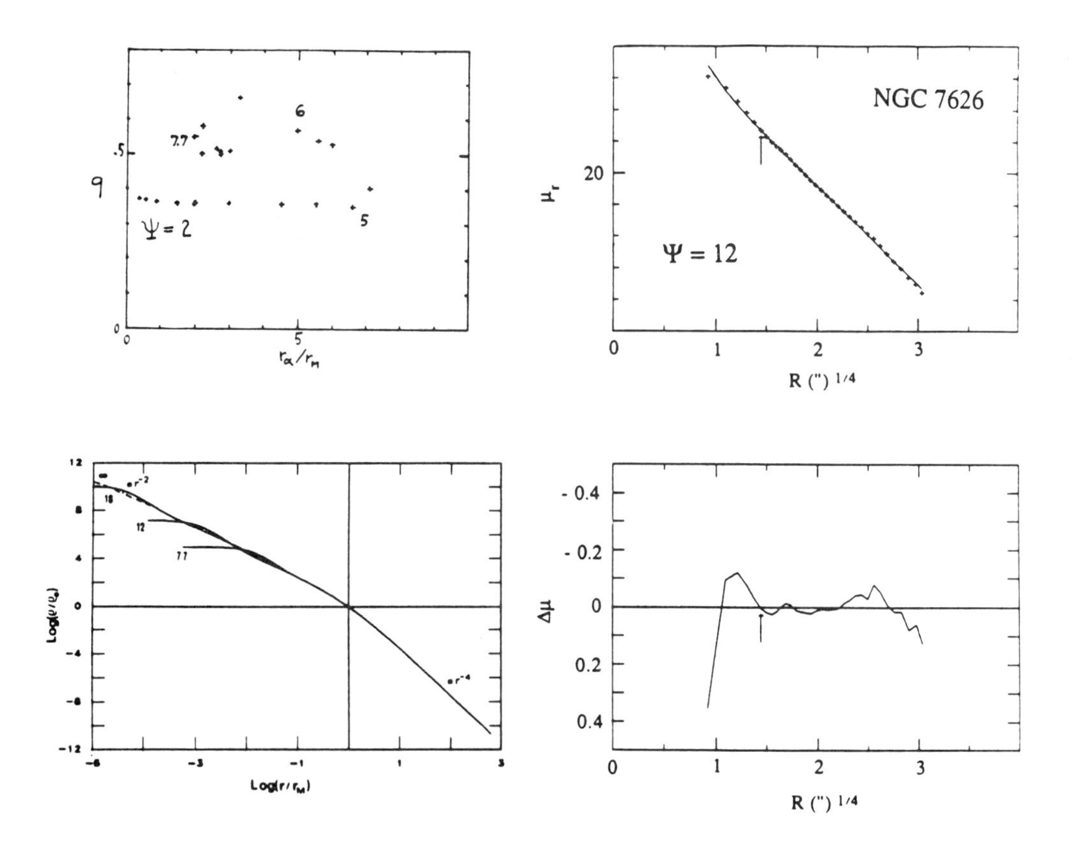

A fit of the angular galaxy 3–point function based on bias expressions

S.A. Bonometto

Dept. of Physics of the University of Perugia - 06112 Perugia. Italy
I.N.F.N. Sezione di Padova - Via Marzolo 8 - 35132 Padova. Italy

I shall report on a study of the 3-point function, based on the analysis of momenta deduced from the Zwicky catalog, indicating that an expression containing a cubic term, besides the usual second degree polynomials of 2–point functions, provides a good fit of angular data. This point is relevant in connection with the attempts to discriminate between models generating Gaussian or non–Gaussian primeval fluctuations. E.g., canonical inflationary models predict a Gaussian spectrum, while models based on topological strings lead to strongly non–Gaussian fluctuations. The method followed is illustrated in Sharp *et.al.* (1984). I shall also very briefly report on analogous results concerning the 4–point function.

The use of the Zwicky sample, to fit a 3–point function expression containing an angular cubic term, is preferable to the use of deeper samples. In fact the values of the angular 2–point function $w(\theta) \propto \theta^{1-\gamma}$ scale with the depth D of the sample, according to the law $w(\theta) \propto D^{-\gamma}$; for the Zwicky catalog they are ~ 1 (for angles of a few primes) and would be smaller at greater depths, making the contribution of a term $\propto w^3$ not so significant when compared with terms $\propto w^2$ and, therefore, likely to be confused in Poisson noise.

The fits are based on a reduction of the data operated restricting the sample to galaxies with $m < 15$, $\delta > 0^o$ and $b > 40^o$. In order to perform error estimates, a split of the whole sample into two subsamples A and B was performed along a line of constant galactic longitude. The subsample A contains 2298 objects with $123^o < l < 303^o$; the subsample B contains 1485 objects with $l < 123^o$ and $l > 303^o$. For the sets A+B, A and B the number $N(\theta_n)$ of galaxies in each ring centered on $\theta_n = \Delta\theta(n - 1/2)$ and of constant width $\Delta\theta = 0^o.2$ was evaluated (n=2,...,20). Correcting for border effects we worked out $w(\theta_n) =< N(\theta_n) > /n\Omega_n - 1$ and the moments $P_2(\theta_n) = < [N(\theta_n) - < N(\theta_n) >]^2 >$ and $P_3(\theta_n) = < [N(\theta_n) - < N(\theta_n) >]^3 >$ (Ω_n is the angular area of the n-th ring). Herefrom, correcting for the discreteness of the sample, we evaluated first

$$U^{(2)}(\theta_n) = P_2(\theta_n)\{n\Omega_n w(\theta_n)[1 + w(\theta_n)]\}^{-2} \quad (1)$$

to be compared with the theoretical expression we shall discuss in the sequel.

In a recent work (Bonometto *et.al.* 1987) we started from the expression

$$\delta^{(3)}P = n^3 F^{(3)}(1 + w_{12})(1 + w_{23})(1 + w_{31})\delta\Omega_1\delta\Omega_2\delta\Omega_3 \quad (2)$$

for the joint probability of finding 3 galaxies in the solid angles $\delta\Omega_i$ (i=1,2,3) of the celestial sphere. Here n is the angular galaxy number density, w_{ij} are 2–point angular correlation functions. Expression (2) applies whenever observed objects are selected as high peaks of an underlying distribution within a *bias* context (see, e.g., Kaiser, 1984; Politzer and Wise, 1984; Bardeen *et.al.*, 1986). $F^{(3)} = 1$ if the distribution is Gaussian. As was shown by Matarrese *et.al.* (1986), a similar relation holds also if the distribution is non–Gaussian, the only difference being that F becomes a known function of $1, 2, 3$.

According to (2) it can be shown that $U^{(2)}$ is expected to take the simple form

$$U^{(2)}(\theta) = [< F > -1]/w(\theta) + f < F > \tag{3}$$

where, at 0–th order in $\Delta\theta/\theta$,

$$f = 2^{\gamma-1}\pi^{-1/2}\Gamma(1-\gamma/2)/\Gamma(3/2-\gamma/2) \tag{4}$$

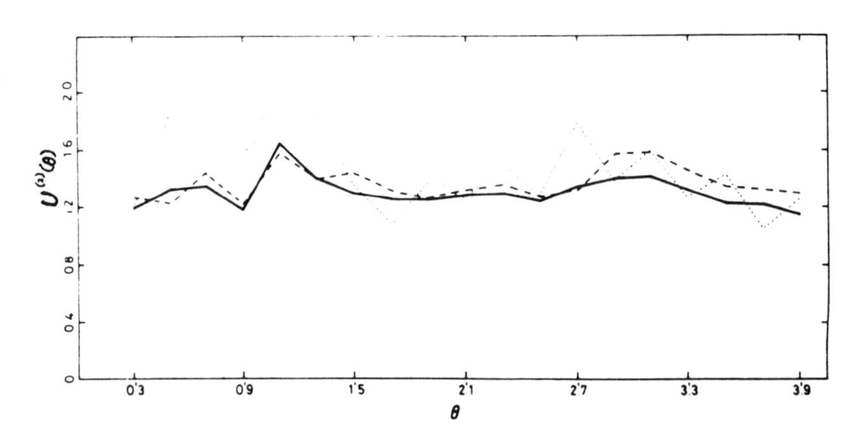

Fig.1 *Observational values of $U^{(2)}(\theta)$. The heavy line refers to the whole set of data. Dashed and dotted lines correspond to the subsets A and B respectively. Data indicate a constant $U^{(2)}$, apart Poisson scatter.*

In Fig.1 the behaviour of $U^{(2)}(\theta)$ is given for the whole sample and its subsets A and B. This yields a direct insight into the fact that the term $\propto w^{-1}$ – which would substantially increase with θ – can only have a negligible weight. In turn this means that $< F >= 1$ and can be interpreted to mean that the underlying matter distribution is Gaussian. A quantitative confirm is obtainable from a least square fit wherefrom γ and $< F >$ can be obtained. The results are given in Table 1.

TABLE 1: *Least square fit of the parameters*

	$< F > -1$	$< Ff >$	γ
whole sample	-1.68×10^{-2}	1.29	1.67
subset A	2.15×10^{-2}	1.27	1.68
subset B	-2.73×10^{-2}	1.45	1.71

A parallel fit aimed to test an expression for the 3–point function without the cubic term was also performed. It leads to discrepancies among the results obtained fitting data coming from the sets A+B, A and B which exceed the ones of the above fit by a factor ~ 7.

A similar fit for the 4–point function was also performed. It was based on the expression

$$\delta^{(4)}P = n^4 F^{(4)}(1+w_{12})(1+w_{23})(1+w_{34})(1+w_{41})(1+w_{13})(1+w_{24})\delta\Omega_1\delta\Omega_2\delta\Omega_3\delta\Omega_4 \tag{2}$$

for the joint probability of finding 4 galaxies in the solid angles $\delta\Omega_i$ $(i = 1, .., 4)$. In spite of the presence of a relevant Poisson noise, this fit also shows that $F^{(4)} \sim 1$, with an error $\sim 12\%$. It should be outlined that this can be considered as the first positive detection of the 4–point function in the Zwicky sample.

The above results concern angular functions. The usual way of relating n–point angular to spatial functions makes appeal to the Limber equations. In this frame, however, the terms $\propto w^s$, with $s > n - 1$, do not yield an analogous spatial term and have no direct spatial interpretation. The way–out from this *empasse* is probably linked to the fact that some of the assumptions required for the validity of the Limber equations – e.g., the independence between space and luminosity distributions – cease to hold in the frame of biased galaxy formation theories. It is however quite relevant that the above fits seem to indicate a Gaussian character for matter density fluctuations.

References

Bonometto, S.A., Lucchin, F., and Matarrese, S., 1987, Ap.J. **322**, in press.
Bardeen, J.M., Bond, J.R., Kaiser, N., and Szalay, A.S., 1986, Ap.J., **304**, 15.
Kaiser, N., 1984, Ap.J. Lett., **284**, L9.
Matarrese, S., Lucchin, F., and Bonometto, S.A., 1986, Ap.J. Lett., **310**, L21.
Politzer, H.D., and Wise, M.B., 1984, Ap.J. Lett., **285**, L1.
Sharp, N., Bonometto, S.A., and Lucchin, F., 1984, Astron. and Astrophys., **130**, 79.

OBSERVATIONS OF UNUSUAL OBJECTS

E. M. Burbidge
Center for Astrophysics and Space Sciences
University of California at San Diego, La Jolla, CA 92093
U.S.A.

INTRODUCTION

The title "Observations of Unusual Objects" suggested to me that you would like me to discuss phenomena observed in QSOs or active galaxies that indicate complex geometries and the asymmetric outflow of matter (relativistic or non-relativistic jets or cones) and beams of non-isotropically emitted radiation.

Much theoretical effort has been, and is being, spent on hypotheses of accretion disks, captures of separate galaxies by larger and more massive objects, and the infall of matter into central massive highly-condensed objects, as mechanisms for producing the energy observed in, for example, the powerful radio lobes in radio galaxies and the energy radiated by QSOs.

Since improved radio resolution has demonstrated the presence of narrow jets of high-energy particles, very well collimated, in some cases right down to VLBI angular dimensions, the production and ejection of collimated jets has also received considerable theoretical attention (e.g. Blandford & Königl 1979; see also the Bridle & Perley (1984) review and references therein). However, there is no general agreement on the radio jet ejection mechanism, nor on the physics of non-spherically-symmetric ejection of bulk matter in general (see, however, paper by W. Saslaw in this Volume).

The observations that I have selected for discussion are two classes of QSOs and one individual, and so far unique, BL Lacertae object:

Broad absorption-line QSOs (BALQSOs)
QSOs with variable emission-line intensity
The BL Lacertae object AO 0235+164.

BROAD ABSORPTION-LINE QSOs

These objects have spectra reminiscent of Type II supernovae in that the permitted emission lines have very broad absorption troughs on their short-wavelength side, indicating outflow of gas at velocities up to, or in a few cases exceeding, c/10. The fraction of all QSOs which exhibit this phenomenon seems to lie between

3 and 10%, as estimated by several workers, and Junkkarinen (1983) has shown that the absorption trough profiles and the way they cut into the adjacent emission lines require that the covering factor of the absorbing material over the emission-line region is ~ 0.1. While this implies that the BAL phenomenon depends on asymmetric outflow of gas and the orientation of this to the line of sight, there appear also to be intrinsic emission-line differences between BALQSOs and non-BALQSOs in that the latter appear to have higher ionization (e.g. N V is unusually strong relative to C IV and Mg II). Can such intrinsic emission-line differences be reconciled with a model in which gas is ejected in a cone oriented close to the line of sight and that orientation to the line of sight is thus the most important factor? The conclusions by reviewers of the subject is that we do not yet know (Weymann et al. 1985). The statistics on the percentage of BALQSOs among all QSOs could probably be refined by looking at surveys reaching to faint magnitudes and covering small areas of sky; for example, Burbidge et al. (1985) found that none of 20 high-redshift QSOs in a small area of sky of about 2 square degrees showed the BAL phenomenon.

Another question that must be asked concerns the length of time that steady outflow will persist. After all, the broad absorption troughs demonstrating gas outflow from supernovae, initiated by a single violent event, dissipate in short timescales (astronomically speaking). The first BALQSO discovered, PHL 5200, has been monitored for 20 years and has shown no change in its smooth, almost but not quite featureless, C IV absorption profile (Junkkarinen et al. 1983 and more recent unpublished data).

I have an unorthodox hypothesis concerning the BALQSOs, and an even more radical extension relating them to QSOs that have bunches of narrow absorption lines at several fairly close values of z_{abs}, where the mean value of z_{abs} is substantially less than z_{em}. Suppose that a BALQSO possesses a strong magnetic field which both controls motion of charged particles and produces non-thermal radiation (the large range of ionization demands a non-thermal continuum extending to frequencies much greater than 1 Rydberg). Let there be violent activity and energy release in the center. Suppose there is a deep gravitational potential well out of which matter "climbs" because it is driven by directed particle flow caused by the large energy release. Let there be a contribution to the redshift from gravitation, a contribution which becomes negligible for matter ejected earliest that has now reached furthest from the central massive object. The material may have large optical depth initially; eventually instabilities in the flow will grow until the gas clumps into clouds or filaments that produce the complex absorption profiles sometimes seen (see e.g. Turnshek et al. 1980, especially 0146+017). What happens to such gas eventually? Is it possible that some of the clumped narrow absorptions such as those around z_{abs} = 1.549 and 1.649 in the BL Lac object 0215+015 (Pettini et al. 1983) might be the eventual result?

In this (or any other) hypothesis, what is the nature of the violent activity that triggers the energy release and initiates the outflow? A conventional hypothesis might be that it is caused by a large infall of matter from an accretion disk into a rapidly spinning black hole. On the other hand, it has been suggested by some that the center of an AGN (active galactic nucleus) might be a supermassive binary object (Begelman *et al*. 1980). If this were true in QSOs, might two spinning massive objects coalesce? Or might one such massive object with very large angular momentum fission? What if a strong magnetic field has a dipole axis not aligned with the rotation axis of one (or two) such massive objects?

Such speculations are triggered in my mind whenever I contemplate BALQSO spectra. I realize they will be generally disliked, especially the hint that a component of redshift due to the gravitational field might be involved.

In any case, to conclude, I believe that the structure of BALQSOs is more complex than that of most QSOs and I believe that charged particle flow and the effect of a magnetic field should be included in attempting to explain the observations. It continues to be an important observational task to monitor BALQSOs at as high spectral resolution as possible, in order to catch any spectral changes that may occur, and to increase, by means of general survey programs, the known number of these fascinating objects.

VARIABILITY OF EMISSION LINES IN QSOs

The broad permitted emission lines in QSOs are generally considered to arise in clouds in a region (the broad-line region or BLR) nearest the energy source, with average densities of 10^9 - 10^{10} particle/cm^3 and surrounded by a lower density extended volume in which the forbidden emission lines are produced. Models based on photoionization give calculated relative line intensities that fit the observations quite well (e.g. Kwan & Krolik 1981).

While the broad Balmer lines of hydrogen in Seyfert galaxy nuclei are well known to vary on a time-scale of the order of one year, and in some cases even in less than one month, the existence of similar variability in QSO spectra, while suspected ever since early work by Burbidge & Burbidge (1966) and Dibai & Esipov (1967), has remained controversial.

A comprehensive study of some 30 low-redshift QSOs was undertaken at UCSD as a Ph.D. thesis project. The objects studied were QSOs known to have variable continuum flux, and were objects for which many years of observational data obtained with the same instrument were available in our data bank.

The result of the study was that 12 QSOs were found to have variations in flux in the broad Balmer lines, especially in Hβ (Zheng 1986; Zheng & Burbidge 1986; Zheng *et al*. 1987; Zheng 1987; Zheng preprint). These objects are listed in Table 1.

TABLE 1
QSOs WITH BROAD EMISSION LINE VARIABILITY

PG 0026+129	1202+281 (GQ Com)
PHL 0054+144	1318+290 (Ton 155)
PKS 0736+017	1612+261 (Ton 256)
PKS 0837-120 (3C 206)	PKS 2128-123
PKS 1004+130	PKS 2135-147
1104+167 (4C 16.30)	PKS 2141+175

The shortest time in which variation was observed was 10 months in 0026+129 (see Figure 1), in which the Hβ flux decreased by 50% between 1977 October and 1978 August. Variations may take place on a shorter timescale than detected in this work, since there was not continuous and frequent monitoring of all the objects. As far as could be inferred from these observations, the H lines appeared to vary in step with the continuum in most objects.

FIG.1. - Observed spectrum of 0026+129, showing decrease in both continuum and H lines between 1977 and 1978 (from Zheng 1986).

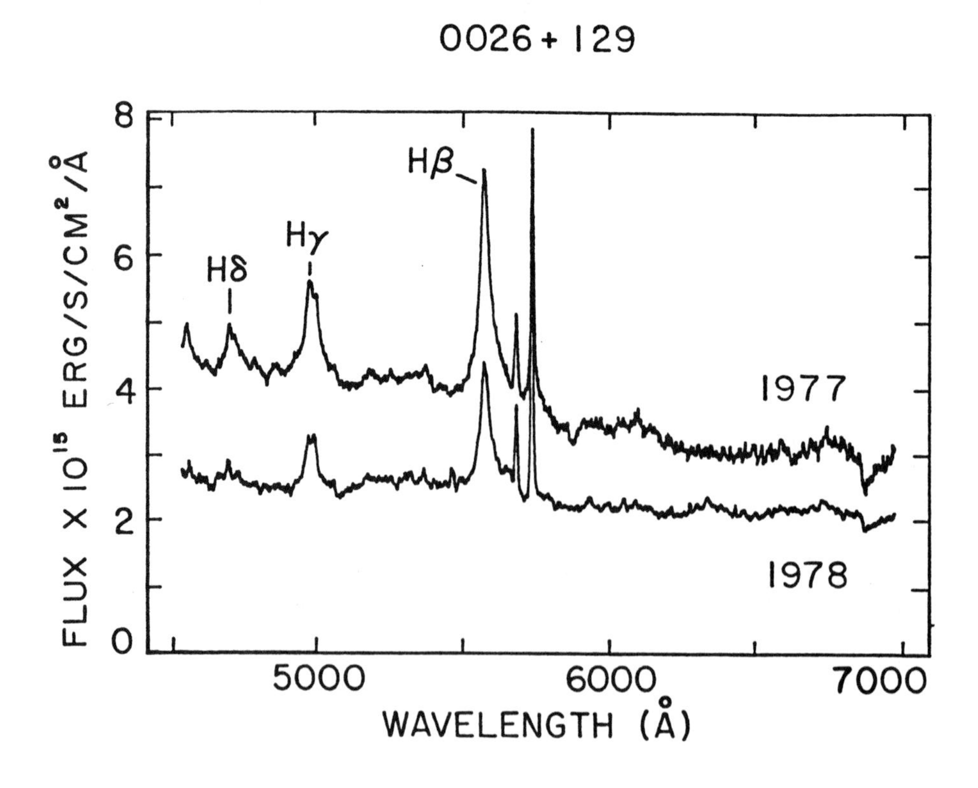

Conclusions drawn from these observations are, first, that the distance of the broad-line-emitting clouds from the energy source may in some cases be as small as one light-year. Consequently, the photon density is higher than previous estimates, which leads, from standard photoionization calculations, to particle densities in these clouds of at least 10^{11} cm^{-3}. Changes in the profiles of the broad emission lines are not yet sufficiently well documented either over long enough time scales or as frequently as necessary, for interpretation by detailed modeling. There is no consensus even on the geometry of the distribution of the BLR clouds, or on the nature of their velocity field which produces the broad profiles.

The interpretations of the data on variability, incomplete as they still are, leave me with the suspicion that the clouds in the BLRs of QSOs may have a more complex distribution than usually modelled. This applies also to radio and Seyfert galaxies, in which double-peaked hyrogen line profiles and variations in these have been known since the early 1970's (e.g. Burbidge & Burbidge 1971; Osterbrock et al. 1975). The idea of two massive compact objects orbiting about each other (Begelman et al. 1980) has been used by several workers (see Peterson et al. 1987 and references given there). Energy release from such a double source might well be non-isotropic - i.e., beamed. Consequences of beamed ionizing radiation have yet to be investigated theoretically.

But I close this section, as I did the section on BALQSOs, with the comment that more observational data are needed, with as high S/N as possible and, in the case of the variable QSOs, on time scales sufficiently short to refine the parameters to be used in theoretical models.

THE EXTRAORDINARY BL LACERTAE OBJECT AO 0235+164

I call this object "extraordinary" because of its extreme variability in radio and optical continuum, its complex and variable 21-cm aborption, and its very unusual optical spectrum. The radio source is very compact, hence its designation as an Arecibo occultation (AO) source from which a precise radio position was derived, yielding an identification with a strongly variable optical object.

Spinrad & Smith (1975) first detected this object's radio and optical variability and, observing no certain spectroscopic lines, either in absorption or emission, they designated it as a faint (~ 19 mag) BL Lac object, on the basis of its lack of emission lines, its rapid variability, and its optical continuum which was non-thermal with a steep slope. They detected a faint luminous extension reaching 3 arcsec south. Soon after these observations, the object underwent in a week or less a dramatic increase in brightness by more than a factor 100, and a similar radio outburst occurred at GHz frequencies. While the object was bright, in 1975, absorption lines were detected in its spectrum at two different redshifts, 0.524 and 0.851, but still no emission lines were seen. Shortly thereafter, 21-cm absorption consisting of several components centered on z = 0.524, the stronger of the two optical absorption systems, was detected and within the next 5 years the 21-cm components were shown to vary in relative intensity. Meanwhile, the faint luminous extension detected by Spinrad & Smith was seen to be a separate apparently stellar object, centered 2-3 arcsec from the active object, in P.A. ~ 185°. Lick Observatory observations of the faint southern object, obtained in unusually good seeing, showed its spectrum to have emission lines at z = 0.524.

References to the work outlined above are given in a paper which describes spectroscopic data obtained from 1975 to 1985 and which yields important new results (Cohen et al. 1987). Firstly, the sum of 9 observations totalling more than 8 hours of integration, obtained at Lick Observatory with the cassegrain spectrograph and CCD detector, showed a very clear broad emission line identified as Mg II λ2800, a clear [O II] λ3727 narrow emission line, and weak emission lines of [Ne V] λ3869 and [O III] λ5007, all at a redshift z = 0.940. The spectra also showed emission lines of [O II] λ3727 and [O III] λ5007 at z = 0.524, and the sum of two KPNO CCD spectra obtained in 1984 in the near infrared showed Na I D-line absorption at 0.524.

AO 0235+164 may thus be described as the object whose spectrum shows two emission redshifts and two absorption redshifts; the two emission redshifts make it a uniquely peculiar object. The two simplest "conventional" explanations of some of the optical spectroscopic data are:

(a) There is an optically-violent variable (OVV) object at z = 0.94 which has at some time ejected material at ~ 14,000 km/sec, and we now see this material, cool enough to display Mg II and Fe II absorption, at z = 0.851 (the difference in redshift between 0.94 and 0.851 corresponds to 14,000 km/sec).

(b) The OVV object is accompanied by galaxies (in a group or cluster with very large velocity dispersion?), one of which is in front of the object and in the line of sight; it produces the z = 0.851 absorption.

These possibilities, however, explain only part of the spectroscopic data. To remain "conventional," we must also postulate an intervening galaxy in the line of sight, unconnected with the OVV object, at z = 0.524. This is an active (maybe Seyfert 2?) galaxy, based on the measured [O II] flux which is approximately the same in spectra centered on the companion object and on AO 0235+164 itself. If uniformly spread over an area of 3 arcsec radius at a cosmological distance corresponding to z = 0.524, the ionized gas would be emitting 3×10^{42} erg/sec in the [O II] line (Cohen et al. 1987), well above that of a normal ScI galaxy. The observed optical continuum emission from the companion object is insufficient to provide the exciting photons for such an extended ionized region, unless the continuum rises dramatically in the ultraviolet.

To postulate that the companion is connected with the OVV object whose redshift is 0.94 would require non-cosmological redshifts, but would, as Cohen et al. point out, provide a convenient source of ionizing radiation for the z = 0.524 emission lines. However, most astronomers would reject such an unconventional hypothesis.

What about the nature of the active object itself? Let us collect the data.

1. It varies violently and apparently non-periodically by at least a factor 100. Variations occur both at optical and radio wavelengths.

2. It is a flat-spectrum very compact radio source, but has a steep optical continuum with $F_\nu \propto \nu^{-3.5}$.

3. It is an X-ray source (Schwarz & Ku 1983).

4. The angle of its radio polarization changes in a time-scale of months in a way that could be interpreted as rotation of some structure within the source (Ledden & Allen 1979).

5. It has an elongated, nearly unresolved, VLBI structure about 1 milliarsec in extent (Jones *et al*. 1984). The VLBI map by Jones *et al*. shows that the central source has an extension in P.A. 7°, approximately 180° from the direction in which the companion object lies.

6. The components of the multiple 21-cm absorption at z = 0.524 vary in relative intensity in times 0.25 to 0.5 yr (Wolfe *et al*. 1982); these authors suggest that such variations might be caused by fluctuations in relative intensity of parts of the active compact source shining through clouds in the z = 0.524 object.

7. It is unique in having two emission-line redshifts in its spectrum.

The BL Lac nature of this remarkable object demonstrated by properties 1-5 might, in the model of Blandford & Königl (1979), be the result of viewing the source along or near the axis of a jet. Altschuler *et al*. (1984) extended this concept to include relativistic ($\gamma \sim 6$-12) expansion of material emitting synchrotron radiation.

There is one further property observed by Cohen *et al*. (1987) which complicates the picture even further. A search through 10 years of data taken at Lick Observatory with lower resolution and lower S/N than that obtained with the CCD spectrograph was made to select observations when AO 0235+164 was near minimum optical brightness and these data were binned and summed. Data taken in 1976 showed a poorly-defined but clearly present weak Mg II emission. In 1980, when the object was even fainter, data with about the same S/N showed no sign of Mg II emission. Thus this feature, which is quite weak compared with its average intensity in a large number of QSOs, is definitely variable on a time scale $\lesssim 2$ yr (in the rest frame), and the line strength is not proportional to the continuum strength.

In conclusion, it can be stated that asymmetric relativistic ejection and beamed radiation occur and can explain some, but not all, of the peculiar features of AO 0235+164. The variable Mg II emission provides a link with Zheng's work on variable H emission lines in some QSOs and strengthens the suggestion that beamed, or at least non-spherically symmetric, radiation may occur in his selection of low-redshift QSOs.

Among the remaining questions are the following: if only cosmological redshifts are allowed and if it is insisted that the material at z = 0.524 is due to an intervening galaxy, what provides the ionizing and exciting photons to account for the z = 0.524 emission line intensities? Is it a coincidence that the companion object is located like a "counterjet" opposite to the much more compact VLBI extension?

As with my previous sections, I end by pointing out that more observations and more frequent monitoring are needed. The object is a prime candidate for Hubble Space Telescope observations, both for that instrument's spatial resolution and its ultraviolet coverage. We hope to discover whether or not there is a luminous bridge between the BL Lac object and its companion, and also to put better limits on the degree of compactness of both, at ultraviolet and optical wavelengths. We look forward also to discovering what Lyman α at the three redshifts (0.524, 0.851, 0.94) looks like, in both emission and absorption.

I wish to acknowledge support for this work by NASA under contract NAS5-29293.

REFERENCES

Altschuler, D.R., Broderick, J.J., Condon, J.J., Dennison, B., Mitchell, K.J., O'Dell, S.L. & Payne, H.E. (1984). Astron. J., 89, 1784-98.
Begelman, M.C., Blandford, R.D. & Rees, M.J. (1980). Nature, 287, 307-9.
Blandford, R.D. & Königl, A. (1979). Astrophys. J., 232, 34-48.
Bridle, A.H. & Perley, R.A. (1984). Ann. Rev. Astron. Astrophys. 22, 319-58.
Burbidge, E.M. & Burbidge, G.R. (1966). Astrophys. J., 143, 271-3.
Burbidge, E.M. & Burbidge, G.R. (1971). Astrophys. J. (Letters), 163, L21-23.
Burbidge, E.M., Smith, H.E., Junkkarinen, V.T. & Hoag, A.A. (1985). Astrophys. J., 288, 82-93.
Cohen, R.D., Smith, H.E., Junkkarinen, V.T. & Burbidge, E.M. (1987). Astrophys. J., 318, 577-84.
Dibai, E.A. & Esipov, V.F. (1967). Astron. Zh., 44, 55.
Jones, D.L., Bååth, L.B., Davis, M.M. & Unwin, S.C. (1984). Astrophys. J., 284, 60-4.
Junkkarinen, V.T. (1983). Astrophys. J., 265, 73-84.
Junkkarinen, V.T., Burbidge, E.M. & Smith, H.E. (1983). Astrophys. J., 265, 51-72.
Kwan, J. & Krolik, J.H. (1981). Astrophys. J., 250, 478-507.
Ledden, J.E. & Aller, H.D. (1979). Astrophys. J. (Letters), 229, L1-3.
Osterbrock, D.E., Koski, A.T. & Phillips, M.M. (1975). Astrophys. J. (Letters), 197, L41-4.
Peterson, B.M., Korista, K.T. & Cota, S.A. (1987). Astrophys. J. (Letters), 312, L1-4.
Pettini, M., Hunstead, R.W. & Murdoch, H.S. (1983). Astrophys. J., 273, 436-49.
Schwarz, D.A. & Ku, W.H-M. (1983). Astrophys. J., 266, 459-65.

Spinrad, H. & Smith, H.E. (1975). Astrophys. J., 201, 275-6.
Turnshek, D.A., Weymann, R.J., Liebert, J.W., Williams, R.E. & Strittmatter, P.A. (1980). Astrophys. J., 238, 488-98.
Weymann, R.J., Turnshek, D.A. & Christiansen, W.A. (1985). In Astrophysics of Active Galaxies and Quasi-Stellar Objects, ed. J.S. Miller, pp. 333-65. Mill Valley: University Science Books.
Wolfe, A.M., Davis, M.M. & Briggs, F.H. (1982). Astrophys. J., 259, 495-521.
Zheng, W. (1986). Ph.D. Thesis, Univ. of California, San Diego.
Zheng, W. (1988). Astrophys. J., in press (January).
Zheng, W. & Burbidge, E.M. (1986). Astrophys. J. (Letters), 306, L67-69.
Zheng, W., Burbidge, E.M., Smith, H.E., Cohen, R.D. & Bradley, S.E. (1987). Astrophys. J., in press (November).

DISCUSSION

G. Chincarini: Did anyone look for a cluster of galaxies with deep imagery around AO 0235+164 ? Could the HI clouds (Arecibo) be related to a cluster gaseous medium?

M. Burbidge: Yes, our two deep photographs with the Kitt Peak 4-meter prime focus showed a few very faint extended objects around AO 0235+164. They have very low surface brightness. We have started a program with the Lick 3-meter CCD cassegrain instrument, both for direct imaging and for low-resolution spectroscopy. The first object we looked at has a tentative redshift less than 0.524.

E. Khachikian: I. Pronik published recently papers in which she shows that [OIII] lines in the Sy galaxy (NGC1275) may have changed during some weeks or months. What is your opinion about it?

M. Burbidge: I think this is a very interesting observation, and suggests [OIII] clouds must be small. I suppose the changes were increases in intensity? Were they in the emission lines at velocity 5200 km/s or 8200 km/s?

A. Treves: In the last years 0026+129 was observed various times in X-rays with Einstein Observatory. No variability was found. A 50% variation between 1981 and 1984 was found with IUE at 1000-2000 Å, but none between 2000-3000 Å. A change of spectrum must have occured.

M. Burbidge: That is very interesting. The Lick results on 0026+129 are part of the paper by Dr. Wei Zheng.

Huge regions of ionized gas around radio galaxies

S. di Serego Alighieri*
Space Telescope European Coordinating Facility
Garching bei München, West Germany

Abstract. The application of improved line and continuum imaging techniques to the study of extended emission line regions around radio galaxies has yielded new results which are summarized here and which shed light on the nature of nuclear activity itself.

1. Introduction

The relationship between active galactic nuclei (AGN) and their environment gives important clues to the properties of the nuclear activity and helps in the understanding of its origin. Radio galaxies are not the most powerful AGN, but are important in this context, because they are found both relatively nearby ($z \leq 0.1$), where the properties of the environment can be studied in detail and compared with that of normal galaxies, and at larger distances where they can be compared directly with bright quasars.

We know that a large fraction of radio galaxies with nuclear emission lines are surrounded by ionized gas extending much further than the classical narrow- line region (see *e.g.* the review by Fosbury, 1986). These extended emission line regions (EELR) have been studied mainly using long slit spectroscopy. These studies heve shown that the EELR extend up to several tens of kiloparsecs from the nucleus, that they show a varying, often high, ionization state and that their velocities relative to the systemic velocity range up to a few hundreds kilometer per second and are generally consistent with normal rotation curves. EELR have also been found around quasars (Stockton & MacKenty, 1987 and references therein) and around Seyfert galaxies.

* Affiliated to the Astrophysics Division, Space Science Department, European Space Agency.

2. An imaging study

We have recently performed an imaging study of radio galaxies with EELR using narrow-band filters and special techniques (di Serego Alighieri, 1987) to obtain line-only (mainly [OIII]λ5007 and Hα) and continuum-only images. The purpose was to study the morphology of the different physical components, *i.e.* ionized gas and stars, to compare it with the radio and X-ray morphology, and to derive line and continuum luminosities and covering factors, thereby putting constraints on the ionizing spectrum and on the mass of the gas.

The results of this study can be summarized as follows. The shape of the stellar continuum is very regular, without deviations from a normal elliptical structure, except for the frequent presence of companion galaxies. On the other hand the morphology of the gas is very complex, with condensations, arms, rings and filaments and with a large range in surface brightness, suggesting that the gas is mostly density rather than ionization bounded. The EELR extend well beyond the stellar continuum up to more than 100 kiloparsec from the nucleus and have a total [OIII]λ5007 luminosity of the order of 10^{41}–10^{42} $erg\ s^{-1}$ ($H_o = 50 kms^{-1} Mpc^{-1}$. These values compare well with those measured for EELR around quasars. The relationship of the gas morphology to the radio one is often not obvious and even in the objects where some of the EELR are aligned with the radio axis, other regions do not show such correspondence. In many cases the morphology of the gas is related to the position of companion galaxies, indicating tidal interaction. In some cases the EELR are aligned with the major axis of the stellar component, suggesting that the gas tends to settle on the principal plane of the ellipsoid. Only in a few cases (mostly broad line radio galaxies) photoionization from a simple power law ($\alpha \sim -1$) extrapolation of the optical nuclear continuum can explain the total line luminosity. Nevertheless the absence of local continuum sources (the case of PKS 2152-69 presented by Tadhunter *et al.* , 1987, is a notable exception) and the line ratios argue in favour of nuclear ionization for most of the EELR. An ultraviolet excess in the nuclear continuum above the simple power law spectrum (Robinson *et al.* , 1987), some anisotropy in the nuclear radiation or a recent large decrease in the nuclear ionizing luminosity (Robinson *et al.* , 1987, in preparation) have been considered in order to solve this problem.

References

di Serego Alighieri, S., 1987, *ESO Messenger*, No. **48**, in press.
Fosbury, R.A.E., 1986, in *Structure and Evolution of Active Galactic Nuclei*, Giuricin *et al.* eds., Reidel, Dordrecht, p. 297.
Robinson, A., Binette, L., Fosbury, R.A.E. & Tadhunter, C.N., 1987, *Mont. Not. R. Astr. Soc.*, in press.
Stockton, A. & MacKenty, J.W., 1987, *Astrophys. J.*, **316**, 584.
Tadhunter, C.N., Fosbury, R.A.E., Binette, L., Danziger, I.J. & Robinson, A., 1987, *Nature*, **325**, 504.

DISCUSSION

G. Burbidge: Have you applied this technique to a sample of non-radio galaxies?

S. di Serego Alighieri: No, our purpose was to study the properties of the known EELR around radio galaxies. But others (e.g. M. H. Ulrich et. al., Philipps et. al., E. Sadler) have searched for gas around radio quiet ellipticals, with a much lower success rate, and the gas when found, is on much smaller scale and lower brightness.

GALAXIES WITH DOUBLE NUCLEI

E. Ye. Khachikian
Byurakan Observatory of the Academy of Sciences
Armenian SSR

At the present time, there are enough data on active galaxies (AG) in order to make some generalizations (see, for example, IAU Symposium No. 121 "Observational Evidence of Activity in Galaxies", 1987, ed. E. Khachikian et al., D. Reidel).

But today I would like to speak about AG with double nuclei in the optical. It is necessary to stress that in addition to double nuclei there are galaxies with three and more nuclei. It is known also that each of the nuclei of double nucleus galaxies can themselves consist of two components.

Therefore, the opinions expressed here concerning the nature of double nucleus AG are relevant to the multinuclei AG as well. It seems unimportant to use the term "multinuclear AG" or to say "central part of AG consists of a number of condensations". The terminology is not important, because I believe that these objects have been formed as a result of division of a single maternal body.

From the time of Kant and Laplace up to the present, the majority of theorists, as well as observers, believe that the Universe develops in a direction from concentrations of diffuse matter to the denser states. Perhaps V. Ambartsumian was the first who declared the opposite point of view and successively put it into practice over the past 40 years. As far back as the end of the 40's, he stated the revolutionary idea that evolution in the Universe goes from the dense condition of matter to the rarefied one.

Unfortunately, very few scientists are attempting the construction of a physical theory for this concept, although there are fairly successful attempts in this field. It seems to me that observational data speaks in favour of this point of view. The existence of double and multinuclei galaxies is the good confirmation of this idea.

It is necessary to note that the notion: "double" or "multinucleus", even "nucleus" itself, needs to be defined. That kind of attempt has been done by Khachikian et al. (1981). The experience shows that any knot or condensation in galaxy considered as a nucleus is connected with some uncertainty. For example, some of the Markarian galaxies (7, 8, 325) consist of several condensations which have similar brightness (Khachikian, 1972, 1984) and which show approximately the same strong emission line spectra. Usually these knots are involved in a common envelope, and it is difficult to distinguish one of them as a nucleus. Even in some nearby galaxies it is difficult sometimes to distinguish the nucleus (for example in LMC).

Since there is no generally accepted definition of a nucleus, it is necessary to adhere to some symbolic criteria such as those which are suggested by Khachikian et al. (1981):

a) High luminosity exceeding the average luminosity of superassociations (SA) (or giant HII region),

b) Almost central position relative to the isophotes of the galaxy,

c) Connection with the structual details of the galaxy (spiral arms, jets, bars, and so forth).

On the basis of these criteria, a number of double nucleus AG's have been detected at Byurakan Observatory. These nuclei show different types of activity (Seyferts, Liners). A number of nuclei with the spectral properties of SA were also detected. In Fig. 1, examples of well-known double nucleus AG's are presented.

Quite recently, the second SA type nucleus has been discovered in NGC 6240, one of the brightest IR sources and a super-starburst AG (Andreasian and Khachikian, 1987).

Here I would like to say some words about the so-called isolated giant HII regions or SA. It is clear now that the majority of them have double-component structure. It is natural to ask what kind of objects they are: a) double nucleus AG's, or b) two isolated SA.

In connection with this, I simply must recall the work which has been done by Dr. Heidmann, our hero of the day Dr. Arp, and myself some years ago (Arp et al., 1974). The paper is about the "twin" objects near Markarian 261 and 262. I speak fairly often about these objects, because of their unique structure and physical properties. The shapes, dimensions, luminosities, radial velocities and spectra of these objects are nearly identical. On the one hand, the "twins" can be considered as a double nucleus galaxy without any stellar environment or common difuse envelope; or on the other hand, as pairs of isolated SAs.

More information about double nucleus AG and their spectra can be found in my article and in the references within it (Khachikian, 1987).

CONCLUSION

The number of known double and multi-nucleus AG is increasing all the time, and it is close to one hundred. No doubt, the majority of them are real

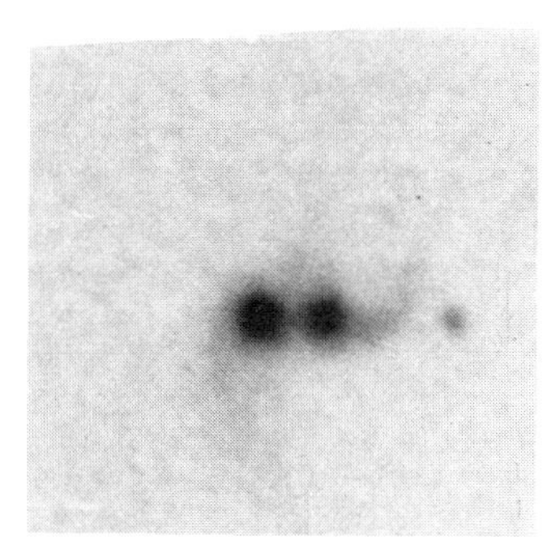

Mark 212
(1 mm ~ 2″;1 ~ 5 kpc)

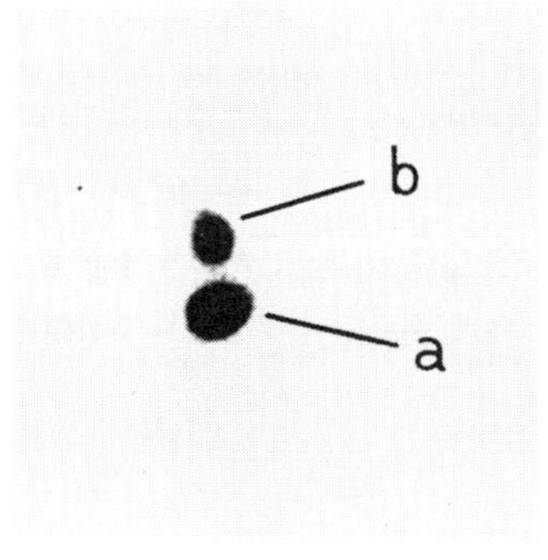

Mark 480
(1 mm ~ 1.4″;1 ~ 3 kpc)

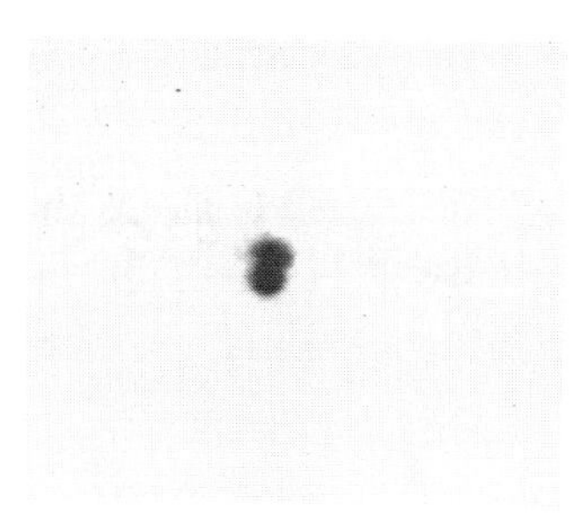

Mark 930
(1 mm ~ 2.2″;1 ~ 1.6 kpc)

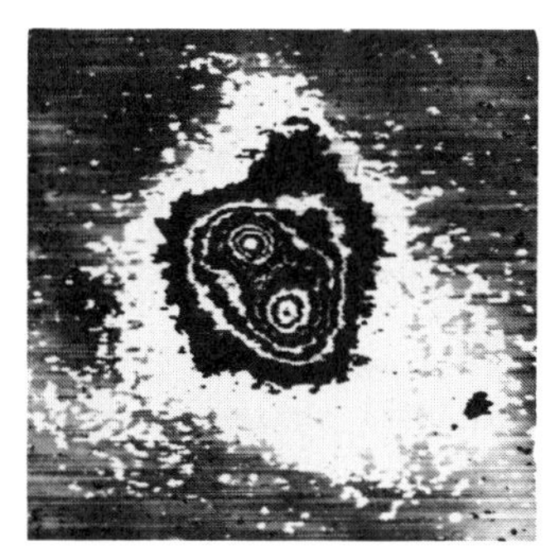

Mark 266
(1 mm ~ 1.7″;1 ~ 6.5 kpc)

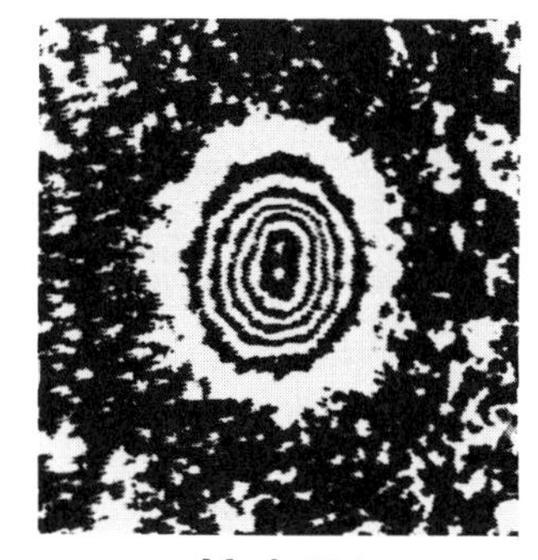

Mark 324
(1 mm ~ 0.6″;1 ~ 0.1 kpc)

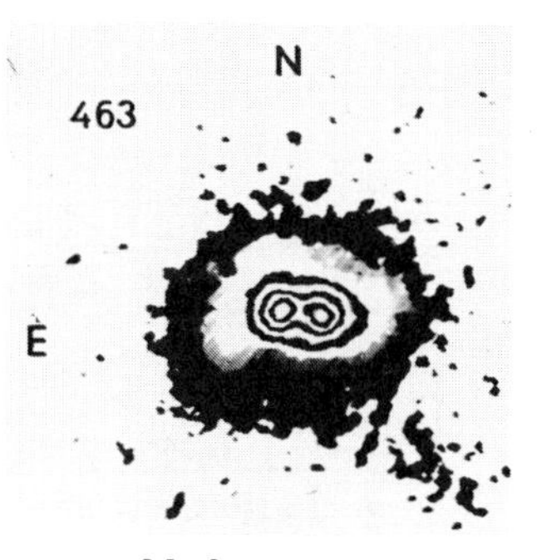

Mark 463
(1 mm ~ 1.3″;1 ~ 4.3 kpc)

Photos and isodenses of some Markarian galaxies
1 is a distance between of nuclei.

double nucleus galaxies and not double galaxies. That is, they are **not a result of interaction of two independent galaxies**. The fact that almost all double nuclei components show either form of activity also speaks in favour of this opinion. It is another question whether double galaxies (close or wide pairs) are formed as the result of dynamical evolution of double nucleus galaxies.

We think that double nucleus AG's are formed from a single maternal body, as a result of its division, in analogy to what takes place in the biological cell.

REFERENCES

Andreasian, N. K., Khachikian, E. Ye. 1987, in "Observational Evidence of Activity in Galaxies", ed. E. Khachikian et al., (D. Reidel; Dordrecht), p.541.

Khachikian, E. Ye. 1972, *Astrofisika*, **8**, 529.

Khachikian, E. Ye. 1984, in "Astronomy with Schmidt-type Telescopes", ed. M. Capaccioli, (D. Reidel; Dordrecht), p.427.

Khachikian, E. Ye. 1987, in "Observational Evidence of Activity in Galaxies", ed. E. Khachikian et al., (D. Reidel;Dordrecht), p.65.

Khachikian, E. Ye., Korovyakovsky, Yu. P., Petrosian, A. R., Sahakian, K. A. 1981, *Astrofisika*, **17**, 231.

DISCUSSION

H. Arp: The genesis of the so-called isolated extragalactic HII regions goes back to Fritz Zwicky. His class of compact galaxies with emission were characteristically double. We discussed this often at that time. It is very significant I feel that this doubling and pairing tendency tends to persist through *all* classes of extragalactic objects, peculiar galaxies, companion galaxies, radio sources, quasars. I agree with Ambartsumian that this has cosmogonic significance which has so far not been confronted.

R. Kraft: How can you explain the presence of strong [SII]/Hα and [OI]/Hα ratios in some of your spectra?

P. Rafanelli: Strong [OI]/Hα and [SII]/Hα ratios, as observed usually in LINERS, can be explained in terms of photoionization by a non-thermal power law source, as suggested by Ferland and Netzer.

M. Burbidge: It is very fashionable at the moment to talk about "mergers" of galaxies to explain unusual morphology, but it seems to me that there are too many of these double nuclei for them ever to be accounted for as "mergers" of previously separate galaxies.

E. Khachikian: I completely agree with you. I am sure that in this case we observe the result of division of one body to two parts.

PECULIARITIES OF THE NGC1275 CIRCUMNUCLEAR REGION
(Is NGC1275 an Arp Type System?)

I. Pronik and L. Metik
Crimean Astrophysical Observatory
USSR

The Seyfert galaxy NGC1275 is the brightest radio and X-ray source in the Perseus cluster. It has two gas systems, whose dimensions are comparable with the galaxy's. The radial velocity of one of them ("high-velocity" system) and of the galaxy differ by about 3000 km/s according to Minkowski (1957). This system contains stars, dust and neutral hydrogen, and can be considered as a late type galaxy. One can suppose that NGC 1275 galaxy is an Arp type system. The recently obtained data permit us to confirm this supposition.

The observations of NGC 1275 have been carried out at the prime focus of the 6-m telescope on January 12-15, 1977. The seeing was 1 -1.5 arcsec. The unwidened spectra of NGC 1275 in the range of $\lambda\lambda$ 3500-7300 Å, with dispersion 95Å/mm, and direct images with glass-filters (λ_{eff} ~4800Å and λ_{eff} ~6650Å) have been obtained. The image scale on the negatives was 17.5 arcsec/mm.

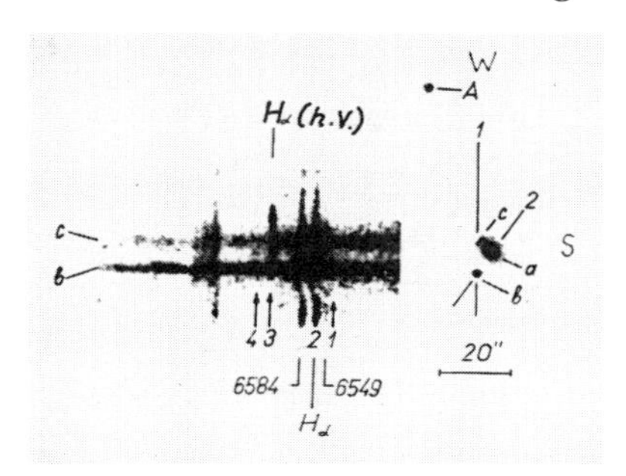

Figure 1– (on the right) the central region of NGC1275 (a copy of an image obtained by the Burbidges (1965). "A"-star, "a"-nucleus, "b", "c"-details in the central region; 1,2- positions of the spectrograph slit. (on the left) the spectra of "b", "c"-details, obtained by 1 position of the slit.

The object "b", is situated 7 arcsec NE from the NGC 1275 nucleus. Minkowski (1957), van den Bergh (1977) and other observers considered it as a star from our Galaxy.

We have been investigating the NGC 1275 central region by multicolor and spectral photometry (Metik & Pronik, 1979, 1985, 1986; Afanasjev & Pronik, 1980) and have obtained a number of results allowing us to suppose that the "b"-object belongs to the NGC 1275 galaxy system.

1. In the blue region of the "b"-object, the spectrum shows the features

of hot stellar radiation, and in the red region - that of the cold stars.

2. In Fig. 1, we present three spectra: "c"-object - at the top, in the middle - the "low-velocity" gas and the "b"-object at the bottom. It follows that "low-velocity" gas in the "b"-object region has a redshift +600 km/s with respect to NGC 1275 galaxy.

3. The emission lines Hγ, Hβ, [*OI*] λ6300 $\AA$ and Hα in the "b"-object spectrum have complicated profiles. In Fig. 1, one can see four components (1-4) which are shifted with respect to NGC 1275 galaxy: 1) at -16$\AA$ (or -700 km/s), 2) at +14$\AA$ (or +600 km/s), 3) at +65$\AA$ (or +3000 km/s), 4) at +107 $\AA$ (or +4500 km/s). The third component corresponds to the "high-velocity" gas. The velocities of components 3 and 4 differ by about 2000 km/s. The results permit us to suspect the existence of a gas stream between "b"-object and the NGC 1275 galaxy.

4. The brightness distribution of NGC 1275 in emission lines, obtained by measurements of unwidened spectra across the dispersion are shown in Fig. 2 (for "low-velocity" gas) and Fig. 3 (for "high-velocity" gas). Labels "b", "c", "d", "e" are the details of the NGC 1275 central region.

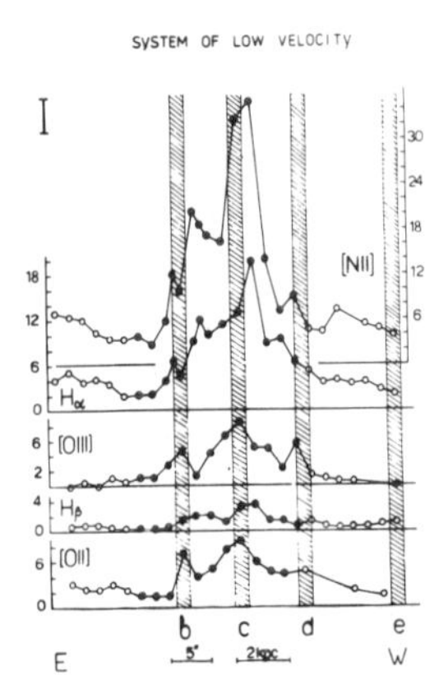

Figure 2

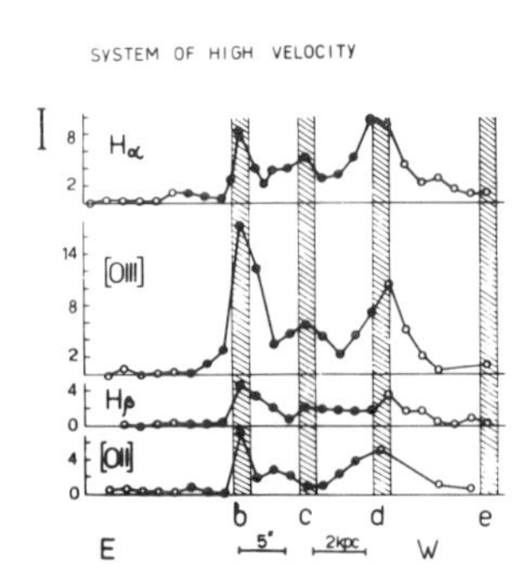

Figure 3

It should be noted that the brightness of the galaxy in emission lines is inhomogeneous. This inhomogeneity is different for "low-velocity" and "high-velocity" gas. The "low-velocity" gas is brighter near "c"-object, and "high-velocity" gas is brighter near "b"-object. Points 2- 4 suggest that "b"-object is connected with "high-velocity" gas system.

5. Fig. 4 shows a bar between the nucleus "a" and "b"-objects, whose direction does not coincide with the "ab"-line. It is reflected by Keel's (1983)

$H\alpha$-picture, published for the "high-velocity" gas.

6. If the "b"-object belongs to the NGC 1275 galaxy system, then the "b"-object velocity should be considered. We attempted to identify the absorption lines in the spectrum, using z = 0.0341. The results are given in the Table.

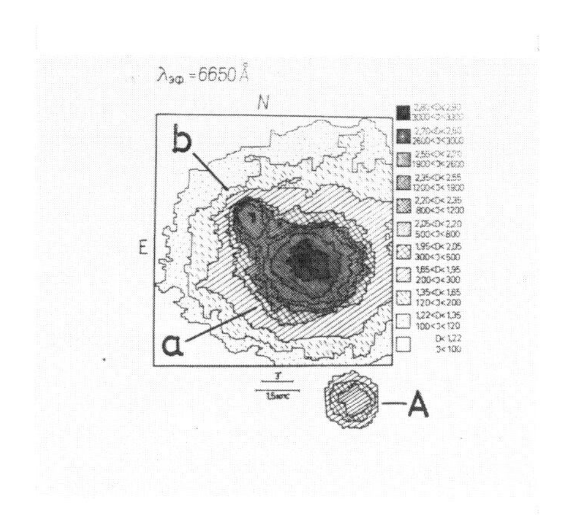

Figure 4– The photometric maps of the NGC1275 circumnuclear region and star "A", obtained in red light (λ_{eff} ~6650Å).

One can see that the lines in column 3 may be adopted as the lines of column 4. In this case, the radial velocity of the "b"-object is about 10000 km/s or 5000 km/s with respect to NGC 1275, and its absolute magnitude M_V= -16.3. (Metik, & Pronik, 1979).

Table 1

Line	λ, Å (z=0)	λ, Å (z=0.0341)	The possible identification of lines having z=0.0341 with the lines at z=0
H_ϵ	3970	4105	H_δ = 4101 Å
H_8	3889	4022	HeI = 4026 Å
H_9	3835	3966	H Ca^+ + H = 3969 Å
H_{10}	3978	3928	K Ca^+ = 3933 Å

The results of our investigations allow us to suggest that the NGC 1275 galaxy, contains objects with different (5000 km/s, 8000 km/s, 10000 km/s) radial velocities, and thus can be considered as an Arp system.

REFERENCES

Afanasjev, V.L. & Pronik, I.I. 1980, *Astrofizika*, **16**, 405.

Bergh van den, S. 1977, *Lick. Obs. Bull.*, **765**.

Burbidge, E. & Burbidge, G. 1965, *Ap.J.*, **142**, 1351.

Keel, W. 1983, *Astron. J.*, **88**, 1579.

Metik, L.P. & Pronik, I.I. 1979, *Astrofizika*, **15**, 37.

Metik, L.P. & Pronik, I.I. 1985, *Astrofizika*, **23**, 451.

Minkowski, R. 1957, in *IAU Symp. No. 4*, ed. H.C. van der Hulst, (Cambridge, University Press), p. 107.

Pronik, I. & Metik, L. 1986, in "Structure and Evolution of Active Galactic Nuclei", eds. G. Giuricin et al., (D. Reidel; Dordrecht), pp. 683-687.

TESTS OF THE DISCORDANT REDSHIFT HYPOTHESIS

Jack W. Sulentic
Department of Physics and Astronomy
University of Alabama
and
Istituto di Astronomia
Universitá di Padova

INTRODUCTION

The controversy over the nature of discordant redshift associations has been with us almost since the discovery of quasars as a major constituent in the universe. The present review is intended to deal with two subjects: *a)* whether Arp's claims for the reality of such associations are supported by the best existing data, and *b)* what possible tests might be made of the Arp hypothesis. We will consider here all of the basic forms of evidence that have been advanced, over the past twenty years, in support of a physical association between galaxies and/or quasars of different redshift. We will not attempt to discuss the body of evidence for the universality of the redshift distance relation. It is our position that this *evidence*, however impressive, should not preclude investigation of alternate viewpoints.

There appear to be three essential elements in the hypothesis developed by Arp: 1) galaxies as the sites of ejection events involving compact objects; 2) the compact ejecta as protogalactic objects; and finally, 3) the redshift as an intrinsic and evolving (non-velocity) property of the protogalactic ejecta. In this view, some first generation galaxies eject compact objects of much higher redshift which we observe principally as quasars. The quasars gradually evolve from dense cores into increasingly less compact forms. At some point, star formation begins and the transformation from quasar to galaxy gradually occurs. At the same time, the redshifts of the ejected objects decrease towards their appropriate Doppler values while, perhaps, other physical properties change as well. This picture is motivated by an apparent observational sequence of forms which suggests a correlation between degree of source compactness and perceived excess (non-Doppler) redshift component in a source spectrum (see Arp 1976, p. 396). Arp (1987a and references) has introduced numerous extensions and refinements to this basic picture. We will not consider them here but, instead, will concentrate upon the evidence which bears on the three basic elements of the hypothesis.

Figure 1 is a summary of the forms of evidence relevant to the Arp hypothesis centering upon associations between objects with large redshift differences. In figure 1, we first distinguish between statistical and direct forms of evidence for such associations between 1) quasars and galaxies (QG), 2) pairs of quasars (QQ), and 3) pairs and groups of galaxies (GG). We also include in figure 1 the evidence for ejection activity in spiral galaxies at both optical and radio wavelengths (upper left) and evidence for smaller redshift discordancies in galaxies (lower left). The former evidence deals with element 1 of the hypothesis, while the latter relates to elements 2 and 3. Galaxies with small redshift discordancies would represent, under the Arp hypothesis, galaxies which still possess a small increment of "intrinsic" (non-Doppler) redshift. We have omitted the evidence for systematic redshift eff-

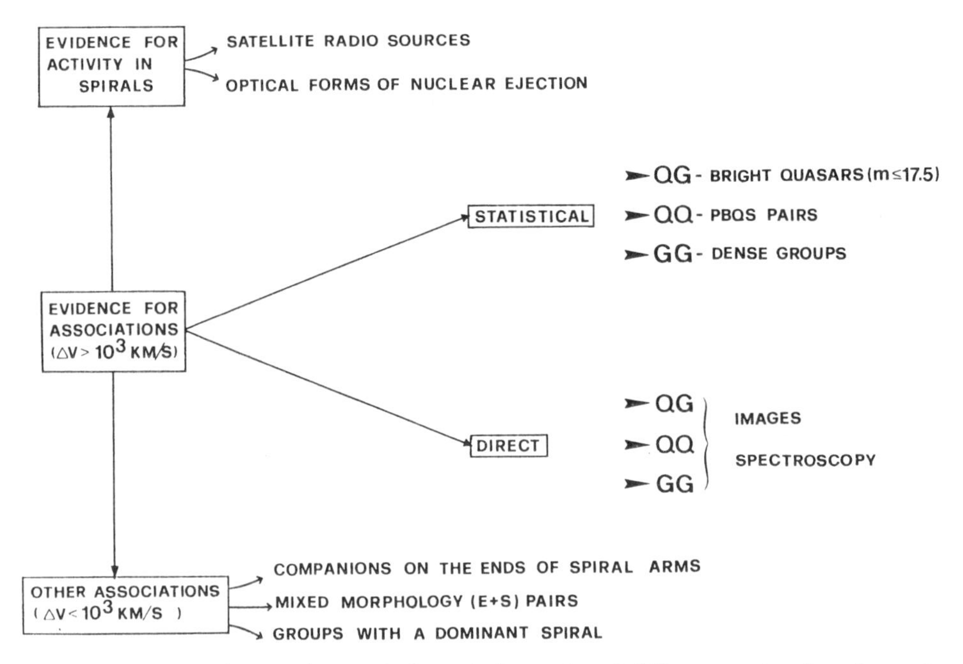

ects in clusters of galaxies (e.g., Tifft 1972), and redshift quantization in pairs (Tifft 1980), which will be discussed separately at this conference. We note that evidence for such systematic effects can be regarded as support for element 3 in the hypothesis. We will consider each entry in figure 1, examining some of the current evidence and suggesting important observations or tests that should be done.

QUASAR-GALAXY ASSOCIATIONS: Statistical Evidence

If some, or all, quasars are local, and if they originate in the nuclei of galaxies, then we would expect some spatial correlation between the two classes of objects. The exact character of the expected correlation is unclear since we know none of the details of the hypothesized ejection process: 1) the magnitude and range of ejection velocities; 2) the types of galaxies that eject quasars; 3) the rate of quasar production by an individual galaxy; 4) the range of initial redshift values that the ejecta possess and 5) whether the formation process can also occur (without ejection) directly from the IGM. We are consequently constrained to the most basic correlation test between quasars and galaxies on a given region of the sky. Although the number of quasars in the literature is now close to 3500, we still suffer from a lack of complete samples over regions of the sky larger than a single 1.2*m* Schmidt plate (25-30 deg^2). This greatly complicates even the simplest association tests between galaxies and quasars because few bright galaxies are present in any one survey field. Despite this restriction to the basic test that could considerably blur any real correlation, an excess of quasars is almost always found when an association test is performed.

Table 1 is a summary of some basic correlation tests that have been performed using the most complete quasar samples. We restrict our summary to 1) the quasar surveys that include significant numbers of quasars brighter than m = 18.5 (or radio-loud) and 2) the brightest galaxies (minus the Local Group) on the sky (generally $m_{pg} \leq 12.8$). The reasoning for these restrictions is that a local hypothesis for the quasars implies that the apparently brightest ones (m $\leq$ 17-18) will be among the most proximate (here we use $m = m_B$ or m_V, assuming average quasar colors). Further restriction would require some knowledge of the intrinsic luminosity function of the quasars. The brightest quasars will presumably be associated with either the Local Group or with some of the other brightest galaxies. Those associated with the Local Group galaxies might be impossible to identify in a correlation analysis. Restriction to the very brightest galaxies is motivated by the assumption that the most luminous and massive galaxies are the best candidates for the kind of ejection activity that is hypothesized. The small magnitude of the quasar excess observed in the Hercules (Burbidge et al., 1985) and Coma (Hoag et al., 1982) clusters suggests that we are either 1) detecting only the most luminous quasars associated with these aggregates, or 2) that galaxies in rich clusters tend to produce few quasars. The former possibility would place most quasars (or at least the fraction that are discordant in redshift) within about 100 Mpc.

Table 1 presents the results of correlation tests reported in the literature, as well as some recent comparisons presented here for the first time. The point of this exercise is not to argue significance levels but to emphasize the general excess quasar density (quasars deg^{-2} in Table 1) that is found near the brightest galaxies. It is difficult to compare the results of the individual tests because different test samples were used and different association tests were performed. The overall positive result obtained in these tests and the difficulty in comparing the different tests motivate the first proposal for a new test of the association hypothesis.

[TEST 1] A multi-variate analysis incorporating the basic forms of data for all known quasars should be performed. The basic correlation test should be performed against a magnitude-limited sample of bright galaxies. The spatial coordinates, redshifts, apparent magnitudes, and (for galaxies) morphological types should be included in this analysis. Other properties, such as X-ray or radio flux measures, might also be included. A principal component analysis is envisioned to determine what, if any, significant correlations exist between these multi-variate data sets. The analysis will be limited by the incompleteness and selection biases in the quasar catalogues. All sky quasar-galaxy correlation tests have been performed in the past with generally positive results (Seldner & Peebles 1979; Nieto & Seldner 1982; Chu et al., 1984). The sample of confirmed quasars has more than doubled since the last of these analyses and a proper multi-variate analysis has never been published.

The final entry in Table 1 presents our own most recent test, restricted to 1) galaxies brighter than $m_{pg} \leq 11.0$ (minus Local Group); 2) quasars brighter than m$\leq$ 17.5; and 3) a large (r = 1 deg) search radius around the galaxies. The search area was chosen arbitrarily large enough to provide a reasonable estimate of the quasar surface density. There are currently 30 bright quasars that fall within r $\leq 1°$ of 26 of our sample of 128 bright galaxies.This implies a surface density ρ(m $\leq$ 17.5) = 0.08 quasar deg $^{-2}$ in regions near the brightest galaxies. This value is compared in figure 2 with other estimates for the quasar surface density to m$\leq$ 21 (see Schmidt & Green 1983; Marshall et al., 1983, for error bars and references). Note that virtually all of the "comparison" samples represented in Figure 2 are contaminated with bright galaxies. Figure 2 suggests that we have already found all

Table 1: STATISTICAL TESTS FOR QUASAR–GALAXY ASSOCIATIONS

Quasar Sample	N_Q	Area (deg²)	Galaxy Sample	Results of Test	Reference to Test
3C Survey Parkes Catalogue (Radio Selected)	93	All Sky	Bright Galaxies Shapley-Ames $m \leq 11.0$	Quasars are 3 – 4° closer to bright galaxies on average than random expectation	Arp (1970)
Osmer & Smith (1980) $m \leq 18.5$	108	340	Bright SA $m \leq 12.8$	Within r=1.4°; ρ_Q=0.48 Outside r=1.4°; ρ_Q=0.27	Sulentic (1981)
Kunth et al. ($m \leq 20$) (Sculptor Field)	19	72	Bright SA $m \leq 12.8$	Within r=1°; ρ_Q=0.40 Outside r=1°; ρ_Q=0.21	Sulentic (1981)
Michigan Surveys (271 Quasar Candidates)	-	515	Bright SA $m \leq 12.8$	Within r=1°; ρ_Q=0.63 Outside r=1°; ρ_Q=0.50	Sulentic (1981)
Schmidt & Green (1983) Bright Quasar Survey (All $m \leq 16.5$)	115	10714	Bright SA $m \leq 11.0$	Within 3° of galaxies observe 7 quasars and expect 4 quasars. Overdensity of factor of 2 within 25° of Virgo cluster center.	present work
Virgo Cluster Center He et al. (1984) ($m \leq 20$)	53-82	25	Galaxies brighter than $m \leq 15.0$	"Overdensity near the brightest 15 galaxies"	He et al. (1984) Arp (1986)
Bright Braccesi Sample Marshall et al. (1983) $m \leq 18.25$	22	37	Bright SA $m \leq 12.8$	Within r=1°; $\rho_Q \simeq 0.75$ Outside r=1°; $\rho_Q \simeq 0.50$	present work
Sculptor Fields + 2 others Monk et al. (1986)	8	13.3	Bright SA galaxies $m \leq 11.0$	ρ ($m_V \leq 17.5$)=0.60 near galaxies ρ ($m_V \leq 17.5$)=0.06–0.1 field	Arp (1987) present work
All Bright Quasars ($m \leq 17.5$)	All Q with $m \leq 17.5$ (very incomplete)	All Sky	Bright SA galaxies $m \leq 11.0$ (Omit local group and M81 group)	Observe 30 Q within 1° of 26 bright galaxies, implies ρ ($m \leq 17.5$)=0.08	present work present work

of the expected bright quasars near our adopted galaxy sample. Yet, it is likely that no more than 20-30 percent of the galaxies in our sample have been systematically searched for quasars in any magnitude or redshift range. We suggest that this most general test predicts a large excess of bright quasars near bright galaxies when more complete samples emerge. Most of the 30 quasars included in this test are more than 0.5 magnitude brighter than the adopted cutoff, which argues against severe contamination of the sample by quasars fainter than m =17.5. Most of the quasars are also more than 20arcmin from the nearest bright galaxy which presents obvious difficulties for an explanation involving gravitational lensing. As the next (?) step in a possible multi-variate analysis, we note that most of the 26 galaxies with bright quasars nearby are Sb-Sc spirals of high luminosity ($M_{pg} \leq -21$).

Several papers have appeared in the past few years claiming no spatial correlation between quasars and low redshift galaxies (e.g., Weedman 1980; Impey & He 1986; Monk et al. 1986). There are two general criticisms that can be directed towards these counter claims: 1) the galaxy samples used in the correlation analyses were often heterogeneous in magnitude (or morphology) or they extended to unreasonably faint magnitude levels, and 2) the comparison quasar density was often taken too high, the result in most cases of galaxy contamination in the control fields .

[TEST 2] The results in table 1 suggest that a controlled bright quasar survey is needed (complete in the range 17.5 to 18.5). Since many of the current quasar surveys are otherwise unconstrained as to field selection, it is surprising that so few are designed to test the general association hypothesis. A compilation of 144 bright galaxy, and an equal number of control, fields (Sulentic et al., 1978;hereafter SA78) provides one possible working list for such a survey.

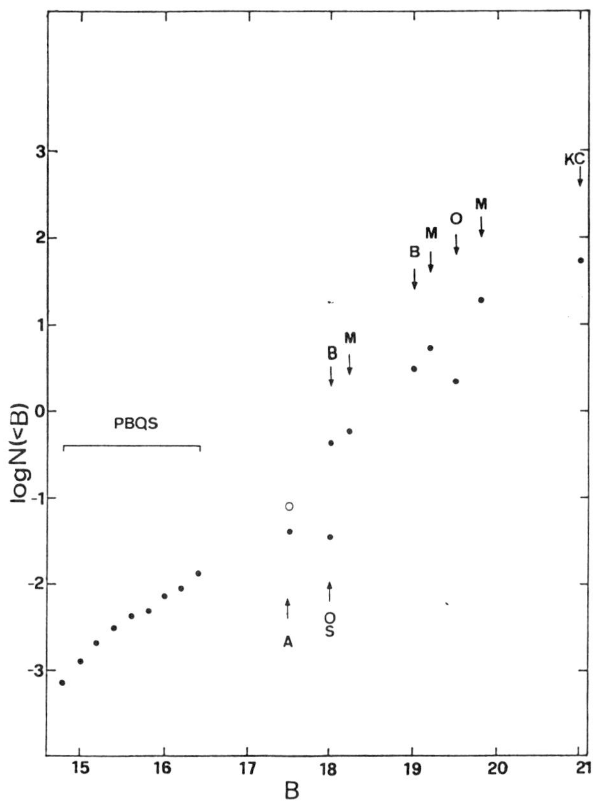

[FIGURE 2] Log of the number of quasars as a function of apparent magnitude. Data points are coded to references in the bibliography. Open circle gives the current density of bright quasars near to bright galaxies.

A possible shortcut for the proposed survey is suggested by the number of serendipitous X-ray quasars that have been found near bright galaxies (e.g., Margon et al., 1985). The density of X-ray sources identified with optical point sources near to a sample of bright galaxies could be compared with one near a sample of globular clusters or other control field sources. The many *Einstein* survey fields centered on galaxies (e.g., Fabbiano & Trinchieri 1985) would provide the test sample. A straightforward comparison of source densities to some predefined optical magnitude or X-ray flux level might provide an indication of any real quasar-galaxy correlation even before complete redshift information became available. Care would have to be taken to avoid confusion with discrete disk sources associated with the spiral galaxies. This would prevent the candidate source density comparison from including objects in the near environs of the galaxies. Still, optical spectroscopy of some of these brighter near-in stellar candidates might reveal quasars such as those found by radio and optical techniques in the disk of NGC 1073 (Arp & Sulentic 1979).

QUASAR-GALAXY ASSOCIATIONS: Direct Evidence

The direct evidence for quasar-galaxy associations has recently been summarized by Arp (1987a,b). Some of the most significant results to emerge in the last few years involve the spiral galaxies NGC 1097, NGC 4319, and 2237+0305. The apparent magnitudes of some of the quasars and/or galaxies involved in these associations would preclude their use in the statistical tests summarized in Table 1, yet, from a direct viewpoint, they argue in a compelling way for physical association between galaxies and quasars. Taken at face value along with the results in Table 1, these associations suggest that the assumed local quasars associated with galaxies span a luminosity range of at least a factor of ~40 ($-12 \geq M \geq -16$)($H_o = 75$). The apparent density excess of quasars near NGC 1097 appears to coincide with the axis defined by the extensive optical jets in that galaxy (Arp et al., 1984). In the second case, a very bright quasar (Markarian 205; $m_V = 15.2$) is linked to the galaxy NGC 4319 by a luminous filament (Arp 1971a; Sulentic 1983a). The final case (2237+0305) would represent one or more quasars still in the center of its "parent" galaxy (Huchra et al., 1985; also Burbidge 1985).

The apparent involvement of X-ray emission with several of the quasars near NGC 1097 serves to reinforce the suggestion made in *[TEST 2]* that a controlled search for point source identifications to X-ray sources near bright galaxies would be useful. We recall that two of the three quasars within 2 arcmin of the elliptical NGC 3842 were discovered as optical identifications to a pair of X-ray sources almost aligned with the nucleus of that galaxy (Arp and Gavazzi 1984).

[TEST 3] We propose high resolution imaging and spectroscopy to clarify the nature of 2237+0305. While part of such a program must await the launch of ST, spectroscopy of the low redshift "host" galaxy might reveal the effects on its interstellar medium and stellar dynamics of the proposed "resident" quasar(s).

Most of the direct evidence for a quasar-galaxy association has centered around NGC 4319/Markarian 205. The connection has now been independently confirmed (Cecil & Stockton 1985) and the suggestion made that it must be an interaction filament associated with the quasar at its appropriate redshift distance. More recently, extensive KPNO 4*m* images and spectroscopy have been analysed (Sulentic 1987a; Sulentic & Arp 1987a,b) and analysis of the new data indicates that: 1) a bright UV knot with an emission spectrum lies on the north edge of the NGC 4319 disk and opposite the galaxy nucleus from M 205; 2) both M 205 and the UV knot appear to be connected to the NGC 4319 nucleus by axisymmetric luminous filaments; 3) the UV knot and gas in its vicinity show outward radial motion with velocity $V \sim 10^2$-10^3 km s^{-1} relative to the nucleus of NGC 4319; and 4) the entire central disk of NGC 4319 exhibits an emission line ratio $H\alpha$/[NII]λ6584 ≤ 0.3, which suggests shock excitation and/or explosive depletion of the disk gas. The evidence that the luminous connection has a high redshift unrelated to NGC 4319 is shown to be seriously affected by image processing artifact (Sulentic & Arp 1987a).

[TEST 4] The evidence summarized above is consistent with a recent ejection origin for Markarian 205. We propose that the best possible spectra are needed for the (20 x 14arcsec) elliptical halo enveloping M 205 (and other quasars) in order to test this idea. If M 205 represents the relatively recent ejection of a quasar that is now rapidly evolving into a galaxy, then an upper limit to its age should be on the order of 10^8 years (see Sulentic and Arp 1987b). It would be important to determine if the halo spectrum, through its implied stellar population, is consistent with this prediction. We expect that spectroscopy of quasar envelopes will reveal them to be characteristically early-type in their spectral properties. Further optical and radio spectroscopic observations of NGC 4319 are also needed to better understand the physical state of the nonstellar component in that galaxy and its possible relationship to M 205. VLA mapping at 21 cm might reveal the location of the higher density gas usually associated with spiral galaxies but apparently absent from NGC 4319.

QUASAR-QUASAR ASSOCIATIONS

The question of the significance of close pairs of quasars has recently been the subject of renewed debate, with Burbidge et al. (1985) arguing that an excess of close pairs ($\theta \leq 120$ arcsec) exist with a radio-loud component. Shaver (1985, 1986;S85 and S86 respectively) has analyzed several different quasar samples and has concluded that no excess of discordant redshift pairs exists in any of them. It was not originally intended to discuss this aspect of the redshift controversy here, but our own analysis of the Bright Quasar Survey (BQS) (Schmidt & Green 1983) and the previous work on this and other samples suggest that the issue is not yet settled. Global analyses of the distribution of quasar separations have the potential to reveal two kinds of evidence relevant to the redshift hypothesis: 1) an excess of discordant redshift pairs of quasars and 2) the tendency for quasars to cluster in the vicinity of low redshift galaxies. The clustering effect, already summarized in table 1, would show up in pairing studies if galaxies eject more than one quasar within the typical timescale for transition from ejected quasar to normal galaxy .

In examining the S85 and S86 studies of quasar pairs from the BQS and other (objective prism and grism) surveys, it was surprising to see a considerable difference between results presented in the form of differential (S85) and cumulative (S86) frequency distributions. The greater uncertainty seems to attach to the differential distributions of QSO pairs, since one is comparing an actual sample distribution of pairs with curves representing the expected random distribution based upon Monte Carlo simulations. The curves are normalized to the *peaks* of the actual distributions, the region where sample edge effects begin to set in. There appears to be considerable latitude in the actual fit because the sample distribution peaks are rather broad (see figure 2 of S85). The slight adjustment of a fit can give the appearance of an excess or deficit of pairs at the small separation end of a distribution. The cumulative distributions of pairs, contrarily, are simply compared to a straight line of slope + 2 (expected random distribution). Surprisingly, the latter distributions reveal a systematic effect in both the BQS and the objective prism surveys that were analyzed (see figure 1 in S86) Close examination of the compressed log-log displays suggest that (n=6) points representing separations $\theta \leq 600$ arcmin in BQS and (n=5 of 6) points for separations $\theta \leq 6$ arcmin in the objective prism results are *systematically high*, while points representing wider pairs in these distributions are *systematically low* relative to the random expectation line.

[TEST 5] We recommend that goodness of fit tests be applied to these distributions to determine at what level the observed distributions deviate from the expected straight line of slope + 2. The grism samples analyzed in S85 and S86 are not large enough in survey area to allow a test of the effect expected from a general quasar-galaxy correlation. It is not clear what, if any, biases might exist against detecting close ($\theta \leq 120$ arcsec) pairs in the grism data. The cumulative frequency distributions for the quasar samples should be *re*determined before doing the above goodness of fit analysis. This is necessary because the Shaver analysis excluded some pairs on the basis of redshift difference. The same quasars rejected in a few close pairs will contribute many (almost twice the sample size) wider pairs. This bias will force the wide separation points in the cumulative distributions higher, compared to the smaller separation values. This will make it more difficult to detect the presence of a close pair excess. We suggest that the most reasonable and unbiased approach is to calculate the cumulative pair distribution for a given sample **without** restriction. The vertical adjustment of the distribution data points to the straight-line expectation should be carried out with a well defined procedure that weights the points relative to the number of pairs that they

represent. A simple (equal weight) "least squares" approach does not seem to be reasonable.If an excess of closer pairs is then observed, it can be analyzed in terms of conventional and unconventional interpretations of the redshift.

The close pair excess suspected from the cumulative pair distribution for the BQS (S86) becomes immediately obvious when the 2-D spatial distribution of the 115 BQS quasars ($\rho = 0.01$ QSO deg^{-2}) is examined. A surprising number (n=16) of pairs are seen in the separation range 0.9 to 3.0 degrees representing an excess over random expectation at the 10^{-4} level. The "significance" level depends strongly on the apparent magnitude limit and resultant quasar surface density that we adopt. The cumulative distribution analysis, of course, does not, which raises the question (quite independantly of the origin of the excess) of why we do not see a more obvious close pair excess in figure 2 of S86. The reason is that Shaver includes in his analysis only 9 of the 16 pairs; 1) 3 pairs were excluded on the basis of redshift difference $\Delta z \leq 0.05$ and 2) 4 quasars were excluded (even from the Schmidt and Green (1983) analysis) on the basis of a redshift-implied absolute magnitude M $\leq$ -23.0. The former exclusion introduces a bias in the slope of the BQS cumulative distribution as discussed above. All three pairs show redshift-implied velocity differences of $\Delta V \geq 7400$ km s^{-1}, suggesting that they are not physical pairs under the conventional redshift interpretation. The latter four rejections exhibit quasar-like spectra and, thus, cannot be removed on other than redshift-dependent grounds. Inclusion of these seven extra pairs would certainly enhance the small separation excess seen in the S86 cumulative frequency distribution of BQS pairs.

As a further corollary to *[TEST 5]*, we recommend a spectroscopic study of close pairs in samples where a significant excess can be demonstrated. A systematic search should be made for possible correlations in spectral properties, including: 1) a search for absorption line systems near the redshift of the higher z components in the spectra of the lower z components and 2) a search for emission and absorption lines at common *wavelengths* in the spectra of the discordant redshift pairs (see e.g., Libby et. al. 1984). An imaging survey would also be of great interest to determine if pair components show similar extended halo structure despite their very different redshifts, **or** if some low redshift components of such pairs exhibit **no** halo structure(or viceversa). The intensity profile widths of the components in BQS pairs, for instance, could be directly differenced. We would expect (conventionally) the low z components to always exhibit the broader profile.

GALAXY-GALAXY ASSOCIATIONS: Statistical Evidence

The redshift controversy can almost be said to predate the discovery of the quasars. It was the detection of discordant redshift galaxy components in the Zwicky Triplet (Zwicky and Humason 1961) and Stephan's Quintet (SQ) (Burbidge & Burbidge 1961) that first attracted the attention of astronomers to a potential problem. The ten years after measurement of the fifth redshift in SQ saw the detection of discordant redshift components in two more of the densest galaxy groups on the sky: Seyfert's Sextet (SS) (Sargent 1970) and VV 172 (Sargent 1968). The improbability of discordant redshifts in such remarkable configurations prompted some interesting discussion at the Vatican Conference on Nuclei of Galaxies (Burbidge & Sargent 1970). It was not only the signs of interaction or calculated improbabilities that attracted so much interest, but the fact that all three of these (relatively rare) groups were unusual even *without* its discordant redshift component.

In the minds of many researchers, the problem of the discordant redshift galaxy quintets was resolved statistically by Rose (1977). It was suggested that these three famous systems were the expected number of optical alignments, based upon a survey that implied more than 400 quartets of galaxies on the sky. It was later shown (Sulentic 1983b): 1) that the actual number of dense groups of four galaxies was a factor of from *4–6 times less* than the Rose (1977) estimate; 2) that the number of quartets with morphological properties similar to VV 172, SQ and SS was a factor of about *10 times less* and, finally, 3) that it was incredible that all of the expected discordant systems had been discovered when very few dense groups had any redshift measures at that time.

A survey by Hickson (1982;hereafter H82), using objective isolation and surface brightness selection criteria, found 100 dense groups ($4 \leq n \leq 8$ members) on the part of the sky covered by the Palomar Sky Survey. The resultant catalogue of dense groups represents an independent confirmation of the results in Sulentic (1983b). Recent galaxy counts within a 1 degree radius of the H82 dense groups allow us to derive an expectation of interloper contamination (n=4 galaxies) for the 100 groups (Sulentic 1987b). Analysis of an incomplete redshift survey for this sample (kindly provided by Paul Hickson) reveals about 17 groups with one or more discordant members. This result suggests that $\geq 25\%$ of the dense groups in the H82 sample will be discordant when the redshift survey is completed. Sulentic (1987b) has proposed that the dense groups are relatively recently formed aggregates of galaxies (ages $t \ll H_o^{-1}$). This suggestion was motivated by several results: 1) the dynamical instability of the dense groups to merger, implying *dynamical* ages $t \leq 0.01\ H_o^{-1}$; 2) the lack of morphological evidence for mergers in progress among the catalogued groups; 3) the typically low density environment in which the H82 groups are found; 4) the presence of substantial diffuse light (attributed to dynamical evolution) in several of the groups (notably, VV 172, SQ and SS) and, 5) the large fraction of discordant redshift components in the dense groups.

[TEST 6] involves several possible tests of the idea that these dense groups are young galaxy aggregates and that the discordant members are physical members of the groups: *(i.)* A southern hemisphere survey corresponding to the already published H82 sample (the combined samples would allow us to estimate the density of such objects within 100–200 Mpc); *(ii.)* Surface photometry measures from calibrated CCD images of the groups (this will provide an estimate of the fraction of diffuse light contained in the groups and a corresponding direct estimate of their dynamical ages); *(iii.)* A spectroscopic survey of the galaxies in the groups with data of sufficient quality to provide a lower limit estimate for the ages of these galaxies (such a survey will also allow us to study discordant redshift components as a class). Burbidge et. al. (1963) and Sandage (1963) provide some relevant discussion of the evidence for and against the existence of young galaxies. The dense groups remain the central element among the evidence for discordant redshifts in galaxies.

GALAXY-GALAXY ASSOCIATIONS: Direct Evidence

There are three kinds of galaxy-galaxy associations that have emerged in the past twenty years: 1) dense groups ($4 \leq n \leq 8$) (H82); 2) isolated triples (Karachentseva et al., 1979); and 3) isolated pairs (Karachentsev 1972)(*CPG*). Redshift measures for the referenced Palomar Sky Survey-based samples, while incomplete, have revealed large fractions (10% to 30%) of wildly discordant components. Table 2 presents a summary of the currently known discordant redshift configurations from these three lists. Statistical analyses (Sulentic 1983b, 1987b)

Table 2: DISCORDANT PAIRS AND MULTIPLETS IN MAJOR CATALOGUES

I. Dense Groups: ΔV(peak to peak)≥1300 km s^{-1}

H#	H#	H#	H#	H#	H#
3	18	36	48	70	92
4	21	41	55	79	93
11	22	45	61	82	

II. Isolated Triplets: ΔV(peak to peak)≥1300 km s^{-1}

KK#	KK#	KK#	KK#	KK#	KK#
5	9	18	27	35	53
6	10	19	29	37	63
7	13	20	30	40	84
8	17	24	32	42	

III. Isolated Pairs: ΔV≥1000 km s^{-1}

K#	K#	K#	K#	K#	K#
2	77	188	315	427	500
15	114	193	328	443	529
27	120	196	338	456	556
30	152	207	357	475	563
35	164	267	360	481	586
45	166	300	391	499	599
48	183	305	421		

involving the first list suggest a large excess over the number of expected interlop ers. Karachentsev (1985) has argued that the number of discordant pairs found in the *CPG* is commensurate with interloper expectations. We would argue that two conflicting effects, a) the selection criteria designed to avoid interlopers and b) the same criteria clearly biased to select dense optical configurations, if they exist in significant numbers, make it difficult to reasonably assess the true interloper fraction without resorting to direct galaxy counts (as in Sulentic, 1987b).

The Table 2 summary is by no means complete since an isolation criterion was used in the compilation of all three surveys. Thus, there is a bias against selecting pairs and multiplets in the near projected environment of bright galaxies. A controlled search suggests a factor of 2 excess in the surface density of such objects near to (r $\leq$ 1 degree) bright low redshift galaxies (SA78). Two striking and well-studied pairs absent from Table 2 are NGC 7603 (Arp 1971b; Sharp 1986) and 4151-46 (in SA78; Sulentic & Arp 1985). Arp (1980) has also discussed several striking southern hemisphere pairs.

[TEST 7] We propose an imaging and spectroscopy survey of objects in Table 2 (see related *[TEST 6]*). The survey would be designed to isolate a sample of discordant galaxy components with common characteristics. Demonstration that the spectroscopic (and surface photometric) properties of the discordant galaxies in Table 2 were completely heterogeneous and "field-like" would argue strongly 1) **against** their interpretation as a class of young galaxies and 2) **in favor of** the interloper hypothesis. It would be useful to extend the Table 2 lists with corresponding surveys in the southern hemisphere and by completing a redshift survey for the 185 interacting systems catalogued by SA78. CCD images will soon become available for all objects in the H82 dense group catalogue (Hickson & Kindl

1987). We expect that the discordant components will be generally more compact, bluer in color, gas-rich, and earlier in average spectral type, than typical field galaxies of similar distance and morphology.

A final program suggested under *[TEST 7]* involves a high resolution 21*cm* survey of objects from Table 2 (or from SA78). Neutral hydrogen measures are a potential monitor of a) the youth of a galaxy through estimation of its fractional hydrogen mass and b) the dynamical state of a galaxy through study of its 21*cm* velocity profile and 2-D spatial distribution. It is not clear what sort of dynamical evidence could be expected in the hypothesized young galaxies and so the primary relevant quantity for these objects might be the measure of gas content. The lower redshift components of these associations, viewed as having almost entirely Doppler redshifts, would provide information both on neutral hydrogen content and dynamical state. The 21*cm* phase of the program is motivated by the observations on SQ which is, perhaps, the **only** discordant system studied in detail at 21*cm*. Recently it was proposed (Sulentic & Arp 1983a) that the (in this case) single low redshift component showed signs of interaction with the remaining members based upon new 21*cm* measures. A similar result was found for the single low redshift component in the quartet NGC 4169,73-5 (Sulentic and Arp 1983b) Detection of a radio continuum (van der Hulst et al. 1981) source and also an X-ray emission centroid (Bahcall et al. 1984) at the interface between the discordant components in SQ is further evidence that all the galaxies are interacting. Much more than detailed observations of single objects however, a survey such as *[TEST 7]* has the potential to clarify the nature of discordant GG associations.

ACTIVITY IN SPIRAL GALAXIES: Radio Evidence

Considerable evidence exists for various forms of nuclear activity in galaxies. There has been much resistance, however, to the evidence that spiral galaxies are sometimes sites of directed ejection activity. We will consider here both radio and optical evidence for such collimated explosive activity in spirals. The relevance of this evidence to the Arp hypothesis is two-fold: 1) most of the QG associations involve spiral galaxies; and 2) models for the ejection of compact bodies from galactic nucleii usually involve a form of collimated activity (see e.g., Saslaw et al. 1974).

Independent investigations (Arp 1967, 1968; Tovmassian and Shahbazian 1976) have pointed toward a possible excess of satellite radio sources near spiral galaxies. The concept was advanced that spiral galaxies might exhibit scaled-down radio galaxy activity but this idea has not met with acceptance because more recent radio surveys have not confirmed the initial claims. Two surveys based on Westerbork observations (Willis 1976; Hummel 1980) have argued against the presence of excess radio sources near spirals. We suggest that the question is still open and that these two surveys are less than conclusive for several reasons: 1) the optical test samples included many galaxies that are not spiral type; 2) the optical samples were either not magnitude-limited; or covered too broad a range in apparent magnitude for a fixed radius survey; and 3) the surveys were not properly controlled.

Figure 3 presents an analysis of the Willis (1976) result that we carried out about ten years ago and which reinforces our above conclusion. We have plotted the surface density of 21*cm* continuum sources (detected above a flux limit of S $\geq$ 100 mJy) from the Willis (1976) bright galaxy sample (W) and a number of possible control surveys (comparison surveys VV,K1-3,and WO are coded in bibliography,K=K1-3). The point of this exercise is to demonstrate that the source density fluctuates considerably between different published samples. Most of these samples are contaminated to varying degrees by bright galaxies in the survey fields. It is clear that an appropriate control sample exists to establish any level of source surface density excess or deficit near spiral galaxies. Figure 3 also shows one test involving 42 spiral galaxies that exist in common between Willis (1976) and SA78. We divided this spiral galaxy sample (AS) into active (29)(AS*) and inactive (13)(AS*) subsamples, based upon the presence or absence of peculiar companions in their neighborhood, respectively (see SA78). We find almost a factor of 2 higher radio source density in our optically pre-defined active class of spirals.

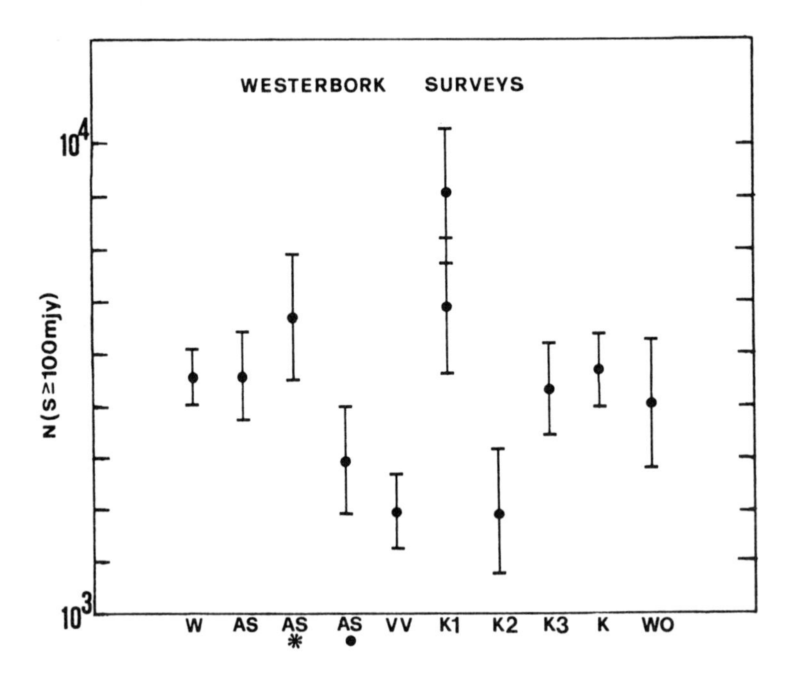

[**FIGURE 3] Comparison of the radio source density (in sources/steradian) for several different Westerbork continuum surveys. The codes in the abcissa refer to references in the bibliography.**

[TEST 8] We propose a radio continuum survey of the near-in environments of a magnitude-limited sample of bright spiral galaxies. The SA78 survey provides a possible working list of (99) bright non-cluster spiral galaxies and control fields. A comparison of radio source surface densities in the two samples should be made both before and after an attempt at optical identification is carried out. Detection of a source excess in all, or a subsample of, the galaxy fields would lead to a search for possible source alignments. The galaxy sample should be chosen in the approximate range $11.0 \leq m_{pg} \leq 13.0$ in order to: 1) avoid the very nearest bright spirals (m-M $\leq$ 27.0), 2) achieve a sample size of at least 100 galaxies (plus a corresponding control sample), and 3) restrict the galaxy sample to a reasonably narrow range in distance. Thus, we will be studying a similar volume of space near all sample galaxies and, with a sample of $n \geq 100$, will be able to detect an effect that might be present in only a small part of the total spiral population. The search could be carried out with either a single dish instrument (e.g. 100m Green Bank

or 300m Arecibo), or with the VLA in the "snapshot" mode. The former choice would facilitate a large area (R $\geq 1°$) search near the galaxies, while the latter would provide maximum sensitivity. The surveys would provide useful statistical information from the resultant source counts, quite aside from their usefulness as a test of the hypothesis.

Evidence has begun to accumulate which suggests that the proposed radio surveys would yield a new insight into forms of spiral galaxy activity. The discovery of the radio "arms" presumably ejected from the nucleus of NGC 4258 (van der Kruit et al. 1972) is a specific example. Valtonen (1977) has found evidence for an alignment of discrete sources in this field with the nucleus of NGC 4258. More extensive observations of Seyfert-class spirals (Wilson & Ulvestad 1983; Ulvestad & Wilson 1984) provides unambiguous evidence for paired radio sources across the optical nuclei. This double-lobed activity, however, appears to be restricted to the optical confines of the spiral galaxies. It would be interesting to see the results of *[TEST 8]* carried out on a subsample of this active class. Most recently, independent observations of three galaxies with spiral structure show, for the first time, 1) radio sources outside of the optical disk, and 2) the sources aligned with the galactic nucleus (Beichmann et.al. 1985; Simkin and Michel 1986; and Sulentic 1986). The latter reference involves NGC 4319, where an attempt has been made (Sulentic & Arp 1987a,b) to relate the proposed radio ejection activity to a cycle of events including the ejection of the quasar Markarian 205.

ACTIVITY IN SPIRAL GALAXIES: Optical Evidence

The number of spiral galaxies with optical evidence for activity continues to increase. The current situation can be summarized by considering the spiral galaxies exhibiting low ionization emission lines in their nucleii (LINER) (see e.g., Keel 1985; Heckmann 1985). There are two general classes of model presently invoked to explain this type of activity: 1) photoionization by an energetic power-law continuum source and/or 2) shock wave heating. The source of the latter phenomenon has been suggested to be "winds" driven by superstarburst activity occurring in some spiral galaxies (Heckmann 1987). Surprisingly, and in spite of the growing evidence, an alternate source of shock excitation — possible explosive ejection activity in spiral galaxies — has not been seriously discussed.

One spectroscopic measure of LINER-type activity is the nuclear emission line ratio Hα/[NII]λ6584, which is often $\leq$1 in these objects. A list of 29 spiral galaxies studied spectroscopically by Keel (1983) includes 8 objects with a stellar synthesis corrected ratio of Hα/[NII] $\leq$ 0.45. This restriction is motivated by our observations of NGC 4319 (Sulentic & Arp 1987b), where Hα/[NII] $\leq$ 0.3 throughout the central disk. We find several galaxies in this spectroscopically defined subclass that exhibit either radio and/or optical evidence for directed explosive events: 1) NGC 5194, which has a pair of radio and optical emitting "bubbles" paired across its nucleus (Ford et al. 1985), 2) NGC 4736, which exhibits possible double lobe radio structure (van der Kruit 1971) and optical evidence for an expanding ring of gas (van der Kruit 1974), and 3) NGC 3627, which has recently been shown to have a double-lobed radio disk morphology (Urbanik et al. 1985). The optical features in NGC 4736 have been successfully modelled as an explosive event by Sanders & Bania (1976). NGC 1097, discussed earlier, is also in the Keel (1983) study with a ratio Hα/[NII] = 0.79.

[TEST 9] involves a survey designed to characterize the forms of ejection activity in spiral galaxies and assemble a catalogue of active objects. It would be followed by a search for quasars possibly related to this activity. We

suggest that galaxies such as NGC 4319, NGC 1097 and, perhaps, NGC 4736 can provide the clues necessary to inaugurate such a study. The criteria for sample selection then would be a) double- or triple-lobed continuum radio morphology, b) an unusual ratio of Hα/[NII] in and especially outside of the galactic nucleus and also, possibly, c) a morphology similar to that predicted by the models of Sanders & Bania (1976). The latter involves a spiral galaxy with a sharply bounded central region which may (NGC 4319) or may not (NGC 4736) contain spiral structure. We believe that the results summarized here indicate that ejection activity occurs in spiral galaxies. The character and frequency of such activity can only be determined from detection and study of a significant sample of these objects. The final step will be to search for the quasars *[TEST 2]*that the hypothesis predicts were recently ejected from these galaxies. Some of them, apparently, have already been found near NGC 1097 and NGC 4319.

SMALLER REDSHIFT DIFFERENCES

If older galaxies eject protogalactic objects, then we might expect to find several classes of unusual objects in their vicinity. Specifically, we might observe various kinds of objects that represent the transition phases from initial ejecta to normal galaxy. One might hypothesize several classes of objects, including 1) stellar quasars with a wide, but unknown, range of z (e.g.,quasars near NGC 1073, 1097); 2) quasars enveloped in various amounts of luminous "fuzz" with an average z less than that of class 1 (Mark205); 3) emission line "galaxies" with varying degrees of compactness; and 4) galaxies with an early-type absorption spectrum.

It was the observation (Ambartsumian and Shahbazian 1957,8) of a class of blue compact objects in the vicinity of more normal-appearing galaxies that motivated Ambartsumian (1958) to suggest that ejection was an ongoing process for creating new galaxies. An excess of smaller companion galaxies near bright galaxies has already been established (Holmberg 1969; SA78). An excess of such companions along the major axes of (an inclined subsample of) bright spirals also led Holmberg (1961) to propose an ejection origin for their formation. The galaxy pairs and multiplets considered in the next three sections have all been suspected of having small systematic redshift anomalies. As the objects with the smallest redshift excesses, they are regarded as possible examples in the above hypothesized classes 3) and 4), which are also, presumably, related to the classes noted by Ambartsumian (1958) and Holmberg (1969). We will consider three types of such objects that may represent the final stage in the hypothesized formation process. It is not clear how the latter two classes might fit into this picture. We simply propose observations that test the suggested redshift anomalies.

Companions on the Ends of Spiral Arms

The two principal catalogues of peculiar and interacting galaxies (Vorontsov- Velyaminov 1959(V-V); Arp 1966) have independently recognized the unusual class of spiral galaxies with compact companions on or near the end of an arm. In a spectroscopic study of 6 such pairs (Arp 1969), it was noted that 1) the companions tend to be high surface-brightness (or compact) objects with μ_B in the range 20-22 mag arcsec^{-2}; 2) they frequently exhibit emission line spectra; 3) when absorption lines are detected, they tend to be of early type; and 4) companion redshifts tend to be slightly higher ($\Delta V \geq 100 kms^{-1}$) than the larger associated spiral galaxy. -V (1975), in an analysis of 160 "M51 type" systems catalogued by him, arrived at similar conclusions about this class of objects (except element #4). Both authors suggested that the companions represented ejecta from the spiral galaxies and that, perhaps, the spiral structure was created in the ejection process. One further important observation made by V-V was that the companions tended toward MCG types N and Ir. They were found to be classifiable as E or S type galaxies only if the ratio of pair component diameters was near unity (next section). V-V concluded that more than 4% of all spiral galaxies possessed one (in a few cases, two?) such compact companions.

[TEST 10] A survey of the compact companions on the ends of spiral arms should be carried out for at least two reasons: 1) they represent a class of binary galaxy for which very few observations exist and 2) a test could then be made of the hypothesis that they are characteristically young objects with a slight intrinsic redshift component. There would be two principal parts involved in such a survey: 1) development of a catalogue of such objects that is unbiased and preselected and 2) a subsequent spectroscopic survey to measure redshift differences and companion spectroscopic properties. The Catalog of Interacting Galaxies-II (V-V 1975) represents an obvious starting point for the preparation of a catalogue. Additional candidate objects are contained in published lists of "superassociations" in or near galaxies (Petrosian et al. 1983,84). Southern hemisphere sample selection will be much easier since this class of compact companion galaxy can be distinguished from stars on the Southern Sky surveys with greater ease.

The principal criterion for sample selection from among the general class of galaxy pairs should be the ratio of component sizes (see V-V 1975; Sulentic 1976). A CCD imaging survey would then provide a basis for refined classification based upon morphology, color, and surface photometric properties. It is possible that there is more than one class of small companion galaxy that frequently pairs with a larger spiral. The principal selection criterion excludes most of the pairs in the *CPG* from this study. Thus, very little spectroscopic data exists for the class of pairs that we propose to study. The limited data which does exist, however, coupled with the initial work of Arp (1969), justifies a serious and systematic program. Figure 4 presents an interesting comparison of redshift differentials for an optically selected sample of 25 E+S pairs and 13 pairs involving a spiral with a relatively compact companion. The pairs are all included in the *CPG* and all were measured spectroscopically in a $6m$ redshift survey (Karachentsev 1980). The redshift differentials are almost all within ± 300 km s^{-1} and, consequently, these pairs are regarded as physical systems. The first sample reveals an excess of higher redshift spiral components (next section), while the latter shows a higher redshift for the compact components of uneven size ratio pairs. It is the latter effect that will be tested by the proposed spectroscopic survey. Further evidence in support of this survey comes from a number of independent observations of pairs fitting our morphological size ratio criterion.

1) 4030-11 (Sulentic & Arp 1985)is a discordant redshift example of a compact companion on the edge of a theta bar spiral galaxy. The (emission line) companion redshift in this case is about 5000 km s^{-1} higher. NGC7603 (Arp 1971) is the prototypical example of a discordant redshift companion on the end of a spiral arm. In that case however, the companion does not show emission lines.

2) NGC 4597 is a theta bar spiral with a high surface brightness stellar (on the Palomar Sky Survey) object at one end of the "theta" arms. The companion shows a pure emission spectrum with a redshift about 200 km s^{-1} higher than the larger component (Sulentic & Arp, unpublished). V-V has noted the continuity in forms, from galaxies to giant HII regions and superassociations, on the ends of spiral arms.

3) NGC 6636, an SBc galaxy, is paired with a bright companion of compact morphology with an emission line spectrum about 300 km s^{-1} higher than its larger neighbor (Kazaryan 1984).

4) A probable edge-on spiral (Anon 2208-252) is connected to a compact companion by an emission filament. The companion has a redshift about 100 km s^{-1} higher than the larger spiral (Bertola & Zeilinger, private communication). In this case it is clear that the companion is connected to the spiral and lies in its principal plane.

5) Neutral hydrogen mapping of the spiral galaxy NGC797 (van Moorsel 1983) reveals that the gas is concentrated in a compact object on the end of a spiral arm. In this case the companion redshift is slightly lower than that of the spiral.

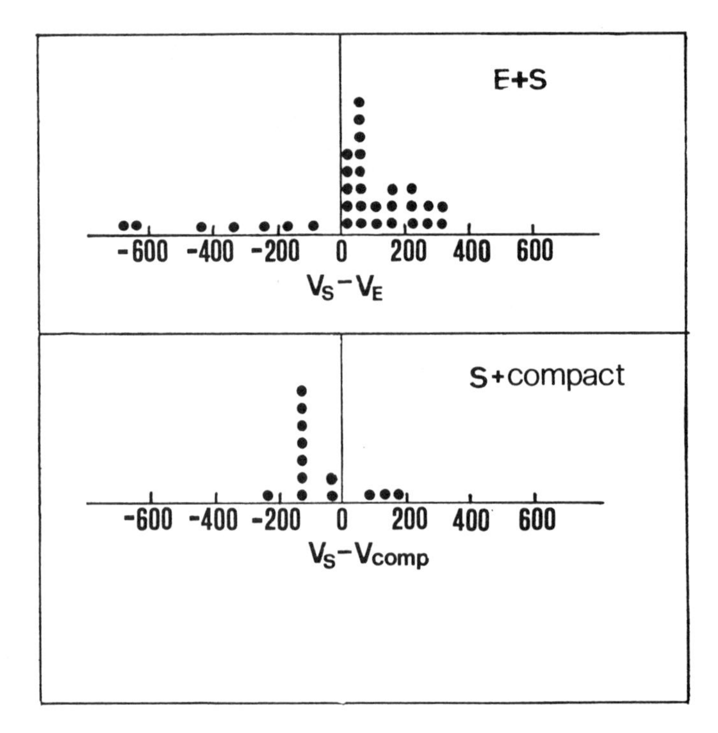

[**FIGURE 4] Redshift differentials for two morphologically refined samples of galaxy pairs. Upper half represents mixed pairs with similar apparent diameters while the lower is for spiral pairs with compact companions.**

Mixed (E+S) Pairs

The upper half of Figure 4 shows an effect that has been debated since Jaakkola (1971) proposed a possible general correlation between redshift and galaxy morphology. If such a correlation exists, then isolated mixed morphology (E+S) pairs of galaxies represent a simple and direct class for quantifying it. V-V (1958) and Burbidge et.al. (1963) have discussed the possibility that the spiral components of such pairs are considerably younger than the elliptical ones. Figure 4 shows a refined subsample of mixed pairs that have 6*m* redshifts (Karachentsev 1980). This optically selected sample shows a clear excess of higher redshift spiral components. The same result was obtained for a larger, but less morphologically refined, sample of such pairs (Sulentic 1982). The Figure 4 subsample is shown in order to emphasize two points: 1) the excess spiral redshift effect seems to become stronger when the analyzed sample is restricted to more certain E+S pairs, and 2) the excess redshift effect reverses when uneven size pairs (spirals with compact companions) are considered.

[TEST 11] We propose a test of the general redshift-morphology correlation using mixed morphology pairs. The test should be composed of two parts: 1) compilation of the largest possible morphologically pure E+S sample, and 2) comparison of component redshifts using existing and new redshift measures. In the northern hemisphere, the *CPG* provides a basis for preliminary sample selection. CCD imagery (Sharp & Sulentic, in preparation) will allow sample refinement based upon morphology, color, and other surface photometric properties. Initial sample selection in the southern hemisphere (Rampazzo & Sulentic, in preparation) will be aided by the higher resolution southern sky surveys and by the *Atlas of Southern Peculiar Galaxies* (Arp & Madore 1987). After morphology, the most important sample selection criterion will be the ratio of component sizes. Two possible techniques might maximize the accuracy of the differential redshift measure: 1) long-slit spectra for the refined sample would permit simultaneous measurement of both components in a pair. The component spectra could be cross-correlated against one another using a Fourier Quotient routine to maximize accuracy of the measure ΔV, and 2) $21cm$ measures could provide an accurate zero point through measures of the spiral redshift. This procedure would also provide an additional check on the morphological purity and dynamical state (see Sulentic & Arp 1983b).

Another independent test involves comparing the mean redshift as a function of galaxy type for members of the same cluster (see papers by Arp and Tifft in this volume).

Groups with a Dominant Spiral

Several reasonably independent surveys suggest that companions in groups show an excess redshift component relative to the most luminous (and usually spiral) members (Sulentic 1984; Arp & Sulentic 1985). Arp (1987b) has suggested that the companions with the most peculiar morphology show the largest redshift excess. If this effect is real, we may be observing relatively recently formed galaxies (Holmberg 1969) or an effect related to conventional dynamics and sample selection (Byrd & Valtonen 1985).

[TEST 12] We propose a test of the suggested excess redshift of group companions. A sample of nearly relatively isolated spiral galaxies should be assembled (see SA78 for a controlled candidate list). Deep images of the near environment of such spirals should allow the creation of a magnitude-limited candidate companion sample in each of the fields. Redshift and morphology measures will then provide the basic information required for such a test. Groups dominated by a single luminous galaxy provide the simplest test sample for such an experiment because only one galaxy is a candidate for the hypothesized ejection activity.

CONCLUDING REMARKS

We have summarized evidence of several kinds that appears relevant to concepts in the Discordant Redshift Hypothesis. We do not claim to understand how all of this evidence can fit together, assuming that it is all significant. It does appear that much of it fits into the single "scenario" summarized at the beginning of this review. Thus, it is not a collection of unrelated curiosities and it does possess limited predictive power. We have used this limited power, and the results previously published, to suggest a series of tests and observations that might clarify (or invalidate) this picture. The tests, in many cases, are specific, but the detailed test procedures have been left deliberately vague. There are many avenues for carrying them out. The only requirement is that the experiments be approached as unbiased tests of the hypothesis.

REFERENCES

Ambartsumian, V. & Shahbazian, R. (1957), *Dokl. Akad. Nauk Arm. SSR*, **25**, 185.
Ambartsumian, V. & Shahbazian, R. (1958), *Dokl. Akad. Nauk Arm. SSR*, **26**, 277.
Ambartsumian, V.A. (1958), in Solvay Conference on *Structure and Evolution of the Universe*, (ed. R. Stoops, Brussels), p.1.
Arp, H.C. (1966), *Ap.J. Suppl.*, **14**, No. 123.
Arp, H.C. (1967), *Ap.J.*, **148**, 321.
Arp, H.C. (1968), *Astrofizika*, **4**, 59.
Arp, H.C. (1969), *Astron. Ap.*, **3**, 418.
Arp, H.C. (1970), *A.J.*, **75**, 1.
Arp, H.C. (1971a), *Ap. Lett.*, **9**, 1.
Arp, H.C. (1971b), *Ap. Lett.*, **7**, 221.
Arp, H.C. (1976), in IAU Colloquium 37, *Decalages vers le rouge et expansion de l'univers*, ed. C. Balkowski and B. Westerlund, (Paris:CNRS), p.396.
Arp, H.C. (1980), *Ap.J.*, **239**, 463.
Arp, H.C. (1986), *J. Astrophys. Astr.*, **7**, 77.
Arp, H.C. (1987a), *Quasars, Redshifts and Controversies*, (Interstellar Media: Berkeley).
Arp, H.C. (1987b), in IAU Symposium 124, *Observational Cosmology*, ed. G. Burbidge, (D. Reidel: Dordrecht), in press.
Arp, H.C. (1987c), *MNRAS*, in press. [**A**]
Arp, H.C. and Sulentic, J.W. (1979), *Ap.J.*, **229**, 496.
Arp, H.C. and Gavazzi, G. (1984), *Astron. Ap.*, **139**, 240.
Arp, H.C., Wolstencroft, R. and He, X-T. (1984), *Ap.J.*, **285**, 44.
Arp, H.C. and Sulentic, J.W. (1985), *Ap.J.*, **291**, 88.
Arp, H.C. and Madore, B. M. (1987), *A Catalogue of Southern Peculiar Galaxies*, (David Dunlap Observatory: Toronto).
Bahcall, N., Harris, D. and Rood, H. (1984), *Ap.J.*, **284**, L29.
Beichmann, C. *et al.* (1985), *Ap.J.*, **293**, 148.
Boyle, B.J. *et al.* (1985), *MNRAS*, **216**, 623.
Burbidge, E.M. and Burbidge, G.R. (1961), *Ap.J.*, **134**, 244.
Burbidge, E.M., Burbidge, G.R. and Hoyle, F. (1963), *Ap.J.*, **138**, 873.
Burbidge, E.M. and Sargent, W.L. (1970), in *Nuclei of Galaxies*, ed. D.J.K. O'Connell, (Amsterdam: North Holland), p.351.
Burbidge, E.M. *et al.* (1985), *Ap.J.*, **288**, 82.
Burbidge, G.R. (1985), *A.J.*, **90**, 1399.
Burbidge, G.R., Narlikar, J.V. and Hewitt, A. (1985), *Nature*, **317**, 413.
Byrd, G. and Valtonen, M. (1985), *Ap.J.*, **289**, 535.
Cecil, G. and Stockton, A. (1985), *Ap.J.*, **288**, 201.
Chu, Y. *et al.* (1984), *Astron. Ap.*, **138**, 408.
Fabbiano, G. and Trinchieri, G. (1985), *Ap.J.*, **296**, 430.
Ford, H.C. *et al.* (1985), *Ap.J.*, **293**, 132.
He, X-T. *et al.* (1984), *MNRAS*, **211**, 443.
Heckman, T. (1985), *PASP*, **98**, 159.

Heckman, T. (1987), in IAU Symposium 121, *Observational Evidences of Activity of Galaxies*, ed. E. Khachikian, K. Fricke and J. Melnick, (D. Reidel: Dordrecht), in press.
Hickson, P.W. (1982), *Ap.J.*, **255**, 382. [**H**]
Hoag, A., Thomas, N. and Vaucher, B. (1982), *Ap.J.*, **263**, 23.
Holmberg, E. (1969), *Ark. Astr.*, **5**, 305.
Huchra, J. *et al.*, (1985), *A.J.*, **90**, 691.
Hummel, E. (1980), *Astron. Ap.*, **81**, 316.
Impey, C. and He, X-T. (1985), *MNRAS*, **221**, 897.
Jaakkola, T. (1971), *Nature*, **234**, 534.
Karachentsev, I.D. (1972), *Comm. Spets. Ap. Obs. USSR*, **7**, 3.
Karachentsev, I.D. (1980), *Ap.J. Suppl.*, **44**, 137.
Karachentsev, I.D. (1985), *Sov. Astron.*, **29**, 243.
Karachentseva, B.E., Karachentsev, I.D. and Shcherbanovskii, A. (1979), *Astrofiz. Issled.*, **11**, 3.
Katgert, P. *et al.* (1973), *Astron. Ap.*, **23**, 171. [**K1**]
Katgert, J. and Spinrad, H. (1974), *Astron. Ap.*, **35**, 393. [**K2**]
Katgert, P. (1975), *Astron. Ap.*, **38**, 87. [**K3**]
Kazaryan, M. (1984), *Astrofizika*, **20**, 24.
Keel, W.C. (1983), *Ap.J.*, **269**, 474.
Keel, W.C. (1985), in *Astrophysics of Active Galaxies and Quasi-Stellar Objects*, Proceedings of the 7th Santa Cruz Workshop on Astrophysics, ed. J. Miller, (Oxford U. Press: Oxford).
Kron, R. and Chiu, L-T. (1981), *PASP*, **93**, 397. [**KC**]
Libby, L.M., Runcorn, S. and Levine, L. (1984), *A.J.*, **89**, 311.
Margon, B., Downes, R. and Chanon, G. (1985), *Ap.J. Suppl.*, **59**, 23.
Marshall, H.L. *et al.* (1983), *Ap.J.*, **269**, 42. [**B**]
Marshall, H.L. *et al.* (1984), *Ap.J.*, **283**, 50. [**M**]
Monk, A.S. *et al.* (1986), *MNRAS*, **222**, 787.
Nieto, J-L. and Seldner, M. (1982), *Astron. Ap.*, **112**, 321.
Osmer, P. (1980), *Ap.J. Suppl.*, **42**, 523. [**O**]
Osmer, P. and Smith, M.G. (1980), *Ap.J. Suppl.*, **42**, 333. [**OS**]
Petrosian, A.R., Saakyan, K. and Khachikian, E. (1983), *Astrofizika*, **19**, 619.
Petrosian, A.R., Saakyan, K. and Khachikian, E. (1984), *Astrofizika*, **20**, 51.
Rose, J. (1977), *Ap.J.*, **211**, 311.
Sandage, A.R. (1963), *Ap.J.*,**138**, 863.
Sanders, R.H. and Bania, T.M. (1976), *Ap.J.*, **204**, 341.
Sargent, W.L. (1968), *Ap.J. Lett.*, **153**, L135.
Sargent, W.L. (1970), *Ap.J.*, **160**, 405.
Saslaw, W., Valtonen, M. and Aarseth, S. (1974), *Ap.J.*, **190**, 253.
Schmidt, M. and Green, R. (1983), *Ap.J.*, **269**, 352. [**BQS**]
Seldner, M. and Peebles, P.J.E. (1979), *Ap.J.*, **227**, 30.
Sharp, N. (1986), *Ap.J.*, **302**, 245.
Shaver, P. (1985), *Astron. Ap.*, **143**, 451. [**S85**]
Shaver, P. (1986), *Nature*, **323**, 185. [**S86**]
Simkin, S. and Michel, A. (1986), *Ap.J.*, **300**, L5.
Sulentic, J.W. (1976), *Ap.J. Suppl.*, **32**, 171.
Sulentic, J.W. (1981), *Ap.J. Lett.*, **244**, L53.
Sulentic, J.W. (1982), *Ap.J.*, **252**, 439.
Sulentic, J.W. (1983a), *Ap.J. Lett.*, **265**, L49.
Sulentic, J.W. (1983b), *Ap.J.*, **270**, 417.
Sulentic, J.W. (1984), *Ap.J.*, **286**, 442.

Sulentic, J.W. (1986), *Ap.J.*, **304**, 617.
Sulentic, J.W. (1987a), in IAU Symposium 121, *Observational Evidences of Activity of Galaxies*, ed. E. Khachikian, K. Fricke and J. Melnick, (D. Reidel: Dordrecht), in press.
Sulentic, J.W. (1987b), *Ap.J.*, Nov. 15 in press.
Sulentic, J.W., Arp, H.C. and di Tullio, G. (1978), *Ap.J.*, **220**, 47. [**AS=SA78**]
Sulentic, J.W. and Arp, H.C. (1983a), *A.J.*, **88**, 267.
Sulentic, J.W. and Arp, H.C. (1983b), *A.J.*, **88**, 489.
Sulentic, J.W. and Arp, H.C. (1985), *Ap.J.*, **297**, 572.
Sulentic, J.W. and Arp, H.C. (1987a), *Ap.J.*, Aug. 15 in press.
Sulentic, J.W. and Arp, H.C. (1987b), *Ap.J.*, Aug. 15 in press.
Tifft, W.G. (1972), *Ap.J.*, **175**, 613.
Tifft, W.G. (1980), *Ap.J.*, **236**, 70.
Tovmassian, G. and Shahbazian, E. (1976), *Astrofizika*, **12**, 201.
Ulvestad, J. and Wilson, A. (1984), *Ap.J.*, **278**, 544.
Urbanik, M., Grave, R. and Klein, U. (1985), *Astron. Ap*, **152**, 291.
Valtonen, M.V. (1977), *Ap.J. Lett.*, **211**, L111.
van der Hulst, J.M. and Rots, A.H. (1981), *A.J.*, **86**, 1775.
van der Kruit, P. (1971), *Astron. Ap.*, **15**, 110.
van der Kruit, P. (1974), *Ap.J.*, **188**, 3.
van der Kruit, P., Oort, J. and Matthewson, D. (1972), *Astron. Ap*, **21**, 169.
van Moorsel, G.A. (1983), *Astron. Ap. Suppl.*, **54**, 1.
van Vliet, W. *et al.* (1976), *Astron. Ap.*, **47**, 345. [**VV**]
Vorontsov-Velyaminov, B. (1958), *Sov. Astron.*, **1**, 9.
Vorontsov-Velyaminov, B. (1959), *Atlas and Catalog of Interacting Galaxies-I*, (MGU:Moscow).
Vorontsov-Velyaminov, B. (1975), *Sov. Astron.*, **19**, 422.
Vorontsov-Velyaminov, B. (1977), *Astron. Ap. Suppl.*, **28**, 1.
Weedman, D. (1980), *Ap.J.*, **237**, 326.
Willis, A.G. (1976), *Astron. Ap.* , **52**, 219. [**W**]
Willis, A., Osterbaan, C. and de Ruiter, H. (1976), *Astron. Ap. Suppl*, **25**, 453. [**WO**]
Wilson, A. and Ulvestad, J. (1983), *Ap.J.*, **275**, 8.
Zwicky, F. and Humason, M. (1961), *Ap. J.*, **133**, 794.

DISCUSSION

J. P. Vigier: Two remarks: a) One should utilize systematically non-parametric statistics instead of the usual evaluations. b) The QSO-Galaxy association is certainly established rigorously in Prof. Burbidge's paper in A&A (Chu *et al*, **138**, 408, 1984). The sample is big enough. One should try to explore completely random empty regions in the sky to control the non-parametric statistics.

J. Sulentic: Agreed, we need some meaningful blank-field bright quasar surface densities to use as control in the quasar-galaxy association tests.

M. Burbidge: Just a comment on your remarks concerning cases where the intensities of [NII]/Hα $\geq$1.0. Some years ago M. Peimbert showed, at least to my satisfaction, that our observation of this in the center of M51 was due to a real overabundance of nitrogen in the center (presumably from enrichment in nitrogen ejected from evolved stars.

J. Sulentic: Perhaps the radio and optical ejection activity seen in M51 are not directly related to the unusual emission line ratio. Any apparent spatial coincidence of such an anomalous line ratio with the optical "lobes" however would argue for a direct relationship. In my opinion, it is suggestive that so many of the (few known) galaxies exhibiting such a ratio *also* show signs of ejection activity.

N. Sharp: Quasar-Galaxy correlation statistics, published by Nieto & Seldner(AA **112**,321,1982) and unpublished by me, seem to show a possible subset of quasars at a fixed *physical* separation from bright galaxies. However this separation is ~10 Mpc, rather larger than the usual scales considered. This anomaly, although still of low statistical significance, may be very relevant to the ejection hypothesis.

J. Sulentic: If all (or a significant fraction) of quasars were formed by ejection from low redshift galaxies, we must see correlation scales out to many degrees. I would expect that such tests would also show correlations on smaller scales. The 10 Mpc signal,if real, may represent the distance at which the maximum correlation occurs although this seems a bit high.

H. Arp: We should acknowledge that these last fine observations on NGC4319 and Mark 205 were done on Geoffrey Burbidge's observing time.

Imaging of the NGC 5296/7 System

N. A. Sharp
National Optical Astronomy Observatories, Tucson, AZ

It was pointed out by Arp (1976) that the small, apparently elliptical, galaxy NGC 5296 (z=0.008), itself a companion of the large spiral galaxy NGC 5297 (z=0.0083), has a much smaller, much higher redshift galaxy (z=0.0863) to its south. Surprisingly, this un-named galaxy showed some sign of being silhouetted against the conventionally much closer NGC 5296, which should not be possible. The system is even more interesting because of the presence of a quasar (z=0.963) to the south-west, and because there seems to be plenty of low surface brightness material reaching from NGC 5297 to NGC 5296 and beyond, stretching in the direction of both the high-redshift galaxy and the quasar.

Images were taken in June 1985 with the #1 0.9m telescope of the Kitt Peak National Observatory, using an 800x800 Texas Instruments CCD ('TI2'), through B, V, and Hα filters. The Hα filter chosen was 'tuned' to the redshifts of the main galaxies. These pictures reveal quite clearly that NGC 5296 is not, in fact, an elliptical galaxy, but has considerable internal structure, and most noticeably a very bright off-center HII region. This is in line with the idea that anomalous redshift galaxies and quasars often appear near disturbed companions of larger galaxies, which would have predicted that something unusual should be found when NGC 5296 was studied in more detail. The current images do not show sufficient absorption in the outer reaches of NGC 5296 near the high-redshift galaxy to support the claim of silhouetting, which may be an illusion due either to contrast effects, or, more probably, to the complex structure of NGC 5296 (which even shows signs of dust lanes).

All three of the original CCD frames were processed with a maximum entropy image restoration algorithm, which improved the apparent seeing by almost a factor of two, although the current implementation of the algorithm has difficulty preserving any structures with low surface brightness. The effect of this processing can be seen in the accompanying pictures of the Hα image, where a low-contrast positive (i.e. white) picture is superimposed on a high-contrast negative (i.e. black) picture, thus revealing both inner and outer structures at the same time. In this red band, the quasar is just visible in the bottom right corner of the unprocessed frame, but did not survive the processing. It can be seen clearly as a blue object in a colour-composite picture.

Arp, H. (1976). Astrophys.J.Lett., 210, L59.

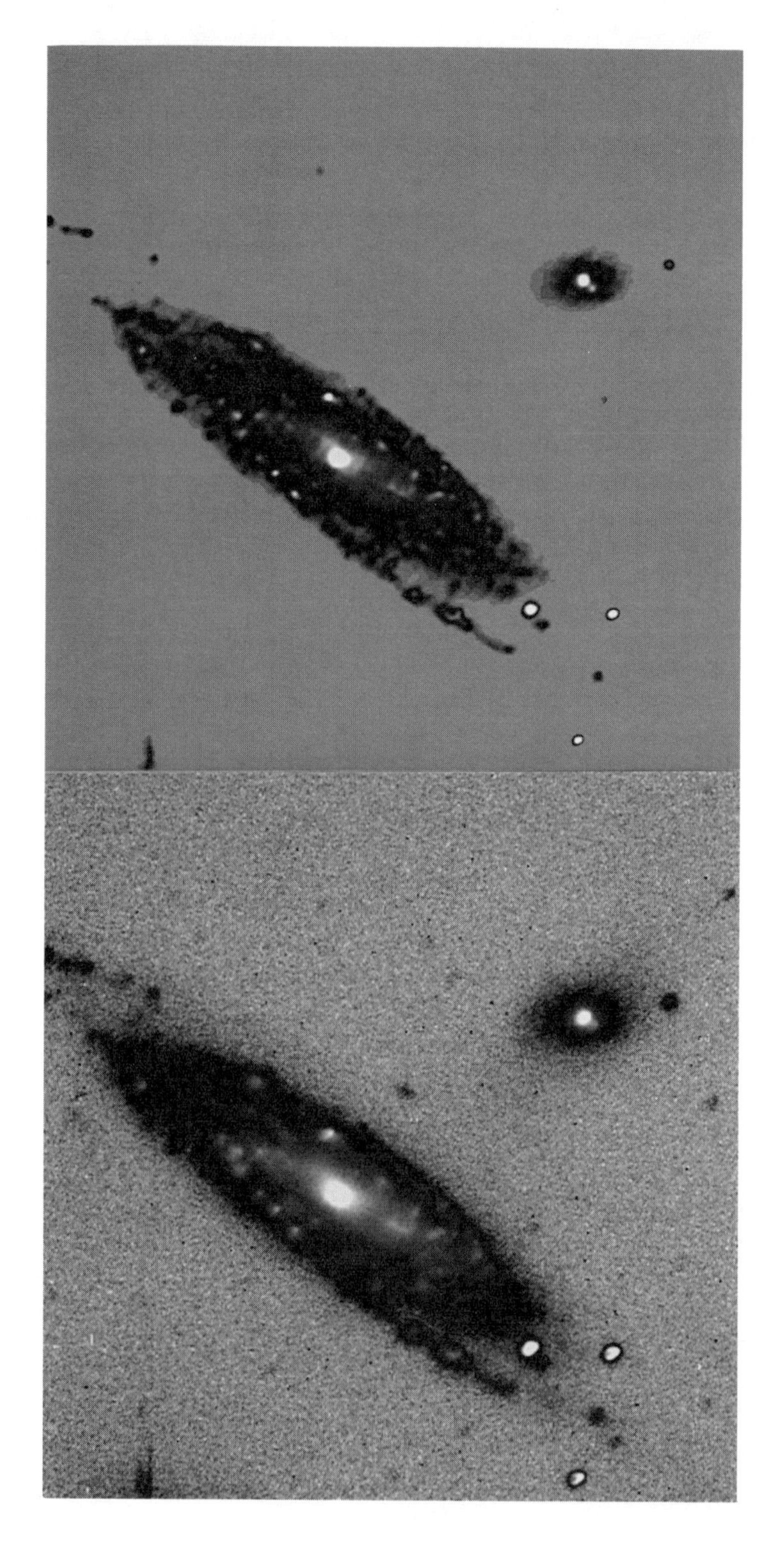

The spiral NGC 5297 and its companion NGC 5296. The original Hα image, at left, shows the distorting effects of slight defocussing and telescope drift. At right, after deconvolution using a maximum entropy algorithm: restoration of stellar images is quite good, although low surface brightness features tend to be removed. Each picture is a superimposition of a negative and a positive, so as to show as large a dynamic range as possible in a single frame.

DISCUSSION

E. Khachikian: Is it real the starlike knot south from the nucleus of companion galaxy on one of the pictures that you showed?

N. Sharp: Yes, indeed! Its strength in the Hα image suggests that this is a very bright HII region just a few arcsec from the center of the companion.

AN OPTICAL JET IN THE STAR-BURST GALAXY NGC 1808

G. F. O. Schnur
Astronomical Institute, Ruhr-University Bochum

J. Kreitschmann
Astronomical Institute, Ruhr-University Bochum

Introduction: NGC 1808 has been classified since 1938 as a galaxy with multiple nuclei but did not arouse the appropriate interest, except by Arp & Bertola (1970), who studied the velocity field of the nuclear region by long slit spectrograms. They concluded that the components of the hot-spot region are not in dynamical equilibrium and are probably moving outwards from the nucleus. Besides a careful study of the general velocity field of NGC 1808 by Burbidge & Burbidge (1968) no other study of its dynamics has been undertaken.

Several papers were published on the physical nature of its central region by Osmer et al (1974), Pastoriza (1975), Alloin & Kunth (1979) and Veron-Cetty & Veron (1985). They agree that NGC 1808 is a typical star-burst galaxy with strong internal absorption.

On plates taken by Arp at the CTIO 4 m-telescope and lend to the authors we detected a linear structure of nearly pointlike sources. As they do not show up in continuum light but only on the Hα plate we predicted them to be H II emission regions. Our aim has been to obtain slit spectra for studying their physical and dynamical properties.

Observations and Reductions

NGC 1808, at a declination of -37.5 deg. has been observed in december 1985 with the ESO 3.6 m-telescope, equipped with a multi-object-spectrograph - OPTOPUS (Lund & Surdej, 1986) - that permitted to obtain up to 45 spectra simultaneously. The size of the entrance aperture of the optical fibres was 2.6 arcsec, the available field size 33 arcmin, and positional accuracy probably better than 0.4 arcsec. The detector was a RCA-CCD with 360 * 512 pixels, aligned such that the 520 lines were parallel to the direction of dispersion. A grating with 172 Å/mm provided a spectral resolution of about 10 Å, and covered the spectral range from 4600 to 7100 Å.

Integration times for our Optopus observations were either 60 or 90 minutes. In addition to the object-frames flat-field and wavelengths calibration exposures were observed for each of the Optopus plates. The data were reduced with the IHAP reduction software package (Middelburg, 1985) involving the following 5 steps: 1. Flat field corrections, to compensate for the varying sensitivity of the CCD chip; 2. Elimination of the cosmic ray events by an appropriate mathematical algorithm; 3. Correction for the different transmission efficiency of the individual fibres; 4. Subtraction of the mean sky brightness from the individual measurements. 5. Flux calibration with Oke standard stars (Oke 1974) . First results including figures of the spectra were given by Schnur (1987).

Observational Results

A map of the Hα intensity distribution is presented in figure 1. The chain of emission regions in the direction to the NW of the nucleus is very obvious. The emission regions are situated along a straight line, which, when extended across the centre to the SE, passes exactly through two other emission regions. Using similar arguments as Bertola and Sharp (1984) for the jet in NGC 3310 we deduce

Fig. 1: The Optical Jet in NGC 1808
Spectroscopic observations are available for the positions 1 to 6 in the jet and counter jet. The very bright central star-burst regions is obvious.

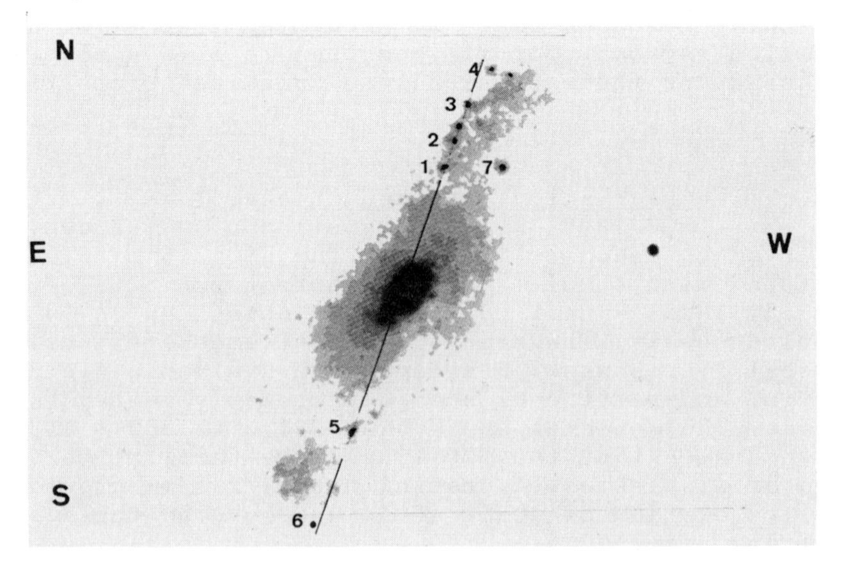

an age of less than $0.5 \cdot 10^6$ years for the Jet-phenomenon in NGC 1808, based on the assumption that any straight feature would get bent due to differential galactic rotation. Spectroscopic measurements are available for each of the indicated positions. In table 1 we present the observed line intensities for the Hβ and Hα lines, the visual absorption as derived from the Balmer decrement (Miller & Mathews 1972), the Hβ fluxes corrected for absorption and the radial velocities relative to the heliocentric velocity of NGC 1808 of 1005 km s^{-1}.

The Hβ luminosities of the knots of the Jet (Tab. 1) are each very much comparable to the Hβ luminosity of the 30 Dor complex in the LMC (Melnick et al., 1987). Since such giant extragalactic H II regions are rarely found, the existence of a chain of such objects is even more unlikely and can hardly be explained by accidental alignment. Furthermore the absorption of $A_V \simeq 3^m$ is very much the same in all the knots of the jet - again an unlikely coincidence.

Table 1: Luminosities, Absorption and Radial Velocities of the Jet in NGC 1808
units: *) erg s^{-1} cm^{-2} $Å^{-1}$
**) log (erg s^{-1})
+) radial velocity difference

Pos.	IH_β*)	IH_α	A_V(m)	log LH_β**)	RV+)
CENTRE	188.8	3442.8	5.43	41.262	0
1	47.64	414.03	3.26	39.663	+ 1
2	32.19	249.10	2.92	39.336	+ 49
3	47.10	397.53	3.17	39.619	+ 48
4	48.05	258.04	1.85	39.017	+106
5	56.74	459.87	3.05	39.645	-112
6	55.27	208.05	0.80	38.598	-146
7	133.16	530.46	1.45	39.205	+172
9	36.32	311.69	3.22	39.528	+167

The velocity field, displayed in figure 2 consists of the observations of the central star-burst region (Schnur, in preparation) complemented by the observations of Burbidge & Burbidge (1968) for the outer regions. The velocity field especially in the NW-quadrant of NGC 1808 indicates a very unusual structure - the velocity 20 arcsec NW of the nucleus drops down to that of the system velocity, a feature normally not found in rotation curves of galaxies.

Fig. 2: The Jet and the Velocity Field of NGC 1808
The filled circles show the velocity field of the central star-burst regions, based on our own observations, while the open circles are from Burbidge & Burbidge (1968). The large crosses refer to the observed relative radial velocities of the jet positions, while the small crosses (positions 7 to 9) refer to positions close to the jet. All velocities are given relative to the system velocity.

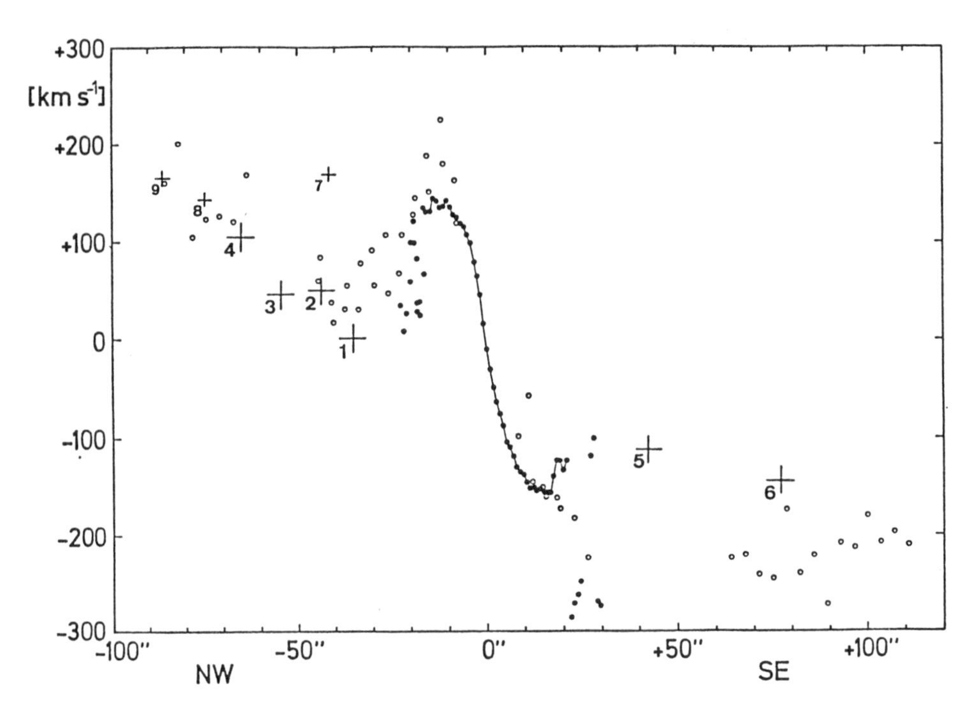

The radial velocities of the knots of the jet seem to agree with this general although disturbed velocity field. The velocities of the jet-knots increase linearly from zero velocity at a distance of 30 arcsec to a velocity of 200 km s^{-1} at a distance of 100 arcsec.

Discussion

The existence of such a chain of giant H II regions, each excited by hundreds of O and B stars and with masses of $5 \cdot 10^5$ $M_\odot$ for the H II gas (Faulkner, 1967), having ages of less than $0.5 \cdot 10^6$ years, stretching over 4 kpc on opposite sides of the nucleus and being perfectly aligned, not only morphologically but also in respect to the radial velocities, needs of course an explanation of a special kind.

The fact that the radial velocities of the knots of the jet agree with that of the general although disturbed velocity field excludes that the radiating material has been ejected from the nucleus. One explanation for the knots could be that a plasma-jet, emitted from the nucleus, has caused by its instabilities condensations in the local interstellar medium, which caused to these giant H II regions. The possibility that we observe an inward streaming of condensed material is improbable, because we observe it on opposite sides of the nucleus.

The effect that the jet is perfectly aligned on both sides of the centre of NGC 1808 however is a clear indication that it originates in the very nucleus. In addition the direction of the jet agrees extremely well with the major structure in the nuclear hot spot region. This confirms Arp's & Bertola's (1970) predictions that there should be ejections from the nucleus of NGC 1808. However, they expected them to be aligned with the minor axis as there is a complex filamentary structure to the NE, while the jet is observed at a different position angle.

NGC 1808, besides being a strong star-burst galaxy, IRAS (1985) source and radio source depicts such peculiar properties that it requires extensive further studies, which will also include an investigation of the surrounding faint galaxies, of which some seem to be related to NGC 1808 and show indications of non-cosmological redshifts.

References

Alloin, D., Kunth, D. (1979). Astron. Astrophys. 71, 335
Arp, H., Bertola, F. (1970). Astrophys. Lett. 6, 65
Bertola, F., Sharp, N.A. (1984). Mon. Roy. Astr. Soc. 207, 47
Burbidge, E.M., Burbidge, G.R. (1968). Astrophys. J. 151, 99
Faulkner, D.J. (1967). Mon. Not. Roy. Astr. Soc. 130, 393
IRAS catalogue: (1985). Explanatory Supplement JPL D-1855
Lund, G., Surdej, J. (1986). ESO Messenger 43, 1
Melnick, J., Moles, M., Terlevich, R., Garcia-Pelayo, J.-M. (1987). Mon. Not. Roy. Astr. Soc. 226, 849
Middelburg, F. (1985). ESO-IHAP Manual
Miller, J.S., Matthews, W.G. (1972). Astrophys. J. 172, 593
Oke, J.B. (1974). Astrophys. J. Supp. 27, 21
Osmer, P.S., Smith, M.G., Weedman, D.W. (1974). Astrophys. J. 192, 279
Pastoriza, M.G. (1975). Astrophys. Space Sci. 33, 173
Schnur, G.F.O. (1987). In ESO/OHP Workshop: The Optimisation of the Use of CCD Detectors in Astronomy, Ed. Baluteau, D'Odorico, 331

DISCUSSION

J. P. Vigier: Have you tried to construct a dynamical model capable of interpreting the jet's red shifts?

G. Schnur: The internal velocity field in NGC1808 is so complicated that we have been unable to derive a reasonable theoretical fit. The observed radial velocities of the knots of the jet are in quite good agreement with that of surrounding galaxy field. Thus it seems that the jet knots have been caused by the interaction of a plasma jet, emanating from the starburst nucleus, and the local interstellar medium. This is confirmed by the already spatially resolved jet knots, which seem to cause the bent ultraviolet arm.

H. van Woerden: You find a jet velocity greater than 2100 km/s from a young age derived from a linear arrangement of HII regions. The velocities of these HII regions differ by less than 100 km/s. Does this imply that the jet, and the chain of HII regions, lies in the plane of the sky (to within a few degrees)? This situation would be quite special.
An alternative would be that the HII regions lie in a plane seen edge-on. In that case the young age does not follow, and no high-velocity jet is required.

G. Schnur: To answer your second question first: Even if the chain of HII regions were arranged in a plane seen edge-on, as you suggest, the extremely small deviation from a straight line would indicate a short age speed of the ejection.
To your first question: We see that some of the HII knots, that are spatially resolved, are observed in the ultraviolet and form an obvious spiral arm structure in the NW area of NGC1808. We assume that these UV-regions are elder remnants of an evolved jet structure ejected several 10^7 years ago, that have started spiral arm formation.

QUASAR PAIRS, ASSOCIATED ABSORPTION AND COSMOLOGY

S. Cristiani[1,2], P. Shaver[1]

1) European Southern Observatory
Karl-Schwarzschild-Str. 2
D-8046 Garching bei München, F.R.G.

2) Istituto di Astronomia della Università di Padova
Vicolo dell'Osservatorio, 5, I-35100 Padova, Italy

Abstract. Associated absorption in quasar pairs is examined in the framework of cosmological versus non-cosmological interpretations of the quasar redshift.

INTRODUCTION

The study of quasar pairs has yielded important results about the origin of absorption lines, the environment in which quasars are located and the size of absorbing regions and their clustering. At the same time it provides an opportunity to carry out a fair comparison of cosmological versus non-cosmological theories of quasar redshifts.

On the one hand, cosmologists should "accept as a proof of the non-cosmological nature of quasar redshift an example of a low-redshift quasar being lensed by a substantially higher redshift object, or showing absorption features from the higher-redshift object in its spectrum" (Birkinshaw 1987). On the other hand proponents of the non-cosmological redshifts should be convinced by a purely logical test, showing that, whenever a case of "associated absorption" is found, it takes place at the lower redshift of the pair (Shaver & Robertson 1983).

This consideration is made more interesting by the relative abundance of suitable objects: at magnitudes brighter than 20th there are roughly 10000 QSO pairs of separation $<$ 1 arcmin over the whole sky (Shaver 1986), more than enough to settle the question.

THE OBSERVATIONS

Some of those ten thousand pairs have already been observed. In addition to 13 useful cases published in the literature, two more are reported here: Q0420-388A,B and KP4,5.

Q0420-388A (R.A. $4^h20^m29\overset{s}{.}9$, Decl. $-38°51'50''$ (1950.0)) is a well-known bright (V=16.9) high-redshift (z=3.12) quasar. Intermediate resolution spectra of it have been published by Atwood et al. (1985), and further spectra were obtained with the ESO Faint Object Spectrograph and Camera (EFOSC, see for a description Buzzoni et al. 1984), at the Cassegrain focus of the 3.6m telescope at La Silla (Fig. 1). Among a number of other systems, a CIV doublet and a Lyα absorption at $z=2.4130\pm0.0003$ are detected. Lyα is blended with other components and therefore was not used in the determination of the redshift. EFOSC was used also to image the field around this QSO and to take a grism slitless frame, which revealed the presence of a possible emission-line object, which was further investigated in the slit-spectroscopy mode at a resolution of 15 Å. In this way Q0420-388B (R.A. $4^h20^m36^s.6$, Decl. -38°50'10", V≃20.8) was discovered and a redshift of 2.403±0.005 assigned to it. The separation on the sky of 2.1 arcmin would "cosmologically" correspond to a separation of about 1.7 Mpc/h_{50}. The difference in velocity between the absorption system and the emission redshift of Q0420-388B is 880 km/s.

For the objects KP4 (0847.6+156A) and KP5 (0847.6+156B), there was no need for a slitless grism frame. First low resolution spectra were taken providing z=2.667±0.002 for KP4, and z=2.041±0.003 for KP5, then echelle spectra were obtained for KP4. No CIV absorption was detected in the spectrum of KP4 around redshift z=2.04 with an upper limit of 1.5 Å equivalent width for the doublet. The separation of the two quasars is 64 arcseconds on the sky, corresponding "cosmologically" to 830 kpc.

THE STATE OF THE ART

The new data were included in Table 1, summarizing the information presently available about 15 quasar pairs. There are five positive detections of associated absorptions within β=0.01 among the higher-redshift members of the pairs, none in the low-redshift group. Application of the Fisher exact probability test to the data of Table 1 shows that the probability of having redshifts randomly distributed with distance is 0.03.

Fig. 1: Intermediate resolution spectrum of Q0420-388A.

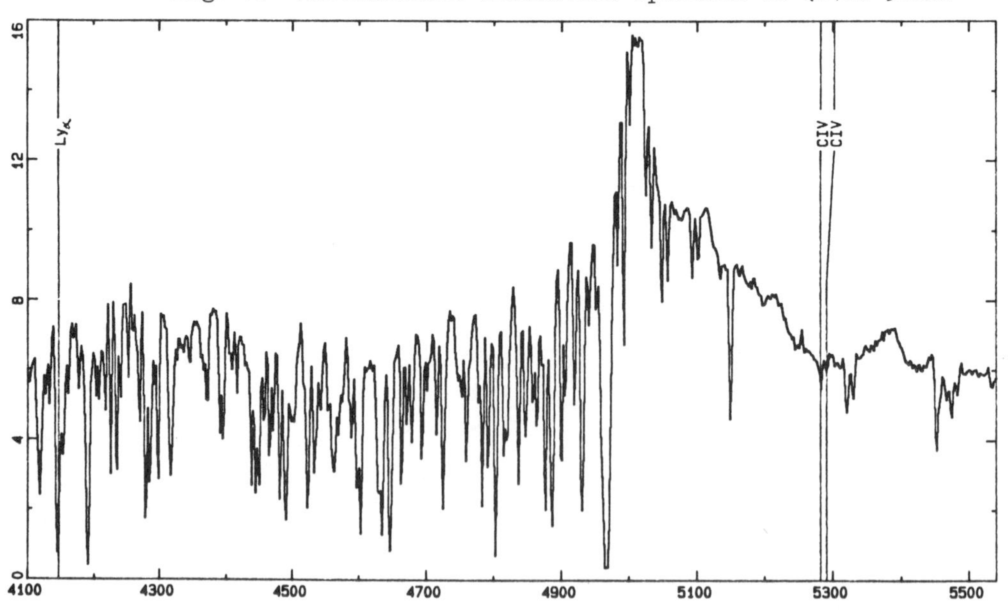

While the present result is not totally conclusive, it does reinforce the cosmological interpretation. It cannot easily be explained by selection effects. In Table 1 cases where confusion is likely (for example, a line falling in the Lyα forest) have been excluded and only CIV and Lyα absorptions have been used. Since the equivalent width of Lyα is greater than CIV 1548 by a factor three on average, both criteria have the effect, if any, of favouring detections of associated absorptions at higher redshifts. For a full use of the information contained in the observations, the angular separation of the objects and the equivalent widths of the detectable absorptions should be included as additional parameters of the statistical non-parametric inference.

Proponents of the non-cosmological interpretation may hope for more negative cases for lower-redshift associated absorptions, or some positive detection at higher redshift. Alternatively, they may attempt to shore up the non-cosmological viewpoint by inventing after-the-fact hypotheses to destroy the symmetry of the test, such as having the higher-redshift quasar embedded in the halo of the lower-redshift quasar. Such hypotheses, like the ancient epicycles, may have to become increasingly complicated to explain more complex situations with

Table 1: Associated absorption in close pairs of QSOs

QSO pair	Separation (arcmin)	$z_{low}-z_{high}$	Associated absorption lower z β	Associated absorption higher z β	Ref
1548+114A,B	0.1	0.436-1.901	*	>	1
0118-031C,B	0.4	1.165-2.112	>	*	2
2359-022A,B	1.0	0.86 -2.81	*	>	3
0307-195A,B	1.0	2.122-2.144	0.0003	>	4
0254-334A,B	1.0	1.86 -1.91	>	>	5
0028,9+003	1.0	1.73 - 2.22	0.0006	>	6
0847+156A,B	1.0	2.041-2.667	>	*	7
0420-388B,A	2.1	2.403-3.12	0.0029	*	7
2206-199A,B	2.1	1.58 -2.55	*	>	3
1623+268,9	2.9	2.52 -2.61	*	>	8
1228+076,7	3.4	1.88 -2.39	0.0067	*	9
1604+176,7	4.4	2.04 -2.32	>	*	3
0118-031A,B	4.8	1.445-2.112	0.0080	*	2
2224-408A,B	4.8	1.95 -2.33	>	>	3
0203-497,8	5.7	1.44 -2.60	>	>	3

> indicates an upper limit in β of 0.01
* indicates an inconclusive spectrum

References:

1) Shaver & Robertson 1985
2) Robertson et al. 1986
3) Shaver & Robertson 1983a
4) Shaver & Robertson 1983b
5) Wright et al. 1982
6) Shaver et al. 1982
7) this study
8) Sargent et al. 1982
9) Robertson & Shaver 1983

associated absorption involving more than two quasars with a variety of redshifts, and the absence of quasar clustering on the sky.

In any case, this approach to the question of the nature of quasar redshifts also provides useful data about the characteristics of the regions where quasars thrive. Indeed, in the cosmological framework, evidence is steadily building up that associated absorption is due in some cases to the cluster of galaxies containing the lower redshift QSO, that the material causing the absorption is patchy (certainly not in the form of a uniform halo surrounding the QSO) and that quasars at larger redshift are located in regions of higher than average matter density.

All these issues should be sufficiently interesting to maintain enthusiasm for gaining the observations needed to honourably settle the "cosmological" controversy, which might be terminated in one night, once a Very Large Telescope becomes operative.

REFERENCES

Atwood, B., Baldwin, J.A. & Carswell, R.F. (1985). Astrophys. J. **292**, 58.
Birkinshaw, M. (1986). In Proceedings of the 119th Symposium of the IAU held in Bangalore, ed. by G. Swarup and V.H. Kapahi, Reidel, p. 472.
Buzzoni, B., Delabre, B., Dekker, H., D'Odorico, S., Enard, D., Focardi, P., Gustafsson, B., Nees, W., Paureau, J. & Reiss, R. (1984). ESO Messenger, **34**, 9.
Robertson, J.G. & Shaver, P.A. (1983). Mon. Not. R. astr. Soc., **204**, 69p.
Robertson, J.G., Shaver, P.A., Surdej, J. & Swings, J.P. (1986). Mon. Not. R. astr. Soc., **219**, 403.
Sargent, W.L.W., Young, P. & Schneider, D.P. (1982). Astrophys. J., **256**, 374.
Shaver, P.A. (1986). In Proceedings of the 119th Symposium of the IAU held in Bangalore, ed. by G. Swarup and V.H. Kapahi, Reidel, p. 475.
Shaver, P.A., Boksenberg, A. & Robertson, J.G. (1982). Astrophys. J., **261**, L7.
Shaver, P.A. & Robertson, J.G. (1983a). Nature, **303**, 155.
Shaver, P.A. & Robertson, J.G. (1983b). Astrophys. J., **268**, L57.
Shaver, P.A. & Robertson, J.G. (1985). Mon. Not. R. astr. Soc., **212**, 15P.
Wright, A.E., Morton, D.C., Peterson, B.A. & Jauncey, D.L. (1982). Mon. Not. R. astr. Soc., **199**, 81.

DISCUSSION

J. P. Vigier: 1) Some QSO's (with flat spectra) could be non-cosmological while steep could be at cosmological redshift (Rowan-Robinson) so your argument is biased. 2) The Fisher test should be compared to results of other non-parametric tests such as Kolmogorov's.

S. Cristiani: 1) Having only a fraction of the quasars with non-cosmological redshift appears a rather "ad hoc" and therefore unsatisfactory, argument. Also in this case the proposed test is useful to reduce more and more the acceptable fraction of non-cosmological redshift until they may eventually vanish. 2) The Fisher test does not fully use the information contained in the data. A statistical inference, using as parameters, also the angular separation and the velocity difference, should further reduce the probability of having redshifts randomly distributed with distance.

F. Hoyle: I think it is important in attempting a disproof of the non-cosmological hypothesis not to make implicit assumptions. The weakness of the non-cosmological hypothesis is that it does not have a clear and generally agreed model for the origin of QSO redshifts. But supporters of the cosmological hypothesis should not play this card more than once. In this matter of absorptions, by assuming that it should be equally likely in a QSO pair to obtain the higher redshift in absorption as the lower, an assumption is being made, which it does not seem to me correct logic to impose on the non-cosmological side, since it is not hard to think of models in which the assumption is not true. For example,in a gravitational model the higher redshift QSO would be the more compact, and this would bias the situation quite appreciably where significantly different redshift values are involved. A gravitational model is also more likely when redshift differences are small to produce double images. By and large it is well known in mathematics that fully decisive disproofs are very hard to achieve.

H. Arp: If the quasars are quite nearby, then the quasars in high and low redshift pairs are physically quite close. It would be natural then for the high redshift quasar to be within the envelope of the low redshift quasar. The low redshift quasar has larger angular extent and is naturally the larger object.

S. Cristiani: Of course models can always be invented to maintain a non-cosmological interpretation of redshifts in spite of the observational evidence. Then the utility of the present test would consist in providing constraints on the "correct" model of non-cosmological redshifts, which should also take into account a number of observed properties, such as clustering on the sky, which cannot be violated.

GALAXY REDSHIFTS

by

Halton Arp

Max-Planck-Institut für Physik und Astrophysik, Institut für Astrophysik

8046 Garching b. München, FRG

ABSTRACT

All possible tests of Sc and later morphological type galaxies show that they are closer than their redshift distances. Sc I galaxies turn out have the largest components of non-velocity redshifts.

Evidence that quasar redshifts are not due to velocity has been most recently summarized by Arp (1986) and Sulentic (this conference). Active, peculiar and companion galaxies also frequently exhibit intrinsic redshifts. It is necessary then to investigate whether more normal kinds of galaxies have components of non-velocity redshifts.

We discuss first the galaxy NGC 262 (= Mark 348) which has a large neutral hydrogen (HI) envelope around it (Heckman, Balick and Sullivan 1978; Morris and Wannier 1980; Heckman et al. 1981). Most recently Simkin et al. (1987) report measures with the VLA and schmidt photography which reveal that it is the "largest known galaxy". Fig. 1 shows how it would appear if placed at the distance of the Local Group. It would easily swallow up a volume containing both M31 and M33! But M31 is by far the largest galaxy for which we have a reliable size. The single key assumption which yields such an enormous size for NGC 262 is the interpretation of its 4700 km s^{-1} redshift as a distance measure.

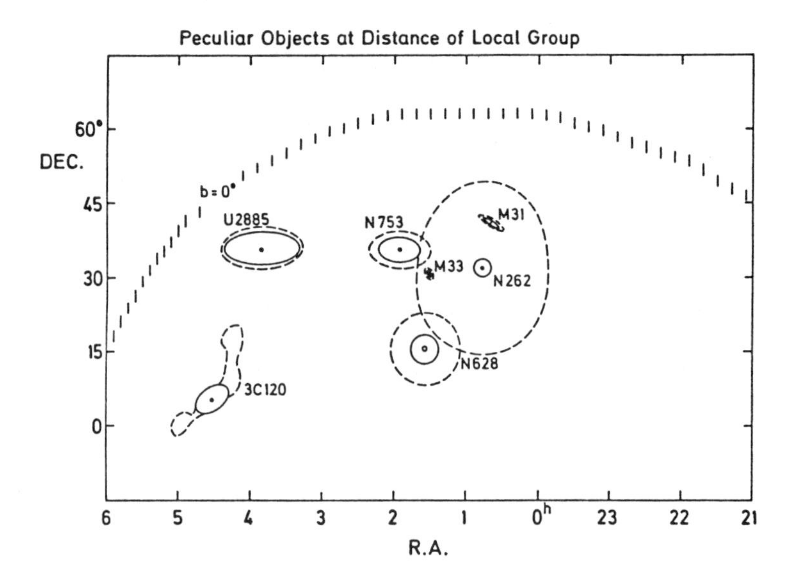

Fig. 1: Galaxies which are very large at their redshift distance are shown as they would appear if moved into the distance of the Local Group. Outer dashed contours represent HI boundaries or (for 3C120) radio continuum boundary, solid boundaries represent optical diameter from Nilsen (1973). Inner points represent apparent angular sizes, that is their absolute sizes if they were really at the distance of the Local Group.

Naturally one such peculiar object cannot define a Hubble relation so there is no actual evidence that NGC 262 is at its redshift distance. Later we will show objects related to this one which violate a linear redshift-distance relation.

An independent distance criterion for NGC 262 should be calculable from the rotational broadening of HI line width. But the HI velocity difference across the galaxy is only 100 km s^{-1}! A large mass should pull on the gas and the only escape from a very small Tully-Fisher distance is to postulate that the galaxy is rotating almost perfectly face on. But the distorted, non-circular optical and HI features argue against such rotation. Except for its redshift the object would clearly be classified as a low surface brightness, hydrogen rich dwarf with an active nucleus.

We next show that dwarfs and late type spirals of higher redshift than the Local Group nevertheless belong to the Local Group and project on the sky in the same close proximity to M31 and M33 as does NGC 262. Fig. 2 here (and Figs. 1 and 2 of Arp 1987a) show that for redshifts as high as 700 to 1000 km s^{-1} most galaxies concentrate in the direction of, and are aligned closely with the distribution of accepted Local Group members along the M31-NGC 404-M33 axis.

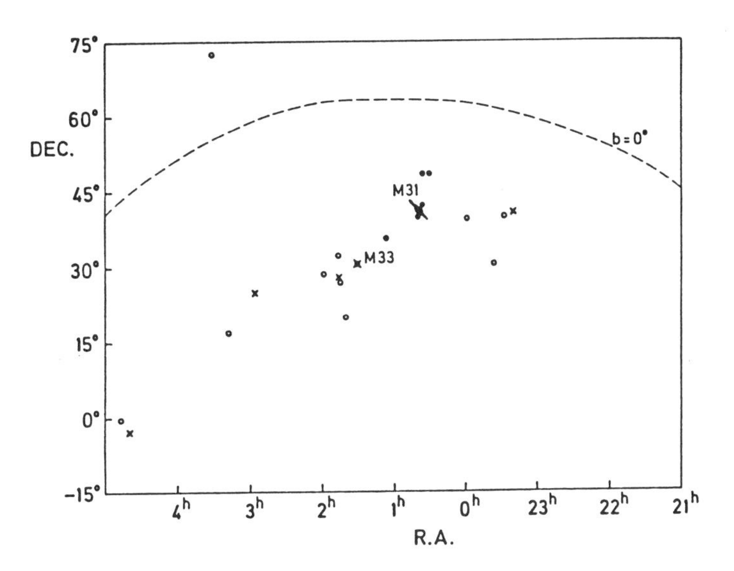

Fig. 2: Conventional members of the Local Group (v_o < 300 km s^{-1}) are plotted as filled symbols. Open symbols (dwarfs) and crosses (spirals) represent all galaxies in the Rood Catalog which have 300 < v_o < 700 km s^{-1}.

These are not the kinds of galaxies that would comprise a background cluster and their inclusion in the Local Group gives the Local Group a range of redshift of constituent galaxies of ≳ 800 km s^{-1}. This is just the range in redshift observed to be characteristic of other groups of galaxies over the sky (Arp 1982; 1986a).

The dependence of excess redshift on morphology is shown in Fig. 3a. In the best known M31 and M81 groups it is seen that the Sb's are the brightest and lowest redshift. Members of all other types have higher redshift. If we were to now add the additional members of the Local Group shown in Fig. 2 in the diagram Fig. 3a, we would see that particularly the Sc and later type members of the Local Group have excess redshifts of Δcz_o ≳ 800 km s^{-1}. The Δcz_o-galaxy type diagram then looks very much like the same diagram for the Virgo Cluster (Fig. 3b). In fact all the richest galaxy clusters that have been investigated in detail show the same characteristics (Figs. 4+5). It cannot be that every major cluster and group which has been carefully examined just accidentally shows the Sb galaxies to have the lowest redshift and the Sc galaxies to have from hundreds to thousands km s^{-1} greater redshift.

Fig. 3:

a) All traditional members of the Local Group (filled circles) and the M 81 group (open circles) are plotted relative to the brightest member of the group.

b) All members of the Virgo Cluster designated V by Sandage and Tamman (1981). Size of symbol proportional to apparent brightness ($B_T^{o,i}$). Full line represents luminosity weighted means and dashed line, number means.

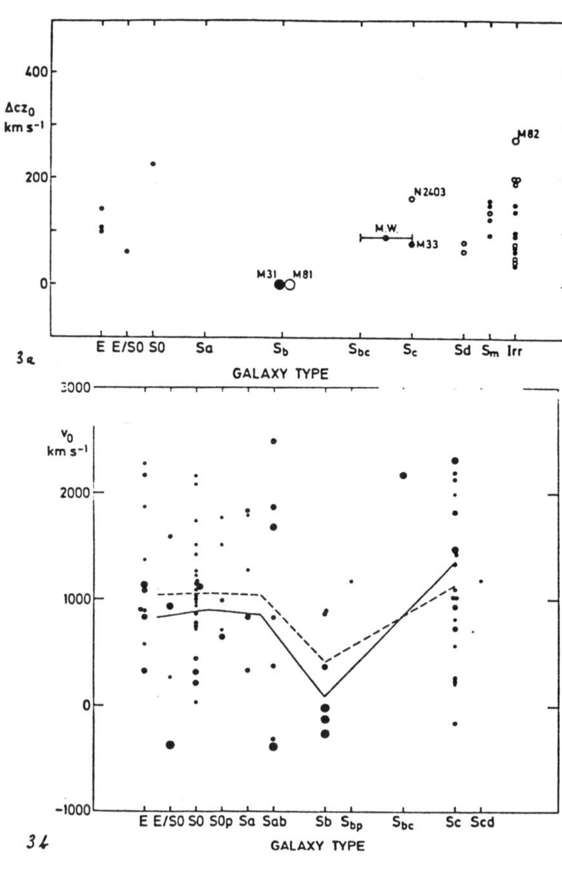

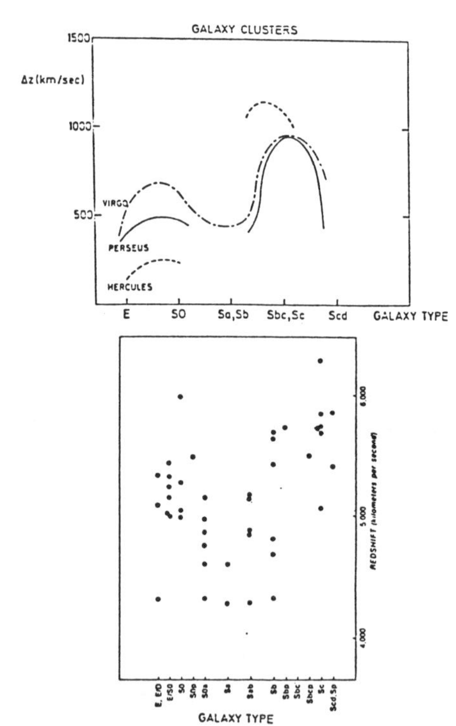

Fig. 4: The redshift-morphology relation for major galaxy clusters. From Giraud (1983).

Fig. 5: Redshift-morphology diagram for the rich cluster of galaxies Abell 262. (From W. Tifft and J. Cocke, private communication).

If we return to the rotational velocity as an independent measure of the (Tully-Fisher) distance to a galaxy we obtain Figure 6.

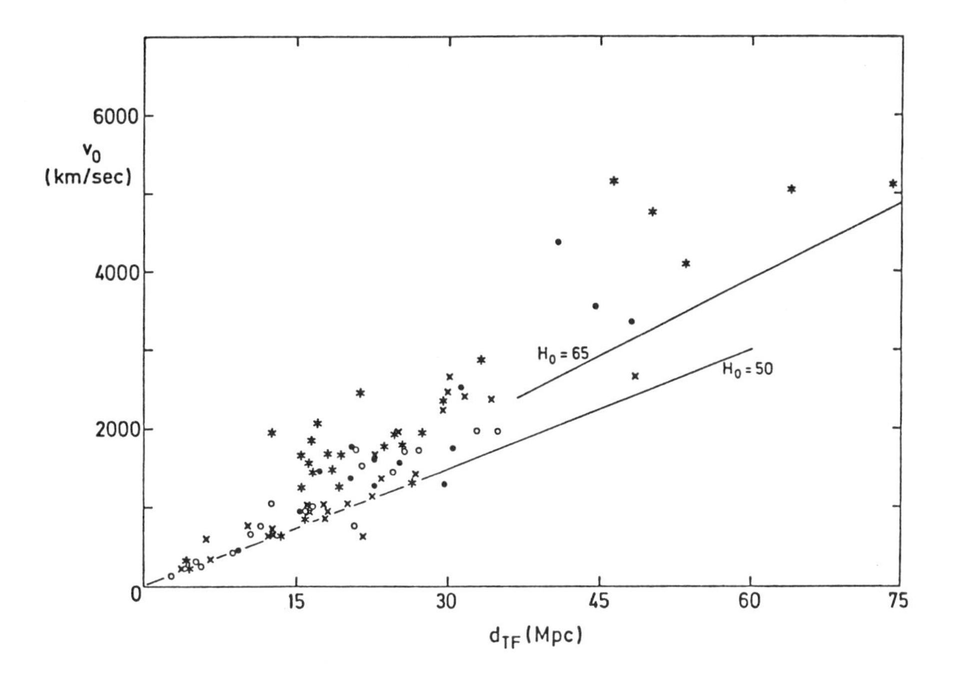

Fig. 6: The redshift-distance plot for Sc spiral galaxies that are listed in the RSA (Sandage and Tamman 1981). The distances are calculated from the Tully-Fisher relation (d_{TF}). Asterisks are from the Local Group region of the sky, filled circles from M81 region, open circles from remainder of northern sky (see Arp 1986a).

It is clear that the galaxies in the broad region of the sky toward the Local Group (asterisks) deviate strongly from a linear Hubble relation. The most severe deviations are by as much as 20-30 Mpc! (Arp 1986a). These galaxies belong to a particular class, ScI, and have redshifts in the ∽5000 km s^{-1} range. Extreme ScI galaxies include NGC 628, NGC 753 and UGC 2885 all of which lie in the general direction of the center of the Local Group as shown in Fig. 1. The latter two, along with NGC 262, are just in this ∽5000 km s^{-1} range which deviate most from the redshift-distance relation.

What do other galaxies in this ∽ 5000 km s^{-1} range show? Fig. 7 shows a complete sky survey (within the Arecibo antenna's $0 < \mathrm{Dec} < 34^{\circ}$ limit) of low surface brightness (LSB) galaxies (Bothun, Beers and Mould; 1985).

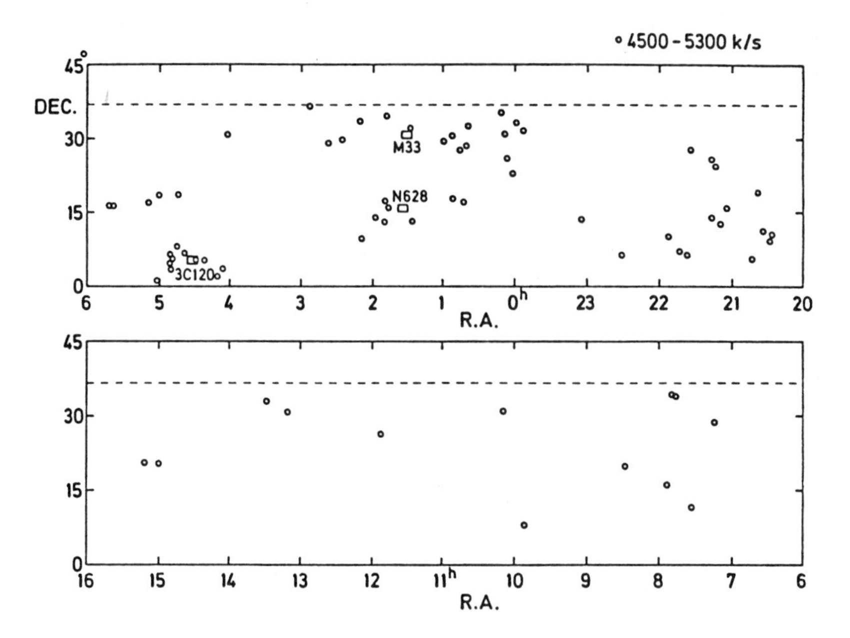

Fig. 7: All low surface brightness, UGC galaxies from Arecibo survey by Bothun, Beers and Mould (1985). In pictured redshift range, 4500 < v_0 < 5300 km s^{-1}, there is strong concentration in Local Group direction and around objects considered to be Local Group objects (see Arp 1986a, 1987b).

The astonishing result is that these LSB galaxies concentrate in the direction of the Local Group! As Fig. 7 shows there are 58 in the half of the sample toward the Local Group and only 13 in the half of the sample in the opposite direction of the sky. Moreover these LSB galaxies clump around previously designated Local Group galaxies M33, NGC 628 and 3C120 (Arp 1986a, 1987b). The latter reference shows that other Local Group constituents such as low redshift HI clouds, large scale luminous filaments, high redshift quasars, and the LSB galaxies concentrate around 3C120 with chances of being accidental of only 6×10^{-4}, 1×10^{-4} and 3×10^{-4}.

This closer distance for 3C120 reduces the superluminal expansion from 6 times the velocity of light to ~ 12,000 km s^{-1}. Since the LSB galaxies in Fig. 1 must contain hydrogen to have their HI redshifts measured, are late morphological type, and in the ~ 5000 km s^{-1} redshift range, they are like the most deviant Sc I's represented in Figs. 3b, 4, 5 and 6. The LSB galaxies in Fig. 7 therefore furnish evidence that the Sc I galaxies of about the same redshift in Fig. 1 are really at the distance of the Local Group. Their redshifts are non-velocity, their sizes are not the enormous outer boundaries depicted in Fig. 1 but instead the very small, pictured inner boundaries which represent their apparent angular extends as observed.

We turn to the only remaining distance criterion that is applicable beyond ~ 5 Mpc, the supernovae. Fig. 8 plots the M_B of supernovae of type II (Tamman, 1982) against the redshift of the host galaxy.

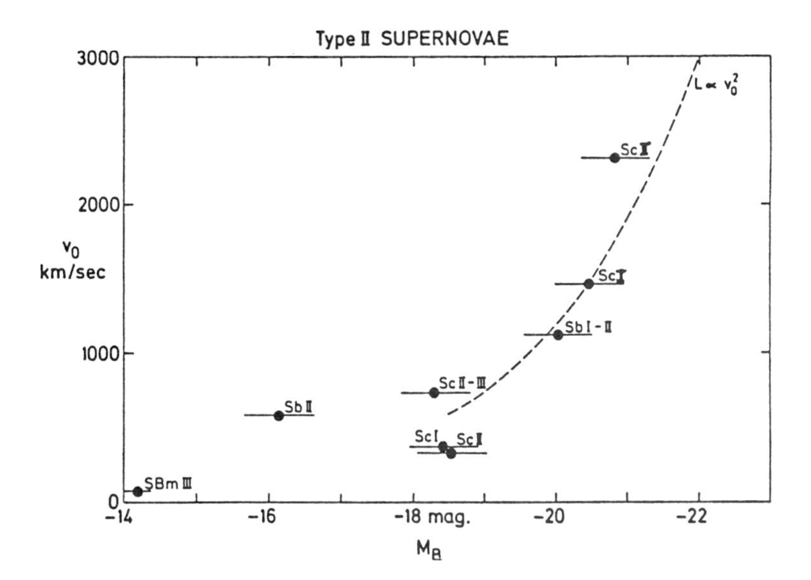

Fig. 8: Absolute magnitude (M_B) of type II supernovae plotted against redshift of host galaxy (v_o). Only the best observed cases with individual reddening determinations are used (Tamman 1982, Table 8). If redshifts of galaxies contain an intrinsic component, the real M_B of the supernovae should be reduced along the curve marked $L \alpha v_o^2$.

There is a luminosity dependence of the supernova on the redshift of the galaxy! The most extreme discrepancies occur in two Sc I systems. If the redshifts of these Sc I systems have an intrinsic redshift excess of a factor of about 4 and 2 respectively, they would slide down the dashed line in Fig. 8 and join the average supernova type II at M_B ~ −18 mag. The most discrepant supernova is in M99. But using stellar models from astrophysical calculations and fitting the observations of the individual supernova outburst, Schurman, Arnett and Falk (1982) derive a fundamental distance to the supernova in M99 of ~ 5 Mpc. This is to be contrasted to the Sandage-Tamman distance to M99 of 21.6 Mpc. We see that accepting intrinsic redshifts for the Sc I's in Fig. 8 actually brings their supernovae into much better agreement with supernovae in much more reliably distanced systems.

Finally we note that the systems which are so giant on the basis of their redshifts would have volumes, as shown in Fig. 1 and Table 1 of 100 to 1000 times that of the best known giant galaxies like M31. The optical apparent diameter of NGC 262 was shown by Simkin et al. (1987) to be almost as large as the HI diameter. Therefore the optical extent would be of the order of 10 times the optical extent of M31. Since the estimated supernova frequency for M31 is one per 21 years (Tamman 1982) the supernova frequency for galaxies like UGC 2885 and NGC 262 should be from 5 to 50 supernovae per year! In anyone doubts that this is a *reductio ad absurdum* it should be an easy matter to produce the observations which establish such a frequency.

The fact that Sc I galaxies have large excess redshifts can hardly be ignored since the four Sc I or SBc I spirals that belong to the Virgo Cluster (V or VR in the RSA) have a mean redshift of $\bar{v}_0$ = 1747 km s^{-1} compared to the redshift of the mean mass of the Virgo Cluster of only $\bar{v}_0^L$ = 863 km s^{-1} (Arp 1987c).

In summary late morphological type galaxies violate every observational test of their redshift distances. The larger their redshifts, the larger the discrepancies. The tests are:

1) Hubble relation (linear redshift-apparent magnitude relation)
2) Tully-Fisher (distances derived from rotational velocities)
3) Type II supernovae (both absolute magnitude and frequency)
4) Association with other objects of known distance.

These are the only criteria possible and all agree that late type galaxies are generally closer than their redshift distances and in particular Sc I's are a low luminosity, nearby class of galaxy.

REFERENCES

van Albada, T.S. and Sancisi, R. 1986, Phil. Trans. R. Soc. Lond. A320, 447.

Arp, H. 1982, Ap. J. 256, 4.

Arp, H. 1986a, "The Hubble Relation: Differences Between Galaxy Types Sb and Sc". Max-Planck-Institut für Astrophysik Preprint No. 236.

Arp, H. 1986b, "Extragalactic Observations Requiring a Non-Standard Approach", IAU Symposium 124, and Max-Planck-Institut für Astrophysik Preprint No. 265.

Arp, H. 1987a, "Additional Members of the Local Group of Galaxies and Quantized Redshifts within the two Nearest Groups", Journal of Astrophysics and Astronomy (India) in press.

Arp, H. 1987b, "3C120 and the Surrounding Region of Sky", Journal of Astrophysics and Astronomy (India) in press.

Arp, H. 1987c, "What is the Mean Redshift of the Virgo Cluster", Astron. & Astrophys., to be submitted.

Baldwin, J.E. 1978, in IAU Symposium 77, Structure and Properties of Nearby Galaxies, ed. E.M. Berkhuijsen and R. Wielebinski (Dordrecht: Reidel) p. 192.

Bothun, G.D., Beers, T.C. and Mould, J.R. 1985, A.J. 90, 2487.

Briggs, F.H. 1982, Ap. J. 259, 544.

Giraud, E. 1983, Docteur these d'Etat, Universite de Montpellier, France.

Heckman, T.M., Balick, B. and Sullivan, W.T. 1978, Ap. J. 224, 745

Heckman, T.M., Sancisi, R., Sullivan, W.T. and Balick, B. 1981, M.N.R.A.S. 209, 15P

Morris, M. and Wannier, P. 1980, Ap. J. 238, L7.

Nilsen, P. 1973, Uppsala General Catalogue of Galaxies.

Roberts, M.S. and Whitehurst, R.N. 1975, Ap. J. 201, 327.

Roelfsema P.R. and Allen, R.J. 1985, Astron. Astrophys. 146, 213.

Sandage, A.R. and Tamman, G.A. 1981, Revised Shapley Ames Catalogue of Bright Galaxies, Carnegie Institution of Washington.

Schurman, S.R., Arnett, W.D. and Falk S.W. 1979, Ap. J. 230, 11.

Simkin, S.M., van Gorkom, J., Hibbard, J. and Su, H.-J. 1987, Science 235, 1367.

Tamman, G.A. 1982, "Supernova Statistics and Related Problems" in Supernova: A Survey of Current Research, ed. M.J. Rees and R.J. Stoneham, D. Reidel Publ. Co. p. 371.

Walker, R.C. 1984, Physics of Energy Transport in Extragalactic Radio Sources, Proceedings of NRAO Workshop No. 9, Greenbank edited A.H. Bridle and Jean Eilek.

TABLE 1

Apparent and Absolute Diameters of Selected Galaxies

Galaxy	apparent diameter optical	apparent diameter HI	distance	absolute diameter optical	absolute diameter HI	Ref.
M 31	3.°3×1.°3	4.°9×.°8	692 kpc	40×16 kpc	59×10 kpc	1
M 33	1.2×.7	2.4×.8	692 kpc	14×8	29×10	1
NGC 628	14'.5×14'	36'×36'	17 Mpc+	73×70 kpc	178×178 kpc	2
NGC 262	1.6×1.5	12×15	94	44×41 ++	328×409	3
NGC 753	3.3×2.1	5.0×–	103	99×63	150×–	4
UGC2885	5.5×2.5	6.2×–	118	189×86	213×–	5
3C120	1'.8×1'.0	5'.2×1'.1*	198 Mpc	104×56 kpc	299×63 kpc	6

+ Larger distances calculated from cz_0 ÷ (H_0 = 50 km/sec/Mpc)

* Extent of radio continuum measurements

++ This is optical diameter listed in Nilsen (1973). Maximum optical diameter as reported in Simkin et al. (1987) is almost as large as HI or about 400 kpc.

1 Baldwin 1978
2 Briggs 1982
3 Simkin et al. 1987; Jahrsbericht Netherlands Foundation for Radio Astronomy 1985, p. 136
4 R. Sancisi (private communication)
5 Roelfsema and Allen 1985; van Albada and Sancisi 1986
6 Walker 1984, Arp 1987b

DISCUSSION

J. Wampler: I know that the number of objects are small but do the "more distant" supernovae that you showed in your plot of SN luminosity versus redshift have different light curves or spectra than the "less distant" ones?

H. Arp: That is the next important step in the investigation, although, as you say, the number of type II supernovae with accurately determined absorptions is small. I am quite sure that it is not due to selection effects, however. The supernova search by Zwicky and collaborators at Palomar–I was able to observe in progress over the years. It is clear to me that supernovae in the bright galaxies shown here would have been discovered easily to $M_V \approx$ -14 mag. (out to m-M$\approx$ 32).

G. Chincarini: The anisotropy you showed between the distribution of galaxies at cz$\simeq$5000 km/s and between RA centered at 0^h and 12^h is naturally due to the presence of the Pegasus- Perseus/Pisces superstructure. This should be interpreted saying that the volume of space up to about 6000 km/s is not yet a fair sample of the Universe, in other words, it is not yet representative and may be close however, to what we may define as an elementary cell. Then, however, having a large volume, I am not surprised to find a large galaxy as NGC262 or NGC2885. It may be a statistical matter due to the luminosity function or distribution in diameters. One has only to see whether or not there are too many so that the bright end of the diameter function, or the probability to find such objects becomes anomalous. I find however the supernova argument extremely interesting and it needs to be carefully studied and tested.

H. Arp: You are right that the Perseus-Pisces superfilament extends over a large angular extent somewhat north of the region I was discussing and is defined by galaxies generally in the range 4000-6000 km/s. There is a severe paradox however, in that at its distance and observed dispersion in redshifts, the dissolution time of the filament is much less than the lifetimes of the galaxies. However in a plot of low surface brightness galaxies which I showed, the sample is limited to $\leq$ Dec=34° limit of the Arecibo radio observations. That cuts out a major portion of the Perseus-Pisces filament. In any case, most of the concentration of 4500-5300 km/s low surface brightness in the sample I showed is in regions around M33, NGC628, and 3C120. It is clear that this large excess in the half of the sky toward RA$\approx$ 12^h is due to galaxies mostly not in the Perseus-Pisces filament. Then the choice comes down to having a gross inhomogeneity on a large scale in the universe or a nearby inhomogeneity on the scale of the Local Group.

QUANTIZATION AND TIME DEPENDENCE IN THE REDSHIFT

W. G. Tifft
Steward Observatory
University of Arizona
Tucson, Arizona

INTRODUCTION AND BACKGROUND

The quantized redshift concept was a direct outgrowth of studies of non-dynamical correlations between redshift, magnitude, and morphology in clusters of galaxies carried out in the early 1970s (Tifft 1974b). Indications that redshift periodicities were present within redshift-magnitude bands were reported in 1973 (Tifft 1974a). An extensive literature review in the mid 1970s resulted in a series of three papers in 1976-77 (Tifft 1976,1977a,b), where the concept was developed in detail and formal predictions were made to guide future work. Central to the concept is the idea that galaxies are composites of specific redshift 'states' dominated by a characteristic redshift interval close to 72 km/s. A program to obtain accurate differential redshifts for double galaxies was begun in the mid 1970s as a critical test of redshift quantization. Double galaxy work first came to fruition in 1979 (Tifft 1980a) when differential redshift data, mostly 21cm, from S. Peterson (1979) was shown to be periodic as predicted. In the early 1980s the Peterson radio sample was improved and extended (Tifft,1982a) and the Steward optical study of close Karachentsev (1972) pairs completed (Tifft 1982b,c). Investigations of both samples confirmed the periodicity as did independent analysis by Sharp (1984). Several studies of redshift differentials in compact groups, including recent work by Schneider et al (1986) further demonstrated quantization (Cocke & Tifft 1983, Arp & Sulentic 1985, Arp 1986).

An important extension of the quantization concept occurred in the late 1970s when it was demonstrated that a solar motion could be found for which redshifts exhibit global quantization effects (Tifft 1978). This work was extended in the early 1980s (Tifft & Cocke 1984) using the full Fisher Tully (1981) redshift sample. Dwarf galaxies with narrow 21cm profiles were shown to be globally periodic in redshift with a well defined period of 24.15 = 72.45/3 km/s. Arp (1986) subsequently derived a similar solar motion to maximize quantization within the Local Group.

The first work to generate a theoretical framework encompassing the observed quantization was carried out by Cocke and appeared in the middle 1980s (Cocke 1983,1985 Nieto,1986). The approach incorporates quantum operators to restrict redshift values. One early contribution was the prediction of a z dependence in the quantization intervals. The results of the global investigations are consistent with such an effect. Two reviews of the early stages of development of the quantized redshift concept exist (Tifft & Cocke 1987, Gribben 1985). Most work prior to about 1985 will not be discussed further here.

The detection of redshift quantization depends upon the quality and quantity of data available (Tifft 1982c, Tifft & Cocke 1984). Serious questions existed in early work concerning the size and nature of both random and systematic error. A major study of 21cm redshifts was begun by Tifft and Cocke in 1984 to study errors and generate new precision data. More than 3000 observations on about 600 galaxies were obtained with the NRAO 300-foot telescope. About a third of the Fisher-Tully catalog was included. Multiple observations demonstrate a very high degree of consistency with better than 1 km/s repeatability at high S/N. Uncertainty rarely exceeds 5 km/s and is now probably dominated by how a redshift is defined rather than by scatter in measurements. Overlap with many sources of data indicates that 21cm redshift uncertainty is generally better than has been stated and is far too small to seriously influence most quantization effects.

QUANTIZATION IN ISOLATED PAIRS OF GALAXIES

One objective of the new work was to obtain a consistent set of accurate data on previously studied radio pairs. Fig 1 shows this data set compared with differentials available in 1982 (Tifft 1982a). Two important conclusions can be drawn. The clumping of differentials is enhanced at the higher precision, and the lowest clump is clearly displaced from zero. The zero deviation had been suspected previously (Cocke & Tifft 1983) but we are now prepared to say with a high degree of confidence that identical redshifts appear to be prohibited within small volumes. This exclusion concept has very important consequences.

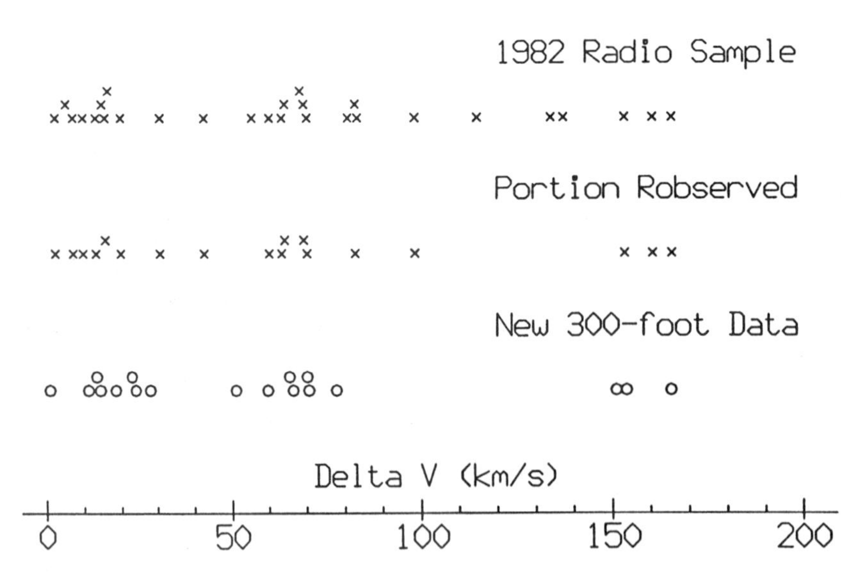

Isolated physical pairs continue to be fundamental in the study of quantization. Three pair samples exist with accuracy sufficient to address quantization, one being wide radio pairs, including the radio sample just mentioned. The second is a new southern hemisphere optical sample by L. Schweizer (1987). Redshift accuracy is about 9 km/s giving differentials with uncertainties near 13 km/s. The second panel in Fig 2 shows the Schweizer differentials for pairs up to about 60 kpc separation. The predicted quantization is clear as is a distinct shift of the lowest clump from zero. The third panel of Fig 2 contains the most accurate subset of the close Karachentsev pairs measured by Tifft (1982b,c). Typical uncertainties are close to 20 km/s. Periodicity in this sample has been previously demonstrated (Tifft 1982b, Sharp 1984). The final panel in Fig 2 is the cumulative distribution. The 72 km/s periodicity with a displaced zero peak is quite striking. The bars and asterisks represent widths and locations of test cells for the old model peaked at zero and the new displaced zero model defined by the radio data. The choice of 24 km/s for the lowest cell is representative only and does not imply that we believe 24 is a unique value present. The three samples cover a range of physical separation from very wide to very close, and involve measurements with both optical and radio techniques. They are the best available and they are clearly quantized.

The deficiency at zero, which is consistent with an exclusion principle in a quantum model, represents a new critical problem for conventional dynamics independent of quantization. Dynamical systems must have a monotonic distribution of differentials peaking at zero (Tifft 1977a). It is of interest in this context to reference the Turner (1976) pair sample which is not accurate enough for direct quantization testing. White et al (1983) note an inexplicable zero deficiency in that sample.

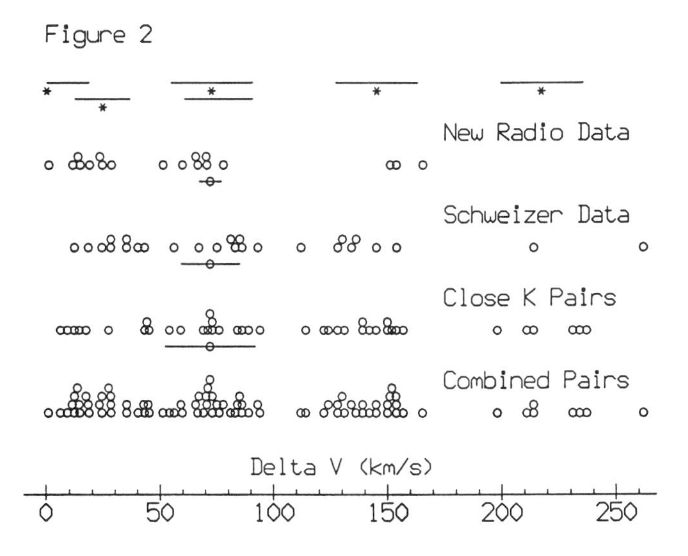

QUANTIZATION AT LOW GALAXY DENSITY, GROUPS

Within the quantization picture an exclusion principle has some interesting testable consequences. In particular, the presence of offsets near 24 km/s must complicate a pattern of simple 72 km/s differentials for groups. A test for this was carried out within the new radio data which contains many objects in loose groups. Each galaxy was paired with the galaxy in the list which had the closest projected physical spacing, up to about 150 kpc, and a redshift differential less than 150 km/s. The resulting 'pairs' are of two types, reciprocal pairs, where each galaxy identifies the other as its closest neighbor, and non-reciprocal pairs, where only one galaxy identifies the other as its nearest neighbor while the neighbor finds that a different galaxy is closer. Reciprocal pairs will include some true isolated physical pairs but since the catalog is incomplete many must be unrecognized non-reciprocal pairs. The non-reciprocal list, however, is unlikely to contain a significant number of real isolated physical pairs. Fig 3 compares the differential redshift distributions (expressed as phase modulo 72.45 km/s) against projected physical spacing (represented as the product of redshift times angular separation). Reciprocal pairs concentrate at closer spacings and about half are consistent with quantum levels associated with physical pairs. The non-reciprocal pairs have larger spacings and, in line with the exclusion concept, show a 2:1 anticorrelation with the quantum levels of physical pairs. They are consistent with a set of 24 km/s offsets from the physical pair levels. A general set of such offsets could easily be the explanation of the overall 24 km/s periodicity observed globally for dwarf galaxies. No restriction on profile width applies to the general pairing study, although dwarves dominate the list.

In order to investigate the distribution of small differentials over very large physical spacings many Fisher-Tully 'pairs' with projected spacings between 200kpc and 1Mpc, and differential redshifts less than

Figure 3

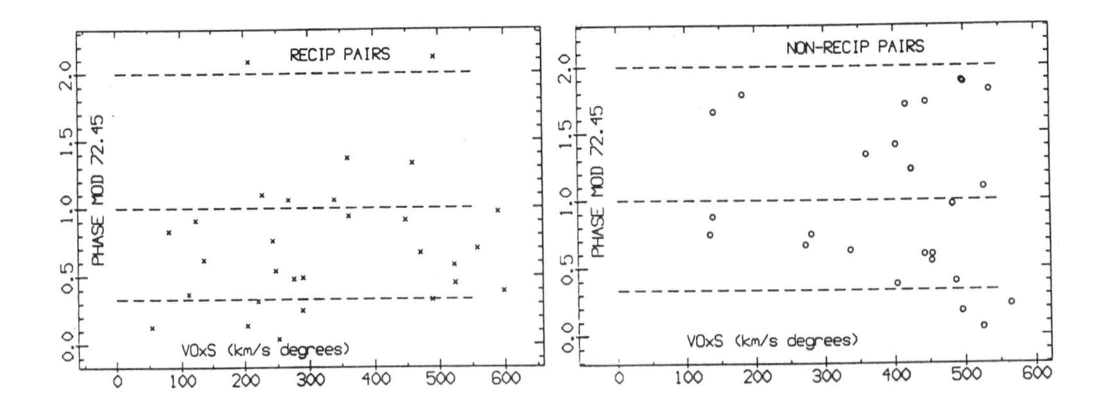

about 50 km/s, were included in the new radio study. Fig 4 shows the distribution of differentials found. The distribution does not peak at zero despite the deliberate bias toward the smallest values possible. The interval 14-28 km/s shows a 1.5 enhancement compared to the 0-14 km/s interval. There is incipient clumping. The asterisk pattern represents the series 72/2, 72/3, 72/4,... as one possible model since intervals near 36 and 24 have been found in previous work (Cocke & Tifft 1983; Tifft & Cocke 1984). The lowest clump from the new Tifft-Cocke radio pairs and the Schweizer pairs is also shown. All the data occupy the same enhanced region. Overall the enhancement has been represented by a cell centered at 24 km/s pending the obtaining of more radio data of very high consistency. The '24' enhancement extends over a spacing range from 10kpc to nearly 1Mpc without detectable change. The region is probably structured and the bias away from zero is quite obvious. The common argument in dynamics that a zero deficiency can be produced by a selective bias against very wide pairs is obviously not supported.

The 24 km/s quantum level certainly extends to VxS=3600 or roughly one Mpc. The same cannot be said for higher quantum levels. The present data suggest that each level is well populated only out to a certain limiting range. Sporadic occurrences may be present at wider spacings but at least a drop in frequency occurs. The transitions occur near VxS of 50, 100, 400, and 3600 for the levels of 217, 145, 72, and 24 km/s. These transition ranges have a square root of R relationship to redshift differentials as shown in Fig 5. Although greater ranges need to be mapped consistently for the larger differentials, this suggests that each differential is strongly supported only out to the range where a standard point mass gravitational potential will support a circular velocity at the level of the differential. The differentials might therefore represent bound levels within a standard gravitational potential. Levels outside the range may have a somewhat different character but are very poorly mapped at present. This is the first observational relationship found which suggests a linkage between quantization and standard gravatational theory.

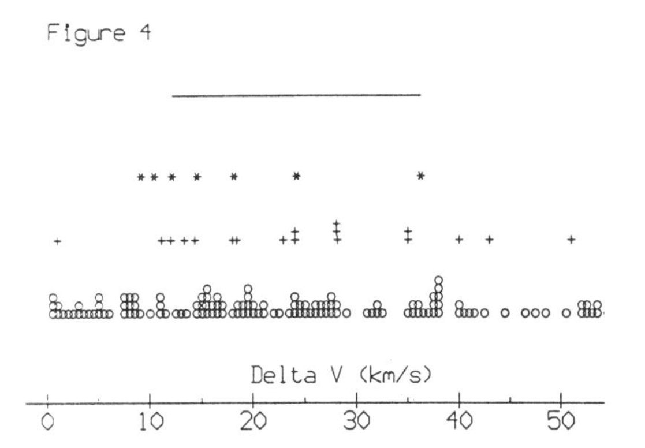

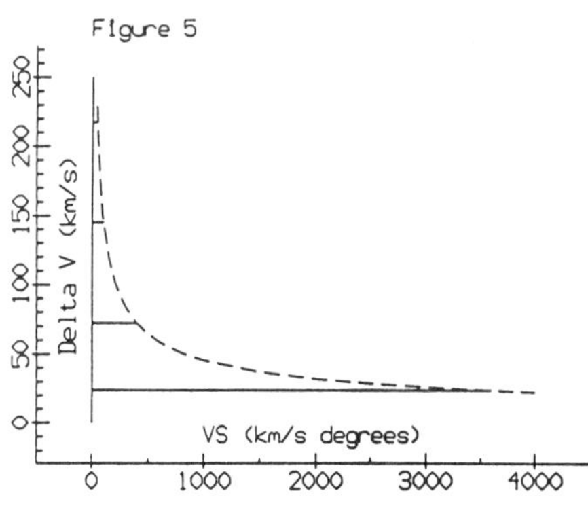

QUANTIZATION AT HIGH GALAXY DENSITY, CLUSTERS

Most of the discussion to this point has dealt with either small numbers of galaxies or low density conditions involving modest numbers of galaxies. Now we turn to more extreme situations. The operation of an exclusion principle within a volume must result in a progressive spreading of redshift as the population rises. The spread in such a case is caused by level degeneracy, not dynamics. In the simplest case we can argue that in a high density region, such as a galaxy cluster, degeneracy would force the population to spread over a wide range of levels in the 72 km/s hierarchy. Many objects within a small region would exhibit a 72 km/s periodicity as a general extension of the concept of a pair, which is a high density region but involves only two objects. One of the early results which led to the formulation of the quantum concept was the demonstration of a general 72 km/s periodicity within the Coma cluster (Tifft,1977a). The availability of modern 21cm data in relatively dense cluster regions permits us to extend this test. As an index of galaxy density surrounding an object we will use counts of galaxies, in the Zwicky CGCG catalog (1963), within 1/2 amd 1/4 degree, expressed as a ratio N.5/N.25. Thus 1/1 represents a galaxy isolated within 1/2 degree, while 50/20 represents a galaxy with 49 companions within 1/2 degree and 19 within 15 arc minutes. If high density forces objects into an ordered 72 km/s periodicity then we should find a strong 72 km/s phase correlation at high density. Table 1 lists the galaxies, in the Chincarini et al (1983) 21cm study of the Coma cluster region, ordered by galaxy density. Above a density index of 4/2 we find 33 of the 39 galaxies associated in half of the phase range. The formal liklihood that such a clumping would occur accidentally in any half phase interval is 0.0005 (Tifft & Cocke 1984). At lower densities the clumping appears to vanish.

Table 1 Phase In Coma-A1367 Ordered by Density Index

Den	In	Out	Den	In	Out	Den	In	Out
54/23	91.21		7/3		87.67	2/2	92.16	
52/19	99.84		7/2	76.86		2/2		86.42
50/18	68.92		6/3	102.18		2/2		87.68
47/5	72.05		6/2		91.57	4/1		93.83
45/13	89.11		5/4	83.91		4/1		87.41
31/10	81.03		5/2	102.23		4/1	94.86	
24/9	74.92		4/3	110.21		3/1	88.98	
17/7	91.18		4/3	81.92		3/1		104.56
16/4	69.19		4/2	123.15		3/1		102.62
15/5	90.28		4/2	90.30		3/1		96.76
14/7		95.49	4/2	76.91		2/1	91.99	
13/2	72/93		8/4D	96.09		2/1	92.91	
12/6	84.11		4/3D	89.99		2/1	110.22	
11/6		99.37	3/2D	96.03		2/1		93.34
11/6		99.45	2/2D	94.90		2/1	86.95	
11/3	76.25					1/1		92.44
11/3	77.00		Hi Density	33:6		1/1		94.58
10/3	113.11		(D=double system)			1/1	94.23	
9/3	97.12					1/1		88.61
9/2	87.32		3/2		89.82	1/1	95.22	
8/3	92.96		3/2		97.63	1/1	99.10	
8/2	95.26		2/2		95.39	1/1	91.20	
8/2		103.61	2/2		99.51			
7/4	97.87		2/2	126.97		Lo Density	12:15	

To illustrate the relationship between galaxy space density and redshift phase more graphically we can invert the problem and use phase to define samples. The left panel in Fig 6 shows the location on the sky of Coma region galaxies which fall in the half phase interval we associate with high density. Asterisks mark cluster centers. Coma lies at 13.0 hours on the right and A1367 at 11.7 hours on the left. Other asterisks identify populous CGCG 'near' clusters which distribute within the Coma-A1367 supercluster structure. Nearly all the phase selected points associate with cluster centers, in other words, high density. Furthermore, the different clusters are in phase with one another; the redshift is globally phased. Redshifts used here have been corrected for solar motion and relativistic effects (Tifft & Cocke 1984). The right panel of Fig 6 shows the location of objects in the other half phase interval which we associate with low galaxy density. The points show no association with specific clusters but do define a continuous filament winding through the cluster complex. The clusters are high density knots on a continuous low density filament, both clearly distinguished by redshift phase alone. Fig 7 is a phase-density plot which contains both the Coma region data and data on another cluster region, A262 (Giovanelli et al 1982), located in a different redshift range and region of the sky, hence with a very different solar motion correction. Similar phase intervals relate to high and low density regions. We tentatively associate the redshift spread in clusters with high density redshift level degeneracy and suggest that it may have little or nothing to do with the conventional idea of hidden mass.

Figure 6

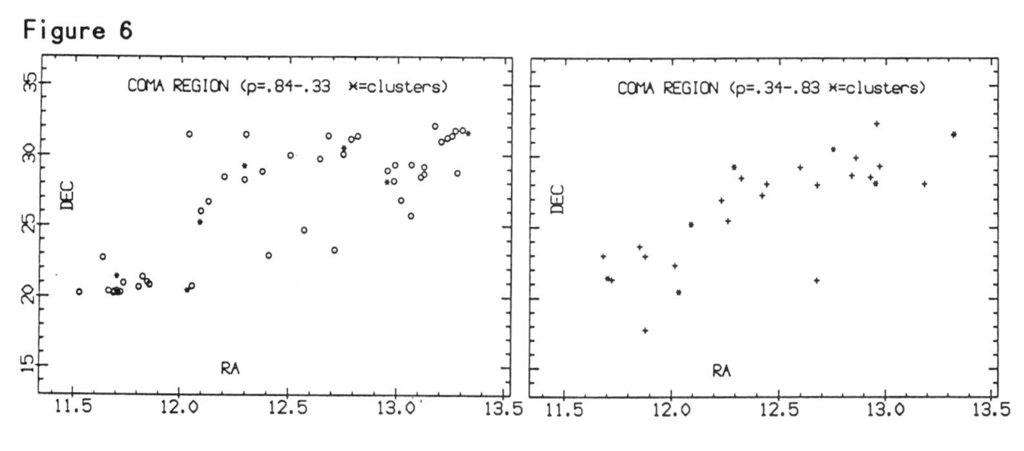

Figure 7 Coma (o) and A262 (x) Galaxies

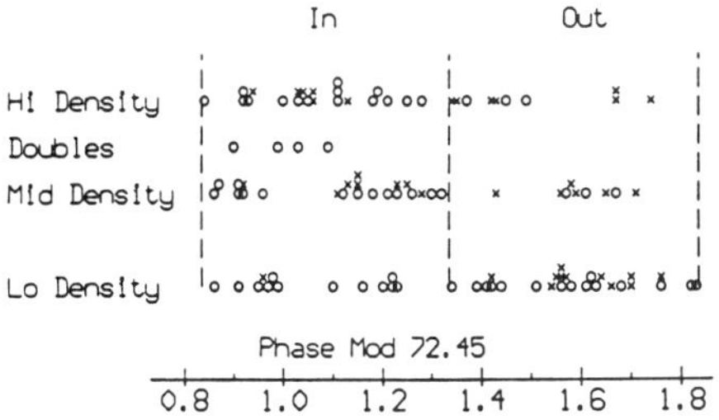

In
Out
Hi Density
Doubles
Mid Density
Lo Density
Phase Mod 72.45
0.8 1.0 1.2 1.4 1.6 1.8

Arp (1987) has also used redshift periodicity in an attempt to define group membership, and especially to associate higher redshift objects. He finds phase shifts between groups which he interprets as due to relative motion. Clearly more studies are required.

QUANTIZATION AND INTERNAL DYNAMICS OF GALAXIES

I will now turn to an aspect of quantization which has great potential but has barely been touched, namely quantization effects within individual galaxies. From the beginning of the quantization concept, galaxies have been viewed as composites of redshift 'states', most likely taking the form of a dipole structure with an inherent redshift discontinuity at the nucleus (Tifft 1976,1977b). For large rotating spirals redshift quantization effects are easily obscured by dynamics. In certain situations, however, quantization effects may be much more important. Two examples concern extreme dwarf systems and elliptical galaxies. For both these categories rotation is smaller than, or comperable to, the quantization, so the quantization can play a significant role in moderating such observables as rotation, dispersion, or 21cm profile width.

The 21cm profiles of extreme dwarves are characteristically single peaked and may have total widths, at 20% of peak, which are less than 40 km/s. There are many with widths of 50-70 km/s. The width of a profile in the dual-redshift-state model of galaxies is necessairly moderated by any intrinsic 'dipole' interval present. Since only specific intrinsic intervals are presumably possible in a quantum picture, when these intervals are comparable to real internal Doppler widths, then width quantization could be observed. The new 21cm program included detailed high resolution studies of a large number of dwarf galaxies. Fig 8 shows the observed width distribution below 80 km/s. The distribution is distinctly non-smooth with several characteristic widths preferred. Pending further refinement in precision on low S/N objects we believe we have detected width quantization.

Figure 8

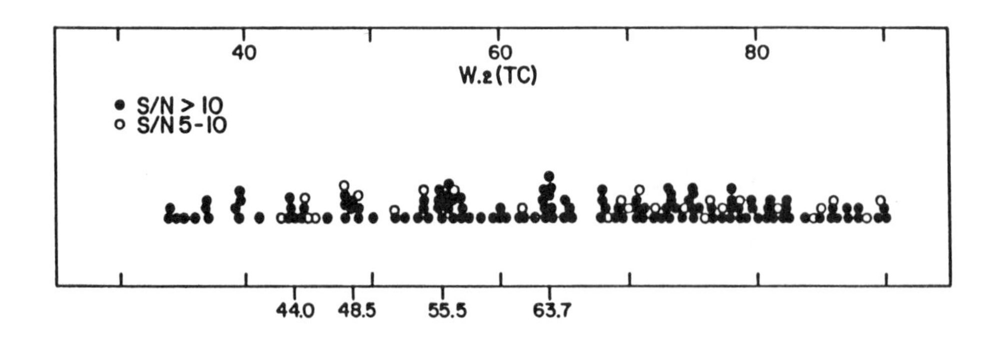

Elliptical galaxies are the antithesis of dwarf irregulars, but they provide an interesting opportunity to detect internal redshift discontinuities. If intrinsic nuclear redshift offsets are present, they will combine with bulk motions and produce effects on observed rotations and dispersions. Unlike conventional forms of motion, the offsets may either enhance or reduce measured velocities depending on the sign of the offset. Fig 9a illustrates the influence of a 72 km/s nuclear discontinuity on measured rotations. Two families of rotation curves are generated, one rising rapidly and one slowly with radius. Fig 9b shows a correlation between galaxy ellipticity and rotation, using data from Davies et al(1983) and Dressler & Sandage (1983). A 'round' object, with no significant rotation in the line of sight, will yield a classical rotation measure of 72/2 km/s due to the offset alone. As rotation and eccentricity increases from zero a two-branched pattern results. One branch, corresponsing to a nuclear redshift discontinuity opposed to the sense of rotation, drops to zero rotation measure at some non-zero eccentricity where the offset and rotation cancel. Beyond that point the rotation measure rises smoothly. The other, 'reinforcing', branch rises smoothly from 36 km/s at zero eccentricity and ultimately parallels the lower branch with a 72 km/s offset. This model is very nicely consistent with available measures and readily explains the fact that two rotational families are observed in ellipticals. It also makes interesting predictions relating to rotation measures on nearly round objects. Dispersion measures, which also show two characteristic patterns with radius, may be similarly analysed (Tifft 1984). Ordinary dynamical analyses have yet to explain the dual forms of rotational and dispersion behavior in ellipticals.

Figure 9a Vc = Conventional rotation measure
Vt = Actual rotation value

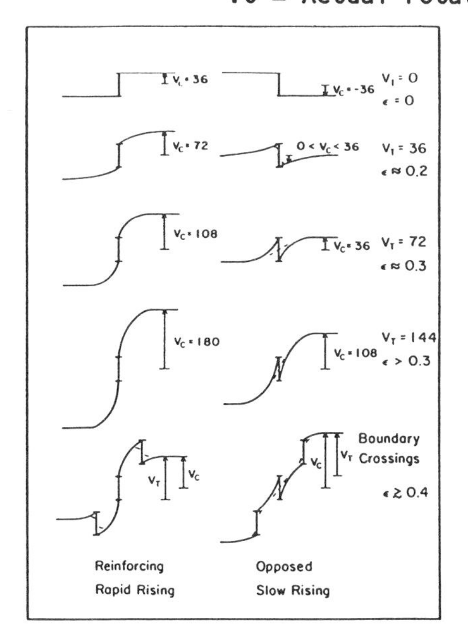

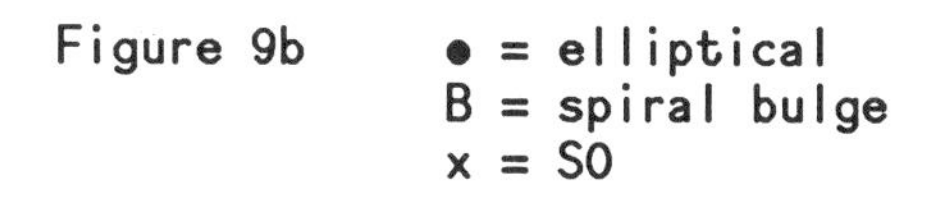
Figure 9b • = elliptical
B = spiral bulge
x = S0

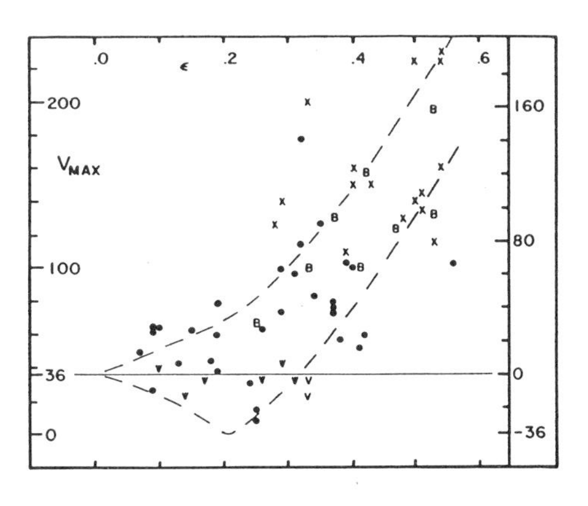

Within spiral galaxies the redshift duality is associated with the two opposed spiral arms which are theorized to differ intrinsically in redshift. Since the arms are interwound, and appreciable rotation is present, the quantum offsets become intermixed with what are usually dismissed as 'non-circular-motions'. An analysis of such deviations played a major role in the original definition of the quantum concept (Tifft 1976, 1977b). Other important manifestations include such phenomenona as warps and extended flat rotation curves. In large open spirals, of which M33 is a good example, the presence of an intrinsic offset, which of course does not project like a velocity, will obviously force a warp into two opposed quadrants as dynamics attemps to accomodate to the offset. Models of M33 show an extreme warp (Rogstadt et al 1976), which is both difficult to produce in this isolated object and even harder to maintain. Such an apparent warp has a natural explanation in a dual redshift model.

Modern rotation curve studies on spirals extend out to distances similar to separations in close pairs. There is an interesting apparent continuity between spiral arms, companions on spiral arms, and companions. Extended rotational levels may assume the same, or similar, quantum levels as companions, for the same underlying reasons. Within a quantum model a degeneracy or exclusion principle can prevent the return to lower quantum levels within the volumes involved. The fact that the rotation curves are flat can be taken to imply that similar quantum rules apply

Although this review can give only a few examples, the overall impact of a quantum model on the interpretation of internal dynamics may be great.

THE VARIABILITY OF REDSHIFT WITH TIME

The final subject to be covered in this review concerns the constancy of redshift with time. The new radio studies had as a prime objective the analysis of uncertainties in redshifts. A large amount of overlap with older studies, especially the Fisher-Tully work, exists.

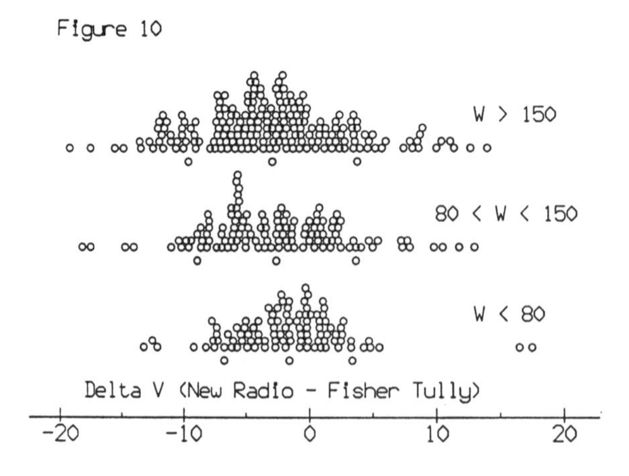

It quickly became apparent that while internal consistency in the new data is extremely good, there are small but clear deviations from older studies. Fisher-Tully redshifts in particular average overall about 2.5 km/s greater than current values. Fig 10 illustrates the deviation for typical subclasses within the new survey. Both the form and the amount of deviation vary somewhat between categories. For the present purposes we will focus on dwarf galaxies with 21cm profiles narrower than about 80 km/s. These are the objects for which the highest redshift accuracy is possible since most have been studied with high resolution. The distribution of deviations is not symmetrical; Tifft-Cocke minus Fisher-Tully deviations are negatively skewed. Further, when one plots deviations against redshift phase (redshift modulo 24.15 km/s) there is a clear correlation. The larger negative deviations associate strongly in phase, as shown in Fig 11. Phase is shown for both Fisher-Tully and Tifft-Cocke epochs to show how specific blocks of data shift with time. The deviation within this homogeneous class of objects is directly related to the redshift quantization phenomenon. It is difficult to conceive of any systematic error which could distinguish by redshift phase. We conclude that real changes in redshift may have occurred in a manner directly related to quantization. About ten years have elapsed between the two surveys.

Other evidence supports the idea that real changes may have occurred. Theoretical work (Cocke & Tifft 1987), where the first investigations have now been extended to a relativistic formulation, include an effect equivalent to Zitterbewegung or high speed 'jitter' present in the relativistic theory of the electron (Dirac 1958). Whereas the electron oscillation frequencies are extremely high, the corresponding time scale of the extragalactic effect is about 10 years. Thus, to the extent that the theory is correct, time variation would actually be predicted.

Figure 11 Phase-deviation diagrams, profile width $40<+<50<\bullet<60<\circ<75$

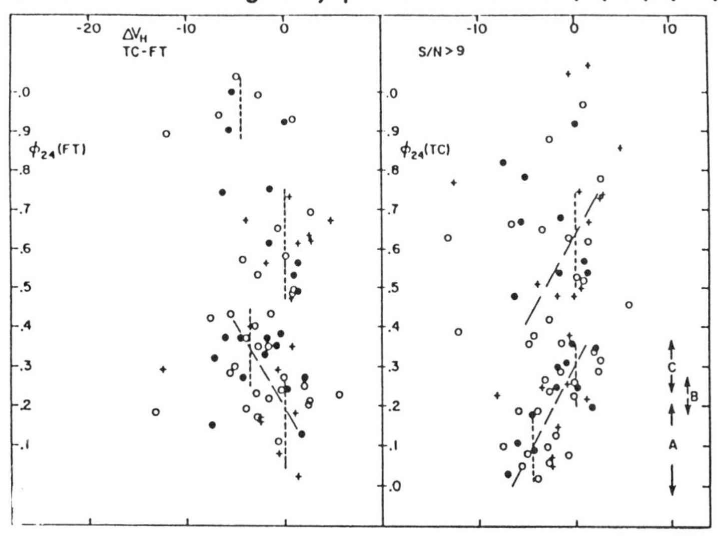

Observational evidence, furthermore, is not limited to the phase correlation. Although large numbers of redshifts predating the Fisher-Tully survey do not exist, some of the best oldest data continue the observed trend. Fig 12 shows a comparison with data from Roberts (1968) which preceeds the Fisher-Tully work by about ten years and shows an average deviation of 12.5 km/s. Also shown in the overall Fisher-Tully deviation distribution, a comparison with Dickel & Rood (1978), and a modern comparison with precision data from Lewis (1983) where the deviation essentially vanishes. A careful analysis of many overlap observations is in progress. Some small changes in profiles have been seen even within the two year time baseline of the new radio study. Such changes are closely related to effects produced by slight pointing errors, however, and it is impossible to distinguish at this time.

Finally, although the internal time baseline is two years or less for most galaxies in the new study, it is possible to examine for differential gradients with redshift phase. Fig 11 contains three phase intervals marked A, B and C. Region A shows the largest change, C very little, and B suggests an intermediate transition interval. Fig 13 shows time gradients, referred to January 1986, of galaxies with multiple observations within the time interval May 1984 to March 1986. Different gradients are suggested, with the largest in region B. Needless to say the effects are very close to our limits of detection. We are presently extending the study to provide a longer time baseline. We now believe, however, that there is a distinct possibility that the redshift is time variable over remarkably short time scales.

Figure 12

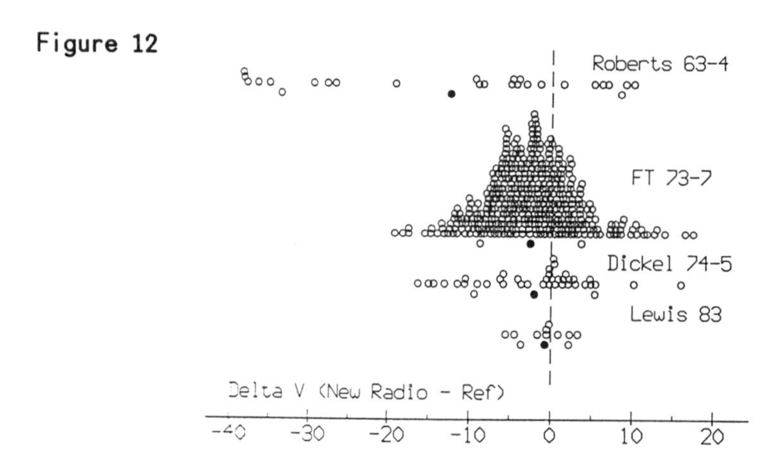

Figure 13

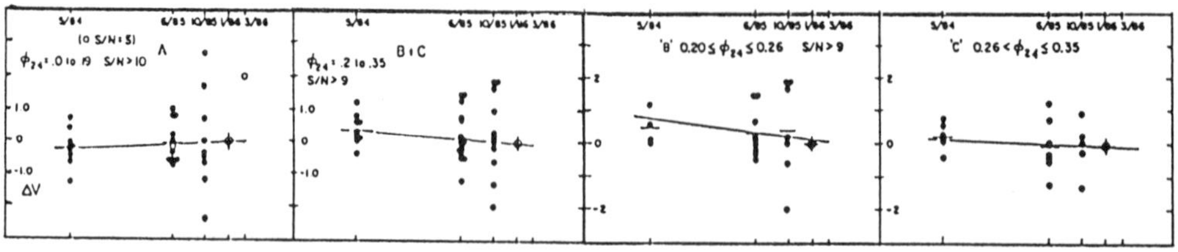

QUESTIONS AND NEW DIRECTIONS

The theme of this symposium is new ideas, something in which the quantized and time-dependent redshift concept certainly abounds. I will close with some questions and thoughts which come to mind if one embraces the concepts. I will begin by asking what relationshps may exist between the concepts and other redshift anomalies. If the redshift is indeed time variable then it requires little effort to associate the rate of variation with luminosity or morphology. Such phenomenona as the redshift excess in dwarf companions, morphologically dependent redshift differences observed in groups and clusters, or redshift-magnitude bands could easily result. I believe there is very clear evidence for such patterns. Fig 14 is a morphology-redshift plot, for the A262 cluster, showing all galaxies down to mp=15.0 in a region centered on the cluster. If such patterns are indeed real, do they arise because time variation in redshift depends upon luminosity and type, or does morphology and luminosity vary as galaxies evolve through redshift space (Tifft 1979)? Redshift patterns in radio and active galaxies (Tifft 1977c, Tifft & Tarenghi 1977) relate to such questions.

A somewhat similar phenomenon concerns redshift-morphology patterns as a function of radius within clusters. Fig 15 shows the pattern in the Coma cluster for galaxies between 0.25 and 1.0 degrees from the center (Tifft 1979). Galaxy density and, if conventional gravitation applies, the gravitational potential, varies with radius. Both factors may be basic in determining levels and populations in a quantum model. Can the general spread of redshift within clusters be explained in terms of a degenerate galaxy gas and thus dispose of the missing mass problem?

Figure 14

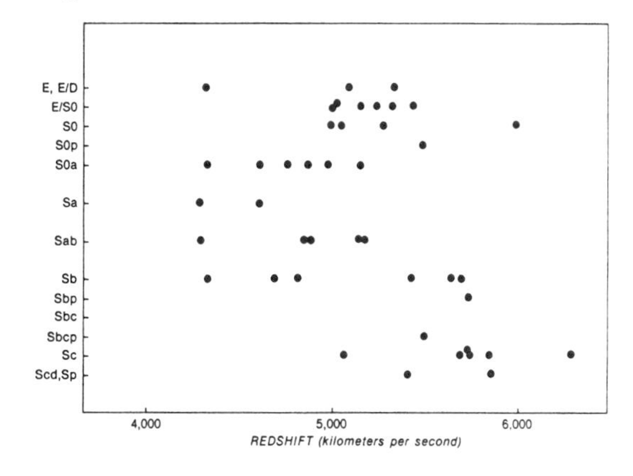

Figure 15

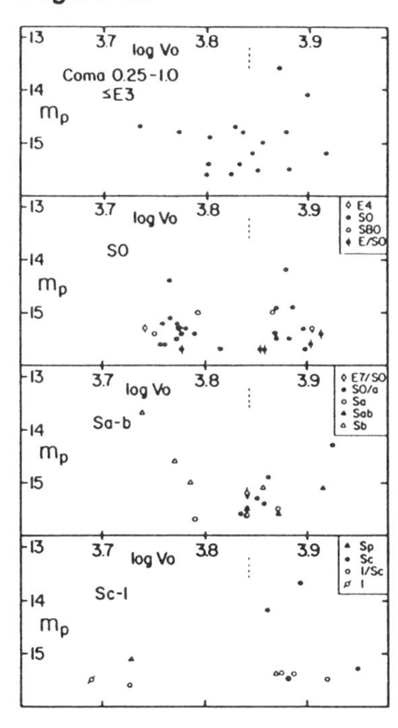

What, indeed, are galaxies themselves? If the nucleus of a galaxy is in fact a redshift singularity, then origin and evolution schemes for galaxies will certainly be quite different from anything we currently consider. The very concept of gravitation, redshift, and the existence of galaxies may reside in nuclei.

Finally, what is the redshift? Is it really a velocity, but subject to quantum rules which, as we go to larger and larger scales, increasingly constrain the average properties to what I have called redshift states? Or, alternately, are redshift differences somehow built in at the subatomic level, intrinsic to the material itself? This requires extraordinary constraints on large scale motion. On the other hand we have no actual hard evidence for real motion on the large scale, and there are some hints of quantization effects at the stellar level (Tifft 1977b). There is evidence that position angle vectors in galaxy pairs are aligned over large distances within superclusters (Tifft 1980b), as shown in Fig 16. This is inconsistent with orbital motion except in very special projections. Are single and multiple galaxies distributed the same way within superclusters, or do they have preferred redshifts or spatial locations? The answer to this question might be interesting.

We cannot even begin to answer most of these questions, or even frame the questions very well. We will not make much progress by ignoring them, however. I sincerely believe that future generations will look back on our present limited view of the redshift much as we regard the Ptolemaic view of the Solar System. We very much need new ideas, and new as well as old Chip Arps! Happy birthday Chip.

Figure 16 Position angle as a function of redshift within a supercluster filament.

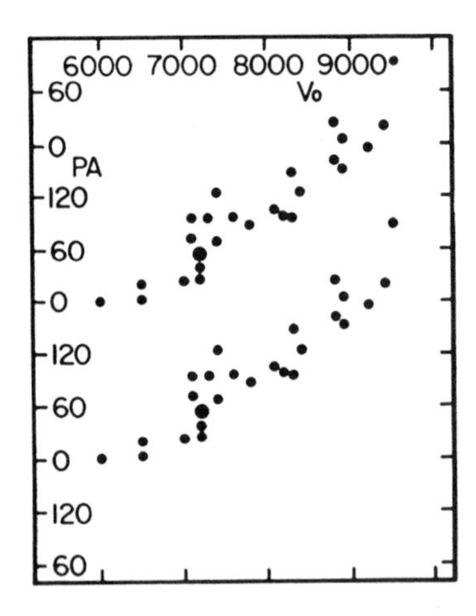

REFERENCES

Arp, H. (1986). Astron.Astroph., 156,207.
Arp, H. (1987). J.Astroph.Astron., in press
Arp, H. & Sulentic, J. W. (1985). Ap.J., 291,88.
Chincarini, G. L., et al. (1983). Ap.J., 269,13.
Cocke, J. (1983). Astroph. Lett., 23,239.
Cocke, J. (1985). Ap.J., 288,22.
Cocke, J. & Tifft, W. G. (1983). Ap.J., 268,56.
Cocke, J. & Tifft, W. G. (1987). in preparation.
Davies, R. L. et al. (1983). Ap.J., 266,41.
Dickel, J. R. & Rood, H. J. (1978). Ap.J., 223,391.
Dirac, P. A. M., (1958). Principles of Quantum Mechanics, 4th ed., London: Oxford
Dressler, A. & Sandage, A. R. (1983). Ap.J., 265,664.
Fisher, J. R. & Tully, R. B. (1981). Ap.J.Suppl., 47,139.
Giovanelli, R., et al. (1982). Ap.J., 262,442.
Gribbin, J. (1985). New Scientist, 20 June.
Karachentsev, I. D. (1972). Comm.Sp.Ap.Obs.USSR., 7,1.
Lewis, B. M. (1983). A.J., 88,962.
Nieto, M. M. (1986). Astroph.Lett., 25,45.
Peterson, S. D. (1979). Ap.J.Suppl., 40,527.
Roberts, M. S. (1968). A.J., 73,945.
Rogstad, D. et al. (1976). Ap.J., 204,703.
Schneider, S., et al. (1986). A.J., 92,742.
Schweizer, L. (1987). Ap.J.Suppl., in press.
Sharp, N. A. (1984). Ap.J., 286,437.
Tifft, W. G. (1974a). IAU Symp., No.58, 239.
Tifft, W. G. (1974b). Ap.J., 188,221.
Tifft, W. G. (1976). Ap.J., 206,38.
Tifft, W. G. (1977a). Ap.J., 211,31.
Tifft, W. G. (1977b). Ap.J., 211,377.
Tifft, W. G. (1977c). IAU Colloq., No.37, 159.
Tifft, W. G. (1978). Ap.J., 221,756.
Tifft, W. G. (1979). Ap.J., 233,799.
Tifft, W. G. (1980a). Ap.J., 236,70.
Tifft, W. G. (1980b). Ap.J., 239,445.
Tifft, W. G. (1982a). Ap.J., 257,442.
Tifft, W. G. (1982b). Ap.J.Suppl., 50,319.
Tifft, W. G. (1982c). Ap.J., 262,44.
Tifft, W. G. (1984). Steward Obs. Preprint No. 513
Tifft, W. G., & Cocke, J. (1984). Ap.J., 287,492.
Tifft, W. G., & Cocke, W. J. (1987). Sky & Telescope, 73,19.
Tifft, W. G., & Tarenghi, M. (1977). Ap.J., 217,944.
Turner, E. (1976). Ap.J., 208,20.
White, S., et al. (1983). M.N.R.A.S., 203,701.
Zwicky, F., et al. (1963-1968). Catalog of Galaxies and Clusters of Galaxies, Pasadena, California Institute of Technology.

DISCUSSION

H. van Woerden: In discussing redshift differences at a level of 1 km/s, precise definitions are of prime importance. What definition do you use? And is it the same definition as used by others? In asymmetric profiles the results will depend on the definition and on the velocity resolution of the observation.

W. Tifft: We measure redshifts as the mean of the two points where the profile crosses 20% of the average height of the central 80% of the profile area. Various definitions are used, as you are well aware, and with modern data such variations in definitions are one of the sources of scatter. Most of the effects I am discussing involve redshift differences much larger than the "definition" scatter. Furthermore *systematic* or phase related *correlated* changes cannot result from such a source of scatter. All the redshifts discussed are based upon profile crossings, not area integrals. Even with area defined redshifts, systematic or phase correlations should not result.

J. P. Vigier: I am worried about the *radial* redshift quantization. Nothing like that exists in quantum theory. (There is no equal interval in Bohr's orbits.) This effect, if true, is a source effect and not a mechanical Doppler effect. One can of course accept redshift variation with time. This is suggested for example by Wolf's experiments in Rochester on source coherence. If it varies with time so does z. One should make further statistics including z error bars to check this quantization's weight.

W. Tifft: The new 21cm data contains detailed S/N and error information as does the best optical data. There is a close relationship between data uncertainty and the detection of periodicities, as has been discussed at length in previous papers. For the best data sets the uncertainties are now quite small.
Although we have used certain analogies with standard quantum theory in discussing the extragalactic effect, it is quite true that there must be distinct differences. At this stage our approach must be almost entirely observational. On the other hand we must start forming a theoretical formulation,however primitive. In the main text of the paper I mention several situations where this primitive theory has been helpful. Lacking a theory must not be a basis for refusing to acknowledge observed effects.

T. Jaakkola: Redshift is not only an effect in the integral spectra, it is also a differential effect observed as redshift fields within galaxies, and as redshift gradients across the disks of external galaxies and the Milky Way. As a universal effect it should be observed and is observed also in the spectra of stars-but certainly not in the size of your z-quanta of tens of km/s. How do these facts fit your quantization picture and how are these interpreted there?

W. Tifft: From the earliest studies in the mid 1970's I have viewed galaxies as consisting of a set of redshift "states". The overall redshift of a galaxy represents the *average* of a blend of states. Within a single state, at the level of stars or lesser parts of galaxies, properties pass over to normal continuous relationships. Only in the grand average is a redshift quantized. The situation is like describing an atomic quantum state. An electron satisfies an overall probability distribution but *within* the distribution it is not constrained. This idea may of course be wrong and there is some evidence that some of the 72 km/s effects are actually seen at stellar levels (paper III in my 1976-77 series). This would suggest that the effect arises at the fundamental particle level. The effect is clear only at the scale of galaxies and above. The details of how to couple the effect to smaller scales is simply unknown.

J. C. Pecker: You observe complex objects. Could not the variation of z with time result from a variation with time of respective *intensities* of some features, at different *fixed* z? A similar observation could have resulted from integrated 21cm lines, as reported by M. Burbidge this morning, referring to the behaviour of multicomponent absorption features of some quasars.

W. Tifft: The spectral features discussed by M. Burbidge almost certainly arise in very small volumes where excitation conditions, cloud locations and other effects can actually change noticeably over short times. The total integrated 21cm emission profile of a galaxy is a different matter. Such changes must be very tiny in the total emission. Further, such changes cannot be connected in any obvious way to the redshift phase information. In some manner galaxies, at a specific phase in the redshift periodicity, shift in unison. I cannot conceive how small scale random processes within a galaxy could possibly produce such coordinated shifts. Of course, the shift may not occur in the galaxy, it may be a change in transit or in our perception upon receiving the light, but it is an integrated shift.

ARE REDSHIFTS REALLY QUANTIZED?

W.M. Napier, B.N.G. Guthrie and Bruce Napier
Royal Observatory, Blackford Hill, Edinburgh, EH9 3HJ.

Abstract. It has been argued that extragalactic redshifts are quantized in multiples of ~72 km s^{-1}. A programme of rigorous statistical scrutiny of this claim is being undertaken and preliminary results are described.

Introduction

We describe here the current status of an attempt to test the hypothesis: are redshifts of galaxies quantized in multiples of ~72 km s^{-1} or, more specifically, multiples in the range 70–75 km s^{-1}? A power spectrum analysis (PSA) technique was used, based on circular transformation statistics (Mardia 1972, Lutz 1985) : a fuller discussion of the technique and the associated confidence level analysis will appear elsewhere. The method was applied to accurate 21 cm observations which have become available for some nearby systems of galaxies. These data were not employed (and indeed did not exist) in the original formulation of the hypothesis (Tifft 1972, 1976 etc.) and thus there is no circularity of argument.

The Arp/Sulentic galaxy groups

Arp & Sulentic (1985) have listed 39 'associations' of galaxies, essentially small clusters comprising a dominant galaxy with a few fainter members. They obtained the differential redshifts of the fainter members relative to each dominant galaxy, in HI, using the Arecibo reflector : 65 differential redshifts were obtained, covering the range −146 to +883 km s^{-1}. A PSA of these data was carried out, periods in the range 20–200 km s^{-1} being searched for at 1 km s^{-1} intervals. The results are shown in Figure 1a. It can be seen that, while there is some tendency for residuals to peak in the range 60–80 km s^{-1}, no one peak stands out in particular.

Figure 1. PSA results for the Arp/Sulentic groups.

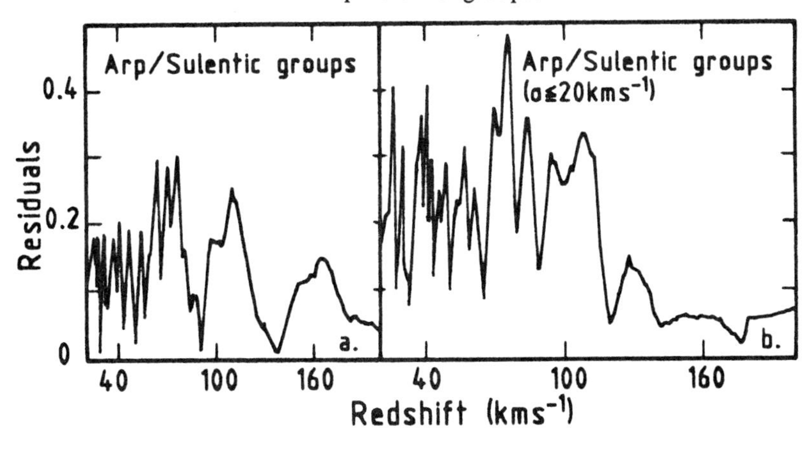

The trial was repeated using only accurately determined redshifts, viz those with standard errors $\leqslant$20 kms^{-1}. The mean s.e. of the 28 redshifts satisfying this condition was 10.6 kms^{-1}. The result of a PSA on these is shown in Figure 1b. A clear peak is now evident, at 74.4±0.6 km s^{-1} (s.e.). A genuine periodicity would have zero phase: that observed is −10 ± 3 km s^{-1}. Confidence levels were assessed by inter alia computing successive PSA's on synthetic data, these being random redshifts in the appropriate range : the probability of obtaining a chance fit to the a priori hypothesis period was found to be $\lesssim$2.10^{-3}, i.e. a confidence level $\gtrsim$ 99.8%.

The M31–M33 group

Arp (1986) has claimed that quantization of redshifts, in multiples of 72.4 km s^{-1}, occurs in a concentration of galaxies (about 60° by 20°) lying along a line between M31 and M33, and that these are Local Group galaxies although not generally recognised as such because of non-cosmological redshifts. Because this group extends 60° over the sky a correction for the (unknown) solar motion must be applied (loc. cit.). The approach adopted was to vary the solar velocity vector $\mathbf{V}_{\odot}$, generating power spectra for each $\mathbf{V}_{\odot}$ and recording the highest residual R_m of each spectrum without regard to its phase or position. PSA analyses were applied to 21 galaxies of this set, the solar speeds chosen being $V_{\odot}$ = 242, 252, 262 and 300 km s^{-1}. In all 5400 power spectra were generated; a representative set of contours is shown in Figure 2a.

Figure 2. (a) PSA contour map in galactic coordinates for the M31–M33 group. (b) PSA using the 'optimum' solar motion.

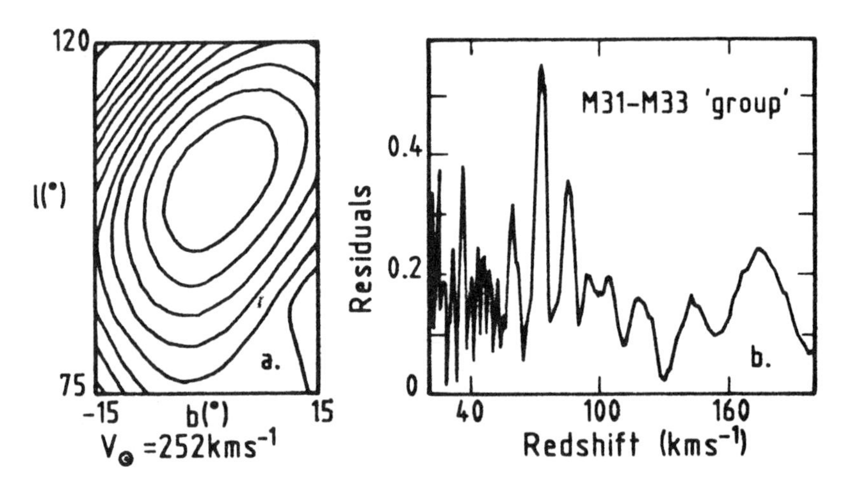

The highest residuals occur around galactic coordinates ℓ = 100°, b = 0°, $V_{\odot}$ = 252 km s^{-1}, with uncertainties ~3° in direction and ~20 km s^{-1} in speed. Figure 2b illustrates the power spectrum for this 'optimum' solar motion. The peak appears at 71.7 ± 0.3 km s^{-1}, and the phase is 0 ± 3 km s^{-1}. If the solar velocity is taken as prescribed, the hypothesis is found to be confirmed at a confidence level $\gtrsim$ 99.9%.

Similar analyses were carried out on the nearby Sculptor group of galaxies, and on the NGC 55 – 300 – 253 group (Arp 1986). No quantization was found in these small groups.

Binary galaxies

PSA's were carried out on assorted data sets of binary galaxies. For 131 binaries with $\Delta V < 1000$ km s^{-1} catalogued by White et al. (1983), no significant periodicities were evident but the accuracy of measurement is generally low (~40 km s^{-1}). From a more recent compilation by Picchio & Tanzella-Nitti (1985), 36 binaries with $\sigma_{\Delta V} \leqslant 20$ km s^{-1} were analysed. A peak at 76 km s^{-1} was evident, although its significance was not particularly high, and when the 21 most accurately measured binaries were analysed ($\sigma_{\Delta V} \leqslant 14$ km s^{-1}), this periodicity vanished. However, when an analysis was carried out on accurately measured wide binaries, i.e. systems with separation $\theta > 10'$ and accuracies better than 25 km s^{-1}, quantization reappeared, at 76±2 km s^{-1} at a confidence level of about 98%.

Discussion

Quantization does not seem to occur in close binaries and compact groups, but seems to be spectacularly present in the more widely spaced systems so far examined. This might lead one to conjecture that the effect is real but goes away if galaxies are crowded together in space. The M31-M33 group would seem at first to contradict this supposition. However a Fisher-Tully analysis applied to 8 galaxies of this set revealed that the group seems to be a chance alignment strung out along a line of sight ~ 10 Mpc deep (Figure 3) : the crowding conjecture could therefore still be valid. That the same data which yield redshift quantization also yield an unexceptionable Hubble constant ~62 km $s^{-1}Mpc^{-1}$ argues that the quantization is a global phenomenon rather than one confined to dynamically coherent groups such as binaries. This resurrects the old spectre of tired light, albeit light which gets tired in discrete jumps.

Figure 3. Redshift-distance relationship for 8 galaxies in the 'group'.

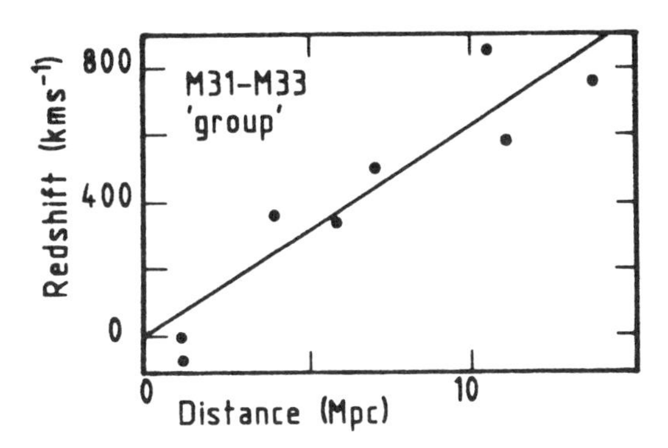

References

Arp, H. (1986). Max-Planck-Institut für Physik und Astrophysik, preprint No. 266.
Arp, H. & Sulentic, J.W. (1985). Astrophys. J., 291, 88.
Lutz, T.M. (1985). Nature, 317, 404.
Mardia, K.V. (1972). Statistics of Directional Data. Academic Press, London and New York.
Picchio, G. & Tanzella-Nitti, G. (1985). Astr. Astrophys., 142, 21.
Tifft, W.G. (1972). Astrophys. J., 175, 613.
Tifft, W.G. (1976). Astrophys. J., 206, 38.
White, S.D.M., Huchra, J., Latham, D. & Davis, M. (1983). Mon. Not. R. astr. Soc., 203, 701.

RECENT OBSERVATIONS OF THE BL LAC OBJECT PKS 2155-304

ALDO TREVES

Dipartimento di Fisica dell' Universita' di Milano, Italy.

Introduction

PKS 2155-304 is one of the brightest BL Lac objects, which has been observed in various spectral regions (e.g. Urry et al, 1986; Treves et al, 1988). Here I refer first on a series of quasisimultaneous multifrequency observations obtained in 1983-1985 when the X-ray satellite EXOSAT was active. Next I concentrate on some recent results on absorptions in the far UV. A complete account of the results will be found in Treves et al (1988) and Maraschi et al (1988).

Multifrequency Observations

The source was observed at 9 epochs with EXOSAT, with a quasi-simultaneous coverage with the International Ultraviolet Explorer. This enabled to measure the flux of the source in five distinct energy bands: medium energy X-rays (1-10 keV) from the EXOSAT proportional counters; low energy X-rays (0.05- 2 keV) from the EXOSAT Channel Multiplier Array; 1200-2000 A, and 2000-3000 A from the two I.U.E. cameras; and the optical band from ESO telescopes and the fine error sensor on I.U.E. The main result from the the examination of the light curves in the various bands (Treves et al 1988) is that the source is variable at all frequencies, with variability increasing with the energy. The minimum observed variability time scales are the following: t(opt) $\sim$ 90 d; t(1500 A) $\sim$ 50 d; t(0.2 keV) $\sim$ 7h; t(3 keV) 2000 sec. No variability has been observed in the radio (Uvelstad and Antonucci 1986). The differences in the values of the time scale strongly suggest that the dimension of the sources responsible for emission in the various bands are different, with the X-ray deriving from a substantially more compact region than the optical and radio. Inhomogeneous models of the source are therefore required. A synchrotron self-Compton model of the source, which accounts for the overall energy distribution was proposed by Ghisellini et al (1986). No relativistic beaming is required, contrary to the results from homogeneous SSC models. On this regard the small value of the variability time scale in medium energy X-rays may be significant. In fact if the source is isotropic and Eddington limited, $M > 10^8\ M_\odot$ and, therefore, the dimension associated with the variability time scale ct becomes close to the gravitational radius (Morini et al 1986). Some beaming could be required, even if much smaller than that requested by the homogeneous SSC models.

Indirect evidence of a jet may be inferred from the presence of an absorption trough at 650 eV, discovered by Canizares and Kruper

(1984). In fact the best explanation is in terms of O VIII Ly absorption from a hot gas with a velocity of 20000 km/s.

UV Absorptions

In view of the discovery of the 650 eV absorption feature, the far UV archived IUE spectra have been examined with special interest on absorption features (Maraschi et al 1988). The averaged spectrum resulting from the superposition of 24 exposures is reported in Fig.1. A number of absorptios are apparent; most of them can be identified with Galactic, and are similar to those observed in the Magellanic Cloud stars. However at 1285 A there is a clear feature, which does not appear of Galactic origin. The proposal is that the feature is due to Ly α absorption by material at z= 0.058. The origin of the material could be the halo of an intervening galaxy, or a H cloud analogous to those inferred from the Lyα forest in high z quasars. Another interesting possibility is a relation with the ejection of material from the source considered by Canizares and Kruper (1984).

References

Canizares C.R. and Kruper J., 1984, Ap. J. Lett. 278, L99.
Ghisellini G. Maraschi L.,and Treves A.,Astron.Ap. 204,212.
Maraschi L. et al 1987, submitted to Ap.J.
Morini M. et al. 1986, Ap. J. Lett. 1986 306, L7.
Treves A. et al 1987, in preparation.
Ulvestad J.S. and Antonucci R.R.J.,1986 Astron. J. 92,6.
Urry C.M.,Mushotsky R.F., and Holt S.S.,1986,Ap.J.305,369.

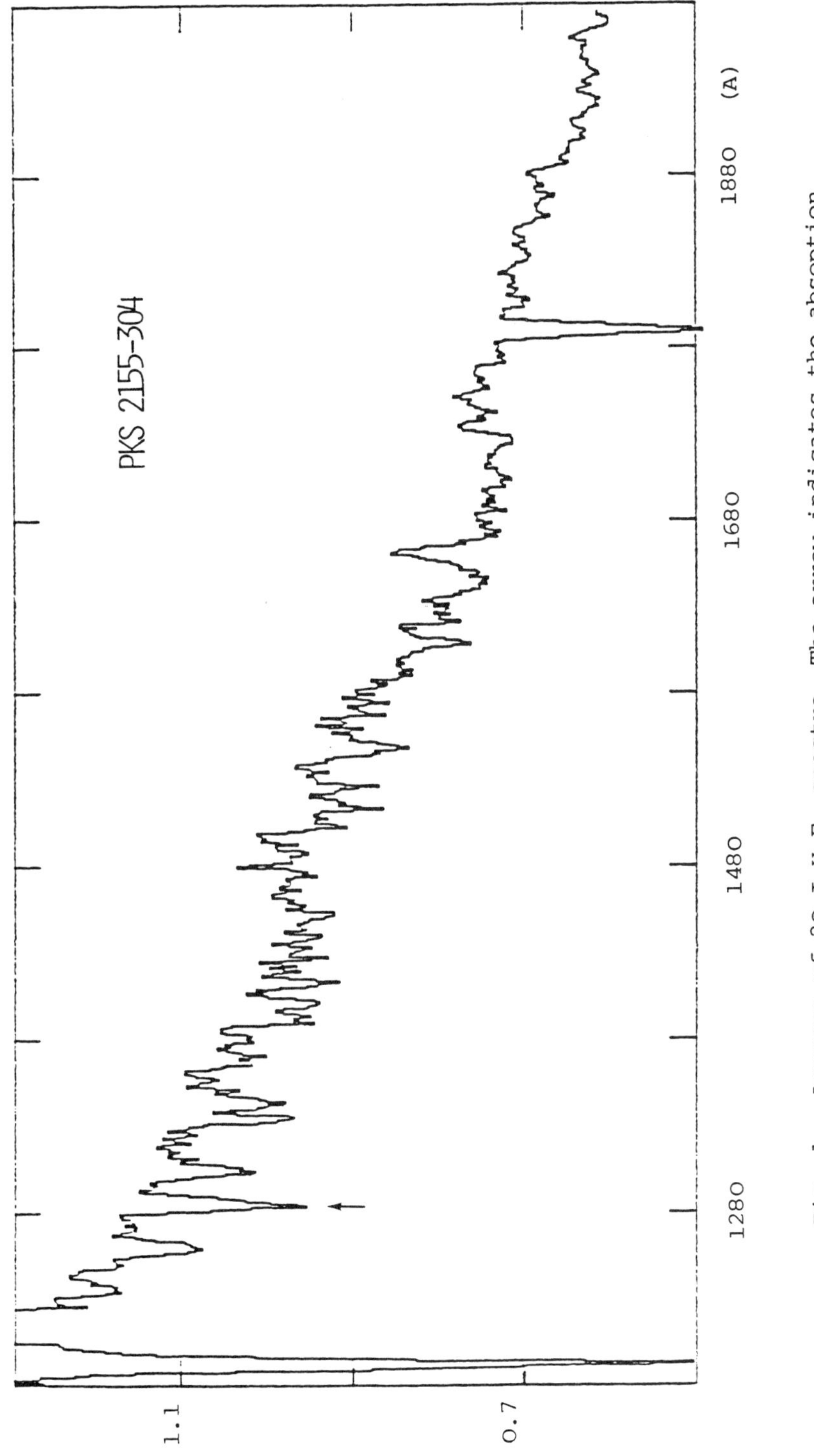

Fig. 1. Average of 20 I.U.E. spectra. The arrow indicates the absoption feature at 1285 A. Ordinates are in arbitrary units. (From Maraschi et al 1987).

EXTRAGALACTIC THEORY AND COSMOLOGY

EJECTION MECHANISMS IN GALAXIES

William C. Saslaw
Department of Astronomy, University of Virginia,
National Radio Astronomy Observatory, Charlottesville,
Virginia, and Institute of Astronomy, Cambridge, England

Abstract

Powerful galactic jets may eject the central engine from the nucleus of its galaxy. Conditions for this rocket effect to occur, and some of its implications, are discussed.

I. Introduction

Many galaxies are in turmoil. They are not the stately, sedate, quasi-static systems that most astronomers believed as recently as twenty-five years ago. During the last quarter century, Chip Arp's observations have been at the forefront of our changing understanding of galaxies. His Atlas of Peculiar Galaxies (Arp, 1966) brought into sharp focus the wonderful variety of galactic activities: jets and plumes, collisions and explosions, starbursts and distorted spirals, tidal disruption and cannibalistic coalescence, galaxies with multiple nuclei and galaxies with no nuclei at all--the list goes on. Arp's detailed photography, spectroscopy, photometry and radio observations of individual galaxies have led us to understand some of this peculiar behavior, and to hope that we might eventually understand even the more bizarre cases. Throughout the development of this subject, Arp's results have forced us to ask whether we need new ideas just as bizarre as the observations they are supposed to explain, or whether the imaginative combination of conventional ideas can produce unexpected explanations.

Among the manifestations of galactic turmoil, some of the most surprising are the galaxies that eject large amounts of matter. These were sufficiently strange and exciting that their discoveries occasionally appeared on the front pages of national newspapers, and often in the popular science magazines. The examples of ejection which are best understood--though far from being well understood--are the radio jets which emerge from the nuclei of giant elliptical galaxies to power the radio lobes often found on each side of those galaxies (for a review see, e.g., Begelman _et al._ 1984). The conventional interpretation of these jets is that they are (possibly relativistic) beams of energetic particles ejected along the rotation axis of an accretion disk around a massive black hole in the nucleus. In keeping with the spirit of the title of this Symposium, I will discuss how this conventional interpretation and simple conventional physics leads to the unconventional idea that the black hole itself can be ejected from the galactic nucleus.

But first, to put the subject in some perspective, I would like to just list the wide range of ideas and observations associated with galactic ejection. Table 1 lists several forms that have been suggested for the central engine and for the more specific ejection mechanism. It also lists the basic observed phenomena. Since each of the engines can combine with several of the ejection mechanisms, and more than one engine may operate at the same time, there is clearly scope for a large number of ideas to explain relatively few observations. Moreover there is no a priori reason why some of these general observed phenomena could not be produced by more than one engine. There are people who prefer unified models of galactic nuclear activity, but I think the observations are too varied for that.

Some of these engines, such as dense stellar systems, may evolve into others such as black holes or supernovae or collapsing gas clouds. Each entry, almost each plausible combination, has become a significant astrophysical industry in the last twenty years. Most of these industries remain viable even today, although some temporarily turn dormant as fashions change. There still does not seem to be direct compelling observational evidence for any particular engine or ejection mechanism; each one has its successes and problems. The most

Table 1: Galactic Nuclear Ejection Phenomena

Engine	Ejection Mechanism	Observed Phenomenon
Single black hole accretion	Accretion jet	Broad emission (absorption) lines
Multiple black hole system	Rocket	Radio jets (lobes)
Spinar	deLaval nozzle	Optical jets (hotspots in lobes)
Dense stellar system	Magnetic flare	VLBI motion
Supernova, pulsar system	Electromagnetic waveguide	Discordant redshift "connections," alignments, concentrations
Colliding or collapsing gas clouds	Gravitational slingshot	
Starbursts	New physics	
New physics		New observations

convincing case is for active galactic nuclei to emit jets of some sort (reviewed by Bridle & Perley, 1984).

II. Rockets

All the ideas of Table I--except one--have been described many times. So in keeping with the title of this Symposium I will not review them here. The idea I'd like to discuss has been mentioned briefly before, but then it dropped almost out of sight in the literature. Perhaps we could at least call it a renewed idea. Shklovsky first mentioned it in 1970 and then again in 1982. In between those years, and subsequently, people hardly thought about it. Shklovsky's (1970, 1982) idea was that a simple consequence of a jet is a rocket. He supposed the engine would be ejected from the galaxy in the opposite direction of the jet.

Perhaps the reason this idea has not caught on is that early impressions of radio galaxies suggested strong symmetry between the two radio lobes. Most models therefore assumed that two-sided intrinsically symmetric jets issued from the active nucleus, either directly from the accretion disk around a black hole or more remotely in a deLaval nozzle found in the gas around the symmetric engine. When more detailed VLA observations began to show many asymmetric jets with intensity ratios of 4:1 or greater, particularly in the more powerful sources, this was generally ascribed to Doppler boosting. There is, however, no direct observational evidence for the necessary relativistic velocities in the outer parts of jets. An alternative explanation is that the energy of the symmetric jet is mostly dissipated on one side, possibly because gas, magnetic fields and turbulence within and around the active galaxy are distributed inhomogeneously.

New observations, however, suggest that in some cases, at least, the jet asymmetry may be intrinsic; what we see is what we get. Asymmetric jets occur both in powerful quasars and in weaker nearby radio galaxies. Examples of one sided jets in quasars are 1007+417 and 1150+497 (both in Owen & Puschell 1984) and 0800+608 (Shone & Browne, 1986). This last case is particularly intriguing since simple models by Shone & Browne suggest that the jet velocity is much too low for Doppler boosting to produce an apparent one-sidedness.

Among nearby asymmetric radio jets, NGC 6251 is usually considered the most dramatic example (Bridle & Perley, 1984). Others are M87 (Arp, 1967; Biretta *et al.* 1983), PKS 0521-36 (Keel, 1986), 1759+211 (Saikia *et al.* 1987) and NGC 3310 (Bertola & Sharp, 1984; Duric *et al.* 1986).

None of these examples are definitely known to be *intrinsically* one sided, since many classes of models which describe them are rather flexible and not easily ruled out. The most straightforward test would be multi-epoch VLBI observations showing a jet flowing out one side of the nucleus. Even here we would have to know that the radio emission represented the overall dynamics, which it might not. Sequences of VLBI maps made over periods of a few years have already shown asymmetric structural variations, as in 4C 39.25 (Shaffer et al. 1987). At present, the main difficulty in interpreting these types of observations is knowing which features are moving and which are stationary. The ambiguity comes in deciding which feature is the engine, which is the galactic nucleus, and which are shocks, obstacles or pinches in the jet. These problems can be resolved by following the features for long enough that their behavior clarifies, or by finding cases whose evolution is quick and lucid. Observing the motion in the nucleus of one radio galaxy relative to a completely different radio source is a promising technique. It has already been done for 3C 345 to determine which component is probably the core, and which are the ejecta (Bartel et al. 1984).

The theory of a single exhaust model has not been greatly discussed. Narlikar and Subramanian (1983) examined how an initial two-sided jet would be affected in quasars ejected relativistically from galactic nuclei by an unspecified mechanism. Their main concern was to quench the blueshift catastrophe which occurs in those models. In their model the strong ram pressure of the intergalactic medium suppressed the forward exhaust. Another intrinsically symmetric model in which the engine was offset in the z direction from the nucleus, resulting in an asymmetric exhaust at larger distances, was discussed by Wiita and Siah (1981). Here the vertical density gradient suppressed the jet on one side.

The models I am concerned with here differ from the previous ones in that the central engine is intrinsically asymmetric. There are several ways this could occur in models where the engine is a black hole surrounded by an accretion disk. First, it may be possible to plug the outflow along one axis, at least temporarily. Although the size and total energy of radio jets are impressive, their mass outflow may be relatively low: considerably less than 1 $M_\odot$ per year in many cases. In a region much less than one light year in radius around the black hole, there is still considerable scope for massive gas clouds or disrupting stars to move into the jet. They may absorb or deflect a considerable amount of the jet's momentum. This could change the shock structure and the resulting flow pattern of the gas around the black hole significantly. Its effect may be to quench the jet on one side of the engine. Then the question becomes whether this intrinsic asymmetry is a new quasi-stable state, or whether the engine would sputter and restart its jet exhaust again. To get real insight into these non-linear three-dimensional relativistic flows requires help from detailed numerical hydrodynamic computations just beyond the present state of the art (cf. Hawley & Smarr, 1986).

Second, it is possible that the engine started off asymmetrically because the initial accretion was predominantly along one side of its rotation axis. This could occur either because the engine was moving through the nucleus--perhaps gravitationally perturbed by large gas clouds or clumps of stars--or because the distribution of fuel was rather one sided. Once a one-sided accretion is set up, the question is again to find the conditions for it to be self-sustaining.

Third, any differential acceleration between the accretion disk and the black hole would tend to produce asymmetry. This acceleration may be produced by tidal forces, or by an initial rocket thrust. To see how the rocket's thrust would produce asymmetry, consider a simplified model. Suppose that most of the thrust is produced by an annular region of mass M_t, in the accretion disk. Mass flows into this annulus where it is accelerated and the reaction moves the position of the annulus ahead of the black hole. Let θ be the angle between the direction from the black hole to the annulus and the direction from the center of the black hole to its equator. Then the gravitational restoring force of the black hole, of mass M_H, on this annulus is $\sim GM_H M_t \theta / r^2$ where r is the distance to the annulus and θ is assumed to be small. The net thrust which accelerates the annulus is $\dot{M}u$ where $\dot{M}$ is the mass loss rate and u is the velocity of the gas leaving the annulus. Let $r_* \equiv r/r_{sch}$ and $\tau_{sch} \equiv r_{sch}/c$ where $r_{sch} = 2GM_H/c^2$ is the Schwarzschild radius and τ_{sch} is the Schwarzschild crossing time. Then the balance of gravity and thrust gives the angle

$$\theta = 2r_*^2 \tau_{sch} \frac{\dot{M}}{M_t} \frac{u}{c} = r_* \frac{\tau_r}{\tau_t} \frac{u}{c} \quad (1)$$

Here $\tau_r \equiv 2r/c$ is the light crossing time of the annulus and $\tau_t \equiv M_t/\dot{M}$ is the mass loss time scale of the annulus. As an example, suppose we take $\tau_t = 1$ year, which is an estimate from the observed time scale of fluctuations, $r_* = 10$, $\tau_r = 10^4$ using $10^8\ M_\odot$ for M_H, and $u/c = 0.1$ if the jet is non-relativistic. This gives $\theta \simeq 3 \times 10^{-4}$ rad $\approx 1'$, which is very small. Possibly even a small angle θ can influence the shock structure of accretion enough to produce a larger asymmetry farther out, but that remains to be seen. The only way to increase θ significantly would be to assume a relativistic jet and a much shorter time scale (days) for gas to flow through the annulus. This might work in some cases, but I am doubtful about it as a general cause of asymmetry. On the other hand, this is still a useful result because it shows that the accretion disk can remain bound to the black hole, despite producing a considerable net thrust. Therefore the accretion disk and the black hole can, together, form a rocket.

III. Escape

For the first question, one might ask whether such a rocket can escape from its galactic nucleus and eventually from its galaxy. Shklovsky simply assumed it would, but Mark Whittle and I have recently looked at the problem in more detail and found that there are several possibilities. If the rocket leaves the nucleus, it is easy to think of many observational implications, and I wonder if Chip Arp may have discovered some examples. First, let us consider some of the simple theory (Saslaw & Whittle, 1988).

The simplest equation describing the velocity, v, of a rocket of mass M_H with thrust $u \dot{M}_H$ moving through a galaxy is

$$M_H v \frac{dv}{dr} = u \dot{M}_H - \frac{G M_H}{r^2} M(r). \qquad (2)$$

Here M(r) is the galactic mass remaining with r. It is assumed to have a spherical distribution, and $u \dot{M}_H$ is assumed to be constant for simplicity. Neither of these assumptions are an essential part of the basic problem and both can be modified easily for more complicated models. Since the mass of the black hole dominates the accretion disk, we can regard M_H as constant where it appears explicitly in equation (2).

The rocket's behavior depends mainly on two non-dimensional parameters. One is the thrust-to-gravity ratio at a reference radius denoted by r_b:

$$T \equiv \frac{u \dot{M}_H r_b^2}{G M_H^2} . \qquad (3)$$

The second is a mass ratio

$$\mu \equiv \frac{4\pi r_b^3 \rho_o}{M_H} \qquad (4)$$

where ρ_o is a density which characterizes the galactic model. The actual values of these quantities will vary considerably from galaxy to galaxy. Just as an illustration to set the scale, consider $M_H = 10^7 M_\odot$, $\dot{M}_H = 1 M_\odot yr^{-1}$, $u = 10^9 cm s^{-1}$, $r_b = 10$ pc and $\rho_o = 10^3 M_\odot pc^{-3}$. Then the thrust-to-gravity ratio $T \approx 2.3$ and the mass ratio $\mu \approx 1.3$. These quantities also define a characteristic time scale of the orbit:

$$t_o = \left(\frac{M_H r_b}{\dot{M}_H u}\right)^{1/2} \qquad (5)$$

which is approximately the time the rocket would take to reach r_b if there were no restoring force. With the values just given, $t_o \approx 10^5$ yr. To describe the galaxy, we may consider a simple model which has four idealized regions: (a) an inner nucleus of radius $r < r_a$ where the stars are bound to the black hole; (b) an outer nucleus $r_a < r < r_b$ where the stars are self-gravitating and have a steep density distribution whose form is taken to be a power law

$\rho = \rho_0 (r/r_b)^{-n}$ within r_b; (c) a density plateau $r_b < r < r_c$ where $\rho = \rho_0$; and (d) a halo beyond r_c with the density decreasing rapidly with radius, perhaps as r^{-2}. The radius r_b of the nucleus and the density ρ_0 of the plateau are thus used as the reference quantities in equations (3)-(5).

Solving equation (2) determines whether there is a radius where v = 0 for which the rocket remains bound to the galaxy. The essential results, however, can be stated simply. Stars in the inner nucleus will respond readily to the motion of a slowly moving low thrust rocket. The density distribution of these stars may well be unrelaxed and is very uncertain. We will assume they have a constant density $\rho \approx \rho_0(r_a/r_b)^{-n}$. Then the radius where the mass of stars equals that of the black hole is $r_a \approx (3/\mu)^{1/(3-n)}\, r_b$. A rocket moving slowly through the stellar distribution produces an enhanced mass $\sim M_H$ at a characteristic distance $\sim r_a$ behind it. So it will feel a gravitational drag force of order GM_H^2/r_a^2. For the rocket's thrust to exceed this drag requires $T \gtrsim (r_b/r_a)^2 \approx (\frac{\mu}{3})^{2/(3-n)} > 1$. So the rocket usually moves into the outer nucleus if $T \gtrsim 1$.

More detailed analysis shows that with typical values of $n \approx 9/4$ and $r_a \approx 0.1\, r_b$, the rocket will usually escape from the outer nucleus if $T/\mu \gtrsim 4$. It is clear that for the range of values of T and μ implied by observational data, we should expect some rockets to remain bound to the nucleus, and others to escape. This is the essential point.

For the rocket to subsequently also escape from the density plateau requires approximately $T/\mu \gtrsim r_c/6r_b$. Thus rockets which can escape from the nucleus will usually escape completely from galaxies with relatively small regions of constant density. Halos provide little restoring force for rockets that can maintain their thrust. Only galaxies with a large density plateau (a few kiloparsecs in radius) are likely to retain very high thrust rockets.

IV. Rocket Redshifts

Next, one can ask for the velocity of a galactic rocket. To maximize this we can imagine, as Shklovsky did, that there is no restoring force and use the standard textbook result for the limiting rocket velocity, v. Now we must take into account the rocket's decrease in mass as it expells its "fuel" through the accretion disk. Assuming, first, that the constant thrust velocity u << c the standard result is

$$\frac{v}{u} = \ln \frac{M_i}{M} \qquad (6)$$

where M_i is the rocket's initial mass and M is the remaining mass after the accreted material is exhausted. Here the best one might hope for is that all the stellar entourage of mass $\sim M_H$ bound to the black hole is processed through the accretion disk. So the maximum initial mass ratio of fuel to rocket $\approx 1/2$ and $v \approx 0.7$ u. This is in the range of the discordant redshifts of interacting galaxies (Arp 1982; 1986) but far from the QSO range. Of course, it would be necessary to show that the discordant companions could actually be related to rockets. In some cases which have been explored in detail they appear to be background galaxies (Sharp 1985).

Can we get higher velocities? Although equation (6) is no longer valid for this regime, it indicates the direction of the answer. Relativistic recoil would require both a relativistic thrust and a very large ratio M_i/M. There are several models of accretion jets which suggest relativistic thrust, so this may not be a great difficulty. It seems harder to find reasons for a ratio $M_i/M \gtrsim 2$ since this value already requires the entire entourage of the black hole to be ejected and converted into thrust. All that seems left is for the black hole itself to evaporate by producing Hawking radiation. To produce a significant effect, this would have to be asymmetric and have a much shorter time scale for a $\sim 10^7\ M_\odot$ black hole than conventional theory suggests. A cluster of mini black holes might work, and should produce other interesting effects, but would have to be very carefully contrived. With $M_i/M = 2$ and $u = c$, we obtain $v=0.6c$ from the relativistic rocket, slightly less than the non-relativistic formula (6) gives. (Because of the inertial effects of the energy, the relativistic analog of equation (6) is $v/c = (M_i^2-M^2)/(M_i^2+M^2)$, on the assumption the rocket's proper acceleration is constant and its thrust velocity, c, is the maximum possible.)

So our conclusion thus far seems to be that discordant redshifts of a few hundred to a few tens of thousands of kilometers per second are possible but that relativistic cases are very unlikely. Examples with non-relativistic discordant redshifts follow from rather conventionally minded models. To learn whether they actually exist, it is necessary to make detailed studies of individual galaxies.

V. General Comments

Rockets are such simple consequences of galactic jets that their complete absence would almost be more surprising than their discovery. If we do not find a rocket associated with a jet, we could essentially conclude that either the engine is intrinsically very symmetric, or that the nuclear restoring force dominates the thrust, or that the basic black hole accretion disk model of the jet is incorrect. Any of these conclusions would be important. If a rocket is discovered, then it provides useful relationships among properties of the galaxy and the jet, as Saslaw & Whittle have discussed.

Many astronomical phenomena are observed long before they are discovered. They may show up in the guise of other phenomena, believed to be better understood at the time. Rockets may be an example. Rockets may, in some cases, masquerade as double galactic nuclei (normally believed to have a capture origin), or as galactic nuclei with a double or peculiar velocity distribution (from the entourage of stars around the black hole), or as compact discordant redshift companions to galaxies (usually thought to be background objects, which they may often be unless redshifts are non-Doppler since the companions are always redshifted relative to the larger galaxy). Absence of a strong jet does not necessarily rule out these possibilities since the rocket may still be escaping from its parent galaxy even though its jet turned off some time ago, possibly after a short but very intense thrust. Fossil remains of the jet or the rocket's motion may be present (Saslaw & De Young, 1972). Presence of a jet would, of course, make the case much more convincing.

For the powerful extragalactic radio jets, it would be useful to look at the highly asymmetric nearby examples for signs of an engine which is off either the optical or the dynamical center of the galaxy. During the lifetime of the jet, typically estimated at 10^6-10^8 years, the rocket may not have moved very far. Therefore this would require rather high precision astrometry; a few examples mentioned earlier have shown it is possible, but very little has been done. More candidates are described elsewhere (Saslaw & Whittle, 1988) along with suggestions for spectroscopic observations. The main observational difficulties are not technical but psychological: an observer must be willing to probe new interpretations of old or unsuspectedly ambiguous data. This is an area where we do not quite know what to look for, or where to look. It is best not to have rigid expectations. Weak radio galaxies, for instance, may once have been powerful and their spent rockets may be far outside their parent galaxy. In this case we might even find a bonus.

So many processes clutter up the nuclei of galaxies that it is difficult to find clear manifestations of the central engine. If we can find cases where the central engine has left its nucleus, emerging perhaps diminished but pristine, we should have a clearer understanding of what drives it.

I am glad to thank Mark Whittle and Alan Bridle, particularly, and many other colleagues more generally, for helpful discussions of this subject. And, naturally, thanks to Chip Arp for showing that galaxies remain full of puzzles.

References

Arp, H. C. 1966. Ap. J. Supp. XIV, No. 123.
Arp, H. C. 1967. Astrophys. Lett. 1, 1.
Arp, H. C. 1982. Ap. J. 263, 70.
Arp, H. C. 1986. In IAU Symposium 124 to be published.
Bartel, N., Ratner, M. I., Shapiro, I. I., Herring, T. A., and Corey, B. E. 1984. In VLBI and Compact Radio Source, IAU Symposium 110, ed. by R. Fanti, K. Kellermann, and G. Setti. (Dórdrecht: Reidel).
Begelman, M., Blandford, R. D., and Rees, M. J. 1984. Rev. Mod. Phys. 56, 225.
Bertola, F. and Sharp, N. A. 1984. MNRAS 207, 47.
Biretta, J. A., Owen, F. N., and Hardee, P. E. 1983. Ap. J. Lett. 274, L27.
Bridle, A. H. and Perley, R. A. 1984. Ann. Rev. Astron. Astrophys. 22, 319.
Duric, N., Seaquist, E. R., Crane, P. C., and Davis, L. E. 1986. Ap. J. 304, 82.
Hawley, J. F. and Smarr, L. L. 1986. In Magnetospheric Phenomena in Astrophysics, ed. by R. I. Epstein and W. C. Feldman (New York: AIP). p. 263.
Keel, W. C. 1986. Ap. J. 302, 296.
Narlikar, J. V. and Subramanian, K. 1983. Ap. J. 273, 44.
Owen, F. N. and Puschell, J. J. 1984. Astron. J. 89, 932.
Saikia, D. J., Wiita, P. J., and Cornwall, T. J. 1987. MNRAS 224, 53.
Saslaw, W. C. and De Young, D. S. 1972. Astrophys. Lett. 11, 87.
Saslaw, W. C. and Whittle, M. 1988. To be published.
Shaffer, D. B., Marscher, A. P., Marcaide, J., and Romney, J. D. 1987. Ap. J. (Letters) 314, L1.
Sharp, N. A. 1985. Ap. J. 297, 90.
Shklovsky, J. 1970. Nature 228, 1174.
Shklovsky, J. 1982. In Extragalactic Radio Sources, ed. by D. S. Heeschen and C. M. Wade, IAU Symposium No. 97, p. 475.
Shone, D. L. and Browne, I.W.A. 1986. MNRAS 222, 365.
Wiita, P. J. and Siah, M. J. 1981. Ap. J. 243, 710.

DISCUSSION

G. Burbidge: Your ideas may fit in well with the local model for QSO's that I published in the proceedings of the Trieste conference (Reidel 1986). There I argued that the rocket process was at work for QSO's and that many BL Lac objects were objects ejected from galaxies and coming towards us. A particularly good example is that of the radio galaxy 3C 66B which in this model is ejecting the BL Lac object 3C 66A.

S. Bonometto: Do you expect also a contribution to z coming from gravitation?

W. Saslaw: Yes, but it may be small because much of the optical radiation may originate from the ensemble of stars that are weakly bound to the black hole. The accretion disk may also produce significant redshifted radiation which would be expected to produce broadened lines. The relative amounts of these contributions depends on detailed models.

R. Sanders: It is presumably the accretion disk which forms and collimates the jet. Thus we would expect the disk to absorb the thrust of the jet. Then the condition that the disk drag the black hole would seem to require that the accretion disk have a mass that is comparable to the black hole. Is this true? If so, then it would seem as though your picture of the engine is more similar to what has been called a "spinar" rather than to a black hole with accretion disk–and therefore subject to all of the instabilities discussed for such objects.

W. Saslaw: The jet's thrust is absorbed by the accretion disk, the black hole and the entourage of stars moving with the black hole, since these are all strongly bound gravitationally. The mass of the accretion disk could be say, 10% of the black holes mass and still provide enough thrust for long enough that the rocket escapes. Even if the accretion disk's mass were somewhat greater, the system would not be a spinar so long as it was dominated by the mass of the black hole.

E. Khachikian: I have found galaxies with a radial optical jet which is detached from the nucleus. Can this be explained by a rocket?

W. Saslaw: The rocket can be at the far end of the jet, and the region of the jet near the nucleus could have cooled, so it will no longer appear to radiate strongly near the nucleus.

COMMENTS ON H. ARP "THE PERSISTANT PROBLEM OF SPIRAL GALAXIES"

Hannes Alfvén
The Royal Institute of Technology, Stockholm
and
University of California, San Diego

Abstract

In his paper "The Persistent Problem of Spiral Galaxies" H. Arp criticises the standard theory of spiral galaxies and demonstrates that introduction of plasma theory is necessary in order to understand the structure of spiral galaxies. In the present paper arguments are given in support of Arp's theory and suggestions are made how Arp's ideas should be developed. An important result of Arp's new approach is that there is no convincing argument for the belief that there is a "missing mass". This is important from a cosmological point of view.

A. Arp's New Approach to the Theory of Galactic Dynamics

In what may be an epoch-marking paper, "The Persistant Problem of Spiral Galaxies" (IEEE Trans. Plasma Sci., PS-14, Dec. 1986), H. Arp draws three especially important conclusions.

1. A careful analysis of the "generally accepted" density wave theories shows that these theories cannot be correct. This is important, because they have long prevented progress in the field of theories of galactic dynamics.

2. Next problem is to fill the void which these theories leave. Arp suggests that this may be done by *plasma theories*. This is likely to be correct. However, his outlines of how this should be done could be strengthened. He bases his arguments mainly on the Chandrasekhar-Fermi paper (1953). With the deepest respect for them, it must be stated that cosmic plasma physics has advanced considerably during the thirty-four years which have passed since its appearance. Especially, it has taken a jump forward with the opening of the X-ray and gamma ray octaves by space research which has given rise to a new model of our cosmic environment, usually referred to as "Plasma Universe".

3. The third conclusion is that there is no evidence for a "missing mass". This will be discussed in D.

B. Plasma Universe

It is fortunate that in the same issue of IEEE Trans. Plasma Sciences in which Arp has published his paper,he can find advice how the new plasma physics should be used for developing his ideas. Indeed, the Special Issue Review paper by Fälthammar with the title "Magnetosphere-Ionosphere Interactions - Near Earth Manifestation of the Plasma Universe" could give much inspiration. The development of Arp's theory may aim at an analogous paper, for example, with the title " Galactic Dynamics as a manifestation of the Plasma Universe".

When preparing the suggested paper it is important to take account of at least three basic processes.

(a) The decisive role of *double layers* (IEEE, the same issue, p. 779). The existence of double layers in laboratory plasmas was studied already by Langmuir in the 1920's. The existence in space of electric fields parallel to the magnetic field was predicted by Alfvén and Fälthammar (1963, Chapter 5.4.2). The theory of double layers in space has been developed by many authors (for a survey see L.Block,1978). One of the first observational indications of electric double layers in space was the discovery of an unexpected electron distribution indicating an electrostatic acceleration in space (McIllwain,1975). Since then massive evidence of electric fields parallel to the magnetic field have been accumulated. (For a review see Fälthammar 1986, see also Alfvén 1981(hereafter CP), Ch. II:6, especially Fig. II:22).

We know now that in the magnetospheres there are double layers with voltage jumps of kV, in the solar atmosphere exploding double layers of MV or GV (as seen from the solar cosmic rays). Carlqvist(1986) has developed a theory of relativistic double layers which in galaxies may accelerate particles up to at least 10^{14} eV (see p.794 in the same issue of IEEE).

(b) Double layers are very often produced by electric currents and their basic role stresses the necessity of clarifying the *circuits in which currents flow*. Space is filled by a network of electric currents (see CP, Ch. II and III). Currents which are homogeneous over large regions exist, but more often currents are "pinched" to thin ubiquitous filaments which are observed almost wherever the resolution is high enough. Surface currents - of the type observed, e.g., in the magnetopause, are also very important. *They give space a general "cellular" structure.* This is relevant for galactic theories. We know from the magnetosphere that the chemical composition of matter may be different on both sides of an interphase associated with the current layer and it seems necessary that similar phenomena exist also in interstellar and intergalactic space. (If such layers also separate different kinds of matter the universe may be symmetric with respect to matter and antimatter, see CP, Ch. IV:9 and especially Ch.VI).

(c) *Critical Velocity Phenomena*, which produce an unexpected high interaction between a neutral gas and a magnetized plasma as soon as their relative velocities exceed a value v_{crit} given by

$$v_{crit} = (2eV_{ion}/M)^{1/2}$$

(V_{ion} = ionisation voltage, M = atomic mass). This means that the kinetic energy of an atom moving with the critical velocity equals its ionisation energy. (See CP, Ch. IV:6 and V).This effect has been decisive for the formation of the band structure of the solar system, and may play an important role also in several galactic phenomena (Alfvén and Arrhenius, 1976).

C. Structure of the Plasma Universe

As shown in Fig.1, space should be divided into two different regions: the *reliable diagnostic region* (out to the reach of spacecraft) where it is possible to build what we mean by a real plasma science, and *regions outside the reach of spacecraft* which basically are fields of speculation. They should be approached by extrapolation of results from the "reliable diagnostics region". Furthermore, it is necessary to distinguish between some different kinds of plasma (Fig.2). Two of them, "magneto-hydrodynamic plasma" and "collisionless plasma", are listed in Fig.2. "Dusty plasma" is a third kind of plasma.

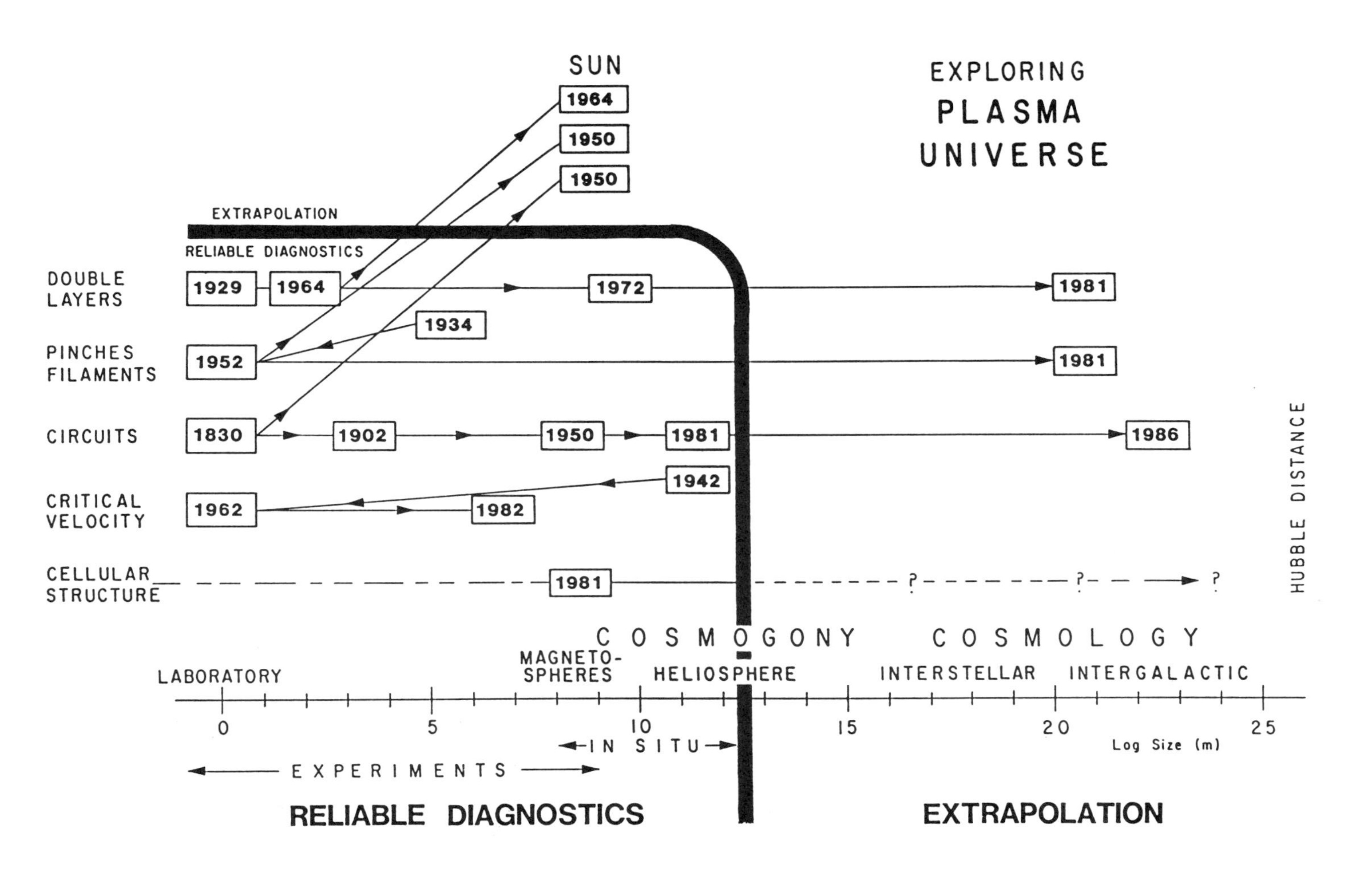
EXPLORING
PLASMA
UNIVERSE
SUN
1964
1950
1950
EXTRAPOLATION
RELIABLE DIAGNOSTICS
DOUBLE LAYERS
1929
1964
1972
1981
PINCHES FILAMENTS
1952
1934
1981
CIRCUITS
1830
1902
1950
1981
1986
CRITICAL VELOCITY
1962
1982
1942
CELLULAR STRUCTURE
1981
?
?
?
HUBBLE DISTANCE
COSMOGONY
COSMOLOGY
LABORATORY
MAGNETO-SPHERES
HELIOSPHERE
INTERSTELLAR
INTERGALACTIC
0
5
10
15
20
25
Log Size (m)
IN SITU
EXPERIMENTS
RELIABLE DIAGNOSTICS
EXTRAPOLATION

D. Symbioses Between Galactic Observers and Reliable Diagnostic Physicists

Because galaxies are outside the field of reliable diagnostics the theory of galaxies must be approached through a symbiosis between galactic observers and magnetospheric-laboratory plasma physicists. Arp's invitation to such a collaboration is certainly very welcome. However when making such extrapolations there are two important caveats.

1. The magnetic field line reconnection approach is misleading and obsolete and must not be used (see p.779 in IEEE Plasma Science Dec.1986, same issue).

2. Whereas it is true that all plasmas are ionized gases, it is not true that what the astrophysicists in general mean by "ionized gases" has very much to do with plasmas (see CP and especially p. 790-792 in the IEEE issue). They use the concept "ionized gases" in a very restricted sense viz to denote a hypothetical medium that does not possess the complex properties of most real cosmic plasmas. Their "ionized gas" is a medium which may be somewhat similar to the quiescent plasma studied in externally heated cesium plasmas in the laboratory Q-machines), but such a medium is observed only in selected regions in cosmical physics.

Furthermore there are strong arguments for the view that a study of plasma regions of different sizes - from the laboratory and the magnetospheres out to the Hubble distance - should be based on the assumption that the basic properties of plasmas are the same within very large variations in the relevant parameters. This means that essential plasma processes in galaxies may be studied by extrapolation from what is known from heliospheric research.

The emission of two or more arms from the galactic nucleus, suggested by Arp, resembles, to a surprisingly high degree, the emission from the sun of high-density streams in the solar wind. The observed flow of ionized material along the longitudinal magnetic field of which Arp speaks is observed also in the heliospheric case. The main difference seems to be that the galactic gravitation plays a much more important role than solar gravitation does. Furthermore, we need not have some exotic mechanism (like black holes) to produce the ejection. A system of electric currents and the production of torsional hydromagnetic waves may be enough (see Belcher, 1971, 1987). These effects can also transfer additional angular momemtum to the arms far out from the center.

E. There Is No "Missing Mass"

The third and perhaps most important conclusion which Arp draws is that there is no support for the claim that galaxies contain the "missing mass" which the Big Bang believers now as always are looking for. Indeed, for decades they have tried to find new and increasingly exotic kinds of invisible mass. A discussion of this is found in Ch.VI of CP, which in some respects is based on works by Arp (1966, 1977, 1978).

Conclusion

As guest editor in the IEEE "Special issue on Space and Cosmic Plasma" Anthony Peratt has opened the dams for the new views on our cosmic environment which space research data make necessary. These will now inundate and drown much of the pre-sputnik geophysics-astrophysics. Arp's new approach to the dynamics of galaxies will probably be an important part of this Great Flood.

Figure 2: PROPERTIES OF MAGNETIZED PLASMAS
From introductory lecture at MIT Plasma Physics Symposium, June 9, 1987–in press

	FLUID PLASMA (Magneto-hydrodynamic)	PARTICLE PLASMA (Collisionless)
General properties	Similar to fluid	An assembly of particles in ballistic orbits
Motion in electric field	Thermal motion superimposed by electric field drift	Ballistic orbits in magnetic and electric field
Velocity distribution	Essentially Maxwellian	Often anisotropic Has a tendency to generate very high energy particles: magnetosphere keV solar atmosphere MeV GeV interstellar space possibly 10^{14} eV or more
Exists in:	Solar, stellar photospheres Ionospheres Comet coma	Solar corona Active regions in magnetospheres Comet tails Interplanetary interstellar intergalactic space
Radiates	Thermal (essentially visual) radiation No X-rays or γ-rays	X-rays, γ-rays (by collisions with residual particles) "Noise" generation, especially in connection with production of high-energy particles
Energy transfer	*Local* theories correct	Only *global* theories correct because currents transfer energy over large distances (often much larger than size of ballistic orbits)
Frozen-in magnetic field	*Yes*	No
Energy release through magnetic merging	Possible	No. This is like Columbus' mistake

References

Alfvén, H. (1981), *Cosmic Plasma*, (D.Reidel: Dordrecht), (referred to as CP).

Alfvén, H. (1986), *Double Layers and Circuits in Astrophysics*, IEEE Transactions on Plasma Science PS-14, 779.

Alfvén, H. and Fälthammar, C.-G. (1963), *Cosmical Electrodynamics, Fundamental Principles*, (Clarendon Press:Oxford).

Alfvén, H. and Arrhenius, G. (1976), *Evolution of the Solar System*, NASA SP-345, (U.S. Government Printing Office, Washington D.C.).

Arp, H. (1986), *The Persistent Problem of Spiral Galaxies*, IEEE Transactions on Plasma Science PS-14, 748.

Belcher, J.W. (1987), *Lecture at the MIT Plasma Physics Symposium*, Cambridge Mass. (In press).

Belcher, J.W. and Davis, L., Jr. (1971), *J. Geophys. Res.*, **76**, 3534.

Block, L.P. (1978), *Astrophys. Sp. Sci.*, **55**, 59.

Block, L.P. (1981), Double Layers in the Laboratory and Above the Aurora, in *Physics of Auroral Arc Formation*, eds. S.-I. Akasofu and J. R. Kan, AGU Geophysical Monograph 25, (American Geophysical Union, Washington D.C.).

Block, L.P., (1983), *Three-Dimensional Potential Structure Above Auroral Arcs*, in Proc. Sixth ESA Symposium on European Rocket and Baloon Programmes and Related Research, Interlaken, Switzerland, ESA SP-183.

Carlqvist, P. (1986), *On the Acceleration of Energetic Cosmic Particles by Electrostatic Double Layers*, IEEE Transactions on Plasma Science PS-14, 794.

Chandrasekhar, S. and Fermi, E. (1953), *Ap.J.*, **118**, 113.

Fälthammar, C.-G. (1986), *Magnetosphere-Ionosphere Interactions -Near Earth Manifestations of the Plasma Universe*, IEEE Transactions on Plasma Science PS-14, 616.

McIllwain, C.E. (1975), *Physics of the Hot Plasma in the Magnetosphere*, Eds. Hultqvist and L. Stenflo, (Plenum, New York).

QUASAR REDSHIFTS

L. Woltjer
European Southern Observatory, Karl-Schwarzschild-Str. 2,
D-8046 Garching b. München

Various kinds of evidence may be considered to ascertain whether the redshifts of quasars have important cosmological or non-cosmological components. These include the following:

	Method	Weight
1.	Associations with galaxies	low
2.	Associations with clusters of galaxies	low
3.	Number counts	medium
4.	Contribution to X- and Gamma-ray background	high
5.	Absorption by foreground objects	medium
6.	Characteristics of surrounding "fuzz"	medium
7.	Physical models	low

The first four methods are purely statistical.

1. Associations between quasars and galaxies of different redshifts have been extensively discussed by Arp, G. Burbidge and others. Since no large, complete, all sky samples exist of quasars with well determined fluxes, there have been some suggestive hints, much controversy and no agreement.

2. The same factor has made the association of clusters of galaxies and quasars very uncertain. There are some cases of quasars in clusters of galaxies with the same redshift, but their statistical significance is far from overwhelming.

3. Most optical number counts of quasars (area density N versus flux S) show a slope m ($N[>S] \propto S^{-m}$) of around 1.8 for the magnitude range B = 15-20. Since this is steeper than the expectation (m = 1.5) for a uniformly filled Euclidean space, it is difficult to reconcile with a distribution of local quasars. Since quasars are distributed rather isotropically (see below) it would imply that we are located just at a hole in the distribution, with the density increasing in all directions. If, on the other hand, the redshifts are of cosmological origin, cosmological evolution can account for the steep distribution in a more natural way.

Recently, Wampler and Ponz (1985) have argued that the optical data are not incompatible with $m = 1.5$, if observational errors are taken into account. While this shows that substantial uncertainties remain, it is also true that the incompleteness and magnitude errors have to be larger than most of us would believe them to be. The radio source counts (dominated by radio galaxies) show trends very similar to those observed in the optical quasar counts. Since these are not subject to such errors, it is clear that cosmological evolution in the radio source population is necessary in any case; a similar evolution of the quasar population is therefore a priori not implausible.

The optical quasar counts will continue to be affected by magnitude errors related to variability and to interstellar absorption and also by various selection effects. The situation is simpler for X-ray source samples, which contain a good fraction of quasars and Seyferts. From the recent discussion by Setti (1987) it follows that if a fit is forced through the data from both the Einstein Medium Sensitivity Survey (MSS) and the HEAO I survey, a value of $m = 1.5$ is obtained. The MSS data alone yield a somewhat larger value. It may be expected that the situation will become clearer when the German ROSAT X-ray satellite will make a deep all sky survey (in 1991 ?) in which about 10^5 quasars should be detected. Not only will this result in a better log N - log S, but also questions of possible associations of quasars and (clusters of) galaxies should be unambiguously settled.

4. Perhaps the most powerful argument about quasar redshifts comes from the brightness and isotropy of the X-ray background (Setti and Woltjer 1973, 1979; Giacconi et al. 1979). The quasars detected by the Einstein X-ray satellite certainly account for more than 10 % of the total surface brightness at 2 kev. Suppose the cosmological redshift component in the spectra of the quasars in these samples were, say, 0.01. Then by the Olbers argument the whole background would be obtained integrating out to less than $z = 0.1$. Since the only cutoff provided by a more or less standard cosmology occurs at redshifts of order unity, the quasars that produce 10 % of it must have a cosmological redshift component not less than 0.1. As the spectral properties of the known quasars and Seyferts are rather different from those of the background (Boldt 1987), it seems highly unlikely that more than 50 % of the background could be due to quasars; the corresponding minimum cosmological redshift is therefore closer to 0.2.

It has been argued that this reasoning can be circumvented by placing most of the quasars in the Local Super Cluster and having large voids till the next clusters. However, with 10 % of the background due to the local quasars, a substantial anisotropy would be produced. The observed anisotropy of the X-ray background, however, is less than about 1 %.

5. If absorption lines produced by object A are seen in the spectrum of object B, then B must be more distant than A. Of

particular interest is the case of PKS 2128-12 with an emission redshift of 0.500. A galaxy with a redshift of 0.430 appears to be responsible for a Mg II absorption at z = 0.4299 in the spectrum of the quasar (Bergeron 1986). In this case, it would seem that most of the redshift of the quasar must be of cosmological origin. The only small doubt about the interpretation might be related to the 64 kpc distance of the absorbing cloud to the galaxy. Larger samples of comparable cases would therefore be important. Absorption of quasars by other quasars, as discussed by Cristiani and Shaver in this volume, is of equal significance.

6. Most quasars with $z < 0.5$ are surrounded by some luminous "fuzz". It was first shown by Kristian (1973) that this fuzz has the characteristics expected if quasars occur at the centers of giant galaxies, a conclusion later strengthened by the observation in some cases that the fuzz has the same redshift as the quasar.

7. Various problems with physical models of quasars at cosmological distances have been discussed in the past. For example, Hoyle, Burbidge and Sargent (1966) discussed the inverse Compton catastrophe which would result if an isotropic relativistic electron distribution would produce the observed quasar continuum by the synchrotron mechanism. However, as shown by Rees (1967) and Woltjer (1966), relativistic expansion may completely change the situation. Similar considerations apply to arguments based on superluminal phenomena ("larger than c expansion"). It would therefore seem doubtful that models could lead to reliable conclusions about the interpretation of quasar redshifts.

In conclusion, it appears that the arguments based on the X-ray background prove conclusively that a substantial part of quasar redshifts must be of cosmological origin and that distances of the order of tens of Mpc cannot be accepted for the majority of quasars. Some of the other arguments we have discussed reinforce this conclusion, but of course this does not prove that there cannot be some quasars with redshifts of a different origin. A scenario with both cosmological and intrinsic redshifts certainly could be envisaged. But in our view its complexity and its introduction of unknown physics makes it unattractive, unless the observational data would unambiguously force its adoption. At present, this does not appear to be the case.

For further information on some of the points discussed here, the reader is referred to the review by Trimble and Woltjer (1986).

References

Bergeron, J. (1986). Astron. Astrophys., 155, L8.
Boldt, E. (1987). IAU Symp. 124, pp. 611.
Giacconi, R. et al. (1979). Astrophys. J., 234, L1.
Hoyle, F., Burbidge, G.R., and Sargent, W.L.W. (1966). Nature, 209, 951.
Kristian, J. (1973). Astrophys. J., 179, L61

Rees, M.J. (1967). Mon. Not. R. Astron. Soc., 135, 345.
Setti, G. and Woltjer, L. (1973). IAU Symp. 55, pp. 208.
Setti, G. and Woltjer, L. (1979). Astron. Astrophys., 76, L3.
Setti, G. (1987). IAU Symp. 124, pp. 579.
Trimble, V. and Woltjer, L. (1986). Science, 234, 155.
Wampler, E.J. and Ponz, D. (1985). Astrophys. J., 298, 448.
Woltjer, L. (1966). Astrophys. J., 146, 597.

DISCUSSION

J. P. Vigier: There is no trouble with energy momentum conservation if you accept a non-zero photon mass. This destroys, as shown by Schrödinger, the Olbers paradox since the potential of X-ray sources decreases like $e^{-\mu r}/r$ and not $1/r$. Combined with Arp's assumption that most QSO's are in our supercluster, then you have no trouble.
As for QSO halos, there are indications (Souriau) that their plane is correlated with the supercluster's equatorial plane.

L. Woltjer: The supercluster assumption leads to problems with the isotropy. Only a rather small number of QSO halos has been detected and I do not believe that the material suffices to make convincing statements about systematic directional effects.

T. Jaakkola: I have a few comments: a) Perhaps your arguments are not valid as such if there exist two populations of QSO's, one at cosmological, the other at local distances. This is suggested by the distribution of radio properties: arguments pointing to cosmological distances are valid for steep spectrum extended sources, those in favor of local objects are valid for compact flat spectrum sources (Jaakkola *et al*, *Ap. Sp. Sci.*, **37**, 301,1975). This division may solve, e.g. the microwave background paradox. b) The steep slope of the optical counts is due to a morphological selection effect (Jaakkola, *Ap. Sp. Sci.*, **88**, 283, 1982). When one adds to the number of bright QSO's the number of the Sy1's, which are not, by definition, included in the samples of QSO's although they are physically identical to them, the counts follow the static steady state prediction $N \propto m^{0.6}$, the law which also the X-ray counts follow closely. c) The argument based on the z-distribution of QSO's with nebulosity is not strict. Size of intrinsic redshift and the compactness are quite plausibly correlated parameters, and hence high z QSO's will not show nebulosity.

L. Woltjer: The X-ray background argument was essentially developed with a $N \propto m^{0.6}$ relation. Most of the background comes from optical quasars and Seyfert's without noticeable radio emission.

F. Hoyle: Does it make a difference to the X-ray background situation if a local model of QSO's is considered on an hierarchical picture, a picture with a local cloud of QSO's and of other more distant clouds?

L. Woltjer: If the cloud-like character is strong enough to significantly reduce the estimate of the background flux, it is also strong enough to produce a marked anisotropy– at least if the distribution of the local cloud is not geocentric, but like the local supercluster.

PROBLEMS OF COSMOGONY AND COSMOLOGY

G. Burbidge
University of California, San Diego, La Jolla, CA 92093
U.S.A.

INTRODUCTION

How do we think that the Universe and its constituents were made, how have they evolved and what do we think will be their fate?

The great problems are associated with the origin and evolution of:

The Expanding Universe; containing the
Microwave Background Radiation, and the
Large-scale Distribution of Visible Matter (galaxies);
The Galaxies, their angular momentum spiral structure and magnetic fields;
Stars;
Chemical Elements;
The Solar System and Life;
Activity in Galaxies;
Quasi-stellar Objects.

What methods do we use to try to understand these problems. We use:

(1) Observations;
(2) the known laws of physics to make theoretical models to compare with observation;
(3) intuition, leading to new creative ideas.

Over the last twenty years or so a conventional picture of the Universe has been put together, and this purports to give answers to some, if not all of the fundamental questions on my list. Let us first look at what this scheme is, what are its strengths, and its weaknesses.

The Universe, The Microwave Background and Galaxies

It is generally believed that the universe originated in a hot big bang and that it has expanded and evolved since this origin. While the discussion of the origin lies outside physics, it is tacitly assumed as an article of faith, that the laws of physics as we know them today have remained unchanged. Thus they either came into being when the universe was created, or they existed before. In either case, they play the role of God, since they make up the only immutable factor in the universe. Everything else is thought to evolve.

The unlikely aspect of this assumption is that it presupposes that we know now, in 1987, all of the laws of physics, i.e. the discovery process has come to an end.

By tracing the history of matter and radiation from the Planck era ($\sim 10^{-43}$ sec) onward, it is argued that galaxies and the large scale structures, clusters, superclusters, chains and voids, must have arisen from density fluctuations at the quantum level.

Undoubtedly the extreme popularity of those ideas and the doctrinaire approach to the hot big bang - it is treated as a proven theory by most astronomers, - starts from the evidence dating from the 1920s that the universe is expanding, and the discovery of the microwave background radiation in 1965. Before 1965 there had been a confused period of a decade or so in which it had been claimed that the counts of radio sources disproved the most extreme alternative to the Friedmann models, the steady-state cosmology, and even earlier, from the 1950s onward, that a sequence of observational results were in conflict with the steady-state model. However, as Hoyle showed in 1968 (Hoyle 1969) none of these latter challenges to a non-evolving universe model had honestly succeeded.

But the fact that the microwave radiation had been predicted to be present, and that it had a black body form (though it took many years before this was established, it was immediately called relict radiation by those who wanted it to be just this, e.g. Zeldovich and many others), led most cosmologists to the conclusion that it was overwhelming proof for a beginning, and thus for evolution.

Studies of the early universe have now dominated all other aspects of modern cosmology for some years. Where has all of this led?

It has led to the general view that various real or imagined problems - flatness, absence of monopoles etc,. require that there was an inflationary phase in the early universe. The preferred inflationary model requires $\Omega = 1$, i.e. it requires that the universe is closed, while to produce the light elements thought to have been made in the big bang, D, He^3, He^4, Li^7, in their correct abundances, requires that the baryon density be considerably less than the closure density. Observationally it has been clear for many years that the mean density of radiating matter gives a value for $\Omega_{baryons}$ of only ~ 0.03, and even if we allow for a reasonable amount of dark matter, $\Omega_{baryons} \simeq 0.1$.

The theoretical prejudice in favor of $\Omega = 1$ has therefore led to a new series of speculations, mostly led or encouraged by particle physicists, in which it is argued that the missing mass is present in the form of non-baryonic matter, for which there is no independent evidence whatsoever!

In a big bang universe the matter and radiation are initially strongly coupled together. Thus the existence of a very smooth radiation background together with a very lumpy distribution of galaxies poses a problem. However galaxies formed, at some level it should be possible to see fluctuations in the microwave background, at a scale of about 10 arc minutes corresponding to primordial fluctuations. Despite the postulation of the existence of non-baryonic matter, or of strings involved in galaxy formation, or of biased galaxy formation, the non-existence of these fluctuations would mean that the idea that galaxies arise from early fluctuations must be wrong, and this would undermine the whole belief in a hot big-bang beginning. However, even if the observations begin to suggest this, there will be the strongest efforts made to fix up the existing view. This is because the beliefs developed over the last twenty years in this area have led to a total imbalance in the direction of research. The observational evidence stemming from the Hubble law and the microwave radiation has led to a single-minded approach based on the hot big-bang. Far-out theoretical ideas are only taken seriously, if they are related to the early universe. There, anything and everything goes. We are told that the unity between particle physics and astrophysics is upon us.

Suppose there was no initial dense state? Suppose that matter and radiation were never strongly coupled together? Suppose that the laws of physics have evolved, as has everything else? The current climate of opinion requires that these questions not be asked.

The Angular Momentum, Magnetic Fields and Spiral Structure of Galaxies

The way that galaxies actually form and the way stars form in them, how galaxies obtain their angular momentum, and how magnetic fields develop, are all origin questions that remain rather unclear. A very early discussion of gravitational collapse and fragmentation was due to Hoyle (1953), and a somewhat more detailed scenario for our own, and presumably other spiral galaxies, was proposed by Eggen, Lynden-Bell and Sandage (1962) and this model has been widely adopted. This was based on the existence of a thin disk, the strong correlation between the metallicity of a star and the ellipticity of its orbit, on the small age spread among the globular clusters, and on the very rapid enrichment of the galactic disk by recycling the elements built by nucleosynthesis in the stars. This model may need some revision if there is no continuous gradient in chemical composition in the halo, as is now argued, but it agrees with our general understanding of the stellar populations, stellar kinematics and chemical composition.

A problem that remains outstanding stems from the fact that no stars with no heavy elements (pure hydrogen and helium stars) have ever been found. Thus the so-called Population III stars have been invoked. These are massive stars which form and evolve early in the life of a galaxy or even before, and produce extreme Population II abundances of the heavy elements. It remains unclear whether or not these Population III stars form and evolve after a galaxy starts to condense, or before.

The origins of the angular momenta of galaxies and their magnetic fields are much more obscure. Long ago Hoyle suggested that the angular momentum arises due to tidal interactions between protogalaxies. This idea was explored further by Peebles, but in my view it is still unclear what the magnitude of this effect is. The very flat rotation curves of spiral galaxies have led to the view, widely believed, that massive halos of dark matter, possibly non-baryonic-matter, if one believes the fairy tales of the particle physicists, or those astronomers heavily influenced by them, are present.

The conventional wisdom about magnetic fields is that there are very weak seed fields present in the early universe that are amplified by shock phenomena and winding up in galaxies. The seed fields presumably arise in current systems present in the early plasma. But magnetic fields and electromagnetic phenomena have not played a significant role in the conventional studies of the early universe. One very difficult unsolved problem is that the strengths of typical magnetic fields in very extended radio sources outside galaxies are quite as strong as the fields in the disks of galaxies like our own, i.e. about 10^{-5} gases. However, the gas densities in these two regions are vastly different.

The existence of spiral forms in a large fraction of all of the galaxies has been known since Lord Rosse first discovered spiral structure in M51 in the middle of the nineteenth century. How such structure is generated, and then maintained, over many periods of rotation, are the two basic problems.

The conventional solution to the origin problem is now claimed to be that spiral structure is generated by spiral density waves, - a view very strongly espoused by Lin and Shu following classical work by Lindblad. It is argued that the spiral density waves are started by shock phenomena in the galactic disk.

While it is possible that this theory may explain some part of the spiral structure that we see, it is clearly not adequate to explain some of the best known features, for example the existence of satellite galaxies on the ends of spiral arms, as in M51, and the existence of magnetic fields which lie along spiral arms. Arp and others (cf. the paper by Alfven in these Proceedings) have argued for ejection from the center as a way of explaining spiral structure.

Stars and the Chemical Elements

Probably the best understood of the major problems of astronomy is the structure and evolution of the stars. While the details of star formation are still not very clear despite the advances in molecular astronomy and in observations in radio and infrared wavelengths, and the spectrum of stellar masses is not understood, we do understand main-sequence structure and the evolution on and partly off the main sequence. The energy sources are understood, and we know why stars with very small masses do not exist. The one nagging problem is the solar neutrino problem, though there is no indication that its solution will lead to a drastic rethinking of stellar physics.

Going along with our understanding of stellar evolution is our understanding of the building of the chemical elements in the stars.

There have been many observational tests of this theory and they all tend to support it. The only major problems in this area are the problems concerning the first generation of stars - Population III - referred to earlier, and also the problem of understanding some of the stars with highly anomalous abundances on their surfaces.

The Solar System and Life

The formation of the solar system is generally and reasonably thought to be similar to the processes of gravitational collapse and condensation of other stars. Thus the concept of a solar nebula collapsing, ejecting matter with angular momentum, and forming planets is invoked. Since a great deal more is known about the solar system than about the environments of any other stars it is perhaps not surprising that many details are not understood.

I am really not qualified to speak about the origin of life. Obviously the establishment point of view extrapolating from the experiments of Miller is that this was a more or less spontaneous event. Those who believe that life will be found elsewhere in the universe obviously believe on that basis that such spontaneous processes cannot be that rare. The alternative, that life exists elsewhere and has spread by the mechanism of panspermia or other mechanisms, is one that a minority espouses. The reader is referred to Hoyle's contribution in this volume.

Active Galaxies and Quasi-Stellar Objects

Among all of the major phenomena listed in the introduction, the existence of active galaxies and QSOs was the last one to be recognized. From the time that galaxies were recognized to be large agregates of stars in the 1920s, up to the 1960s, it was assumed without question that they would only evolve very slowly at a rate determined by stellar evolution. And of course the QSOs were not identified until 1960.* By the early 1960s, galaxies as powerful radio sources, Seyfert nuclei, patterns of gas ejection, and highly luminous non-thermal sources, all showed that the nuclei of galaxies could undergo very violent activity, and the QSOs at their cosmological distances were also seen to be extremely powerful non-thermal sources (Burbidge, Burbidge and Sandage 1963).

In the next few years the orthodox model to explain these phenomena was developed. In this it is supposed that the energy responsible for the activity is gravitational in origin, and that it is derived from matter

---

*Before 1920 we did not know that a Universe of galaxies existed. In the 1980s we all talk as though the only detectable objects in the Universe will be galaxies and QSOs. Are observers thereby foreclosing the possibility of finding something else? The answer is probably "Yes".

falling into massive black holes in the centers of galaxies.

The continuity scheme requires that QSOs are very bright nuclei situated in galaxies and that there is continuity between them and much lower luminosity objects like Seyfert nuclei which lie in spiral galaxies. The existence of fuzz around QSOs at low redshifts has been used as evidence that they lie in galaxies.

The evidence that this approach is not adequate will be the subject of later discussion.

A RECAPITULATION

So far I have given a very brief sketch of some of the major problems of astrophysics and the view that most astrophysicists have of them as we approach the 21st century.

What I have obviously left out, since I have largely followed the establishment view, are the many pieces of observational evidence about the universe that some in this audience, and particularly Chip Arp have discovered. Most of this evidence as it relates to cosmology and galaxies, to the origin of life, and to activity in galaxies and QSOs, does not fit in to the orthodox schemes outlined above. With this evidence have come new and radical ideas and theories which have also been disregarded.

Since, at this meeting we are discussing some of the unorthodox views of the universe, it might be easy to forget that Arp, as well as quite a few of his colleagues present here, has been responsible for work which is also widely accepted and is now part of the main stream of astrophysics. This is especially true in the area of stellar structure and evolution, where the major steps forward were made about thirty years ago. In the 1950s, Arp was considered to be one of the most talented and dedicated of the young observers in the study of color-magnitude diagrams. At the same time other members of the audience here were developing the theory of stellar nucleosynthesis, and showing that stellar abundances agreed with this theory. Some years before this Fred Hoyle with Schwarzschild and others had made the first major step in understanding stellar evolution off the main sequence.

I say all of this as an aside, but given their history it is perhaps not surprising that individuals "with such a track record", as it is often stated in the U.S., are not afraid to advocate that some of the radical solutions I am about to describe are correct and will eventually replace the current views. I discuss these problems under the general heading of the Discordant Observational Universe.

THE DISCORDANT OBSERVATIONAL UNIVERSE

There are many observations of different kinds which cast doubt on some parts of the establishment view of the universe described in the first part of this talk. We approach them starting with those which are merely ambiguous, and which are usually interpreted within the framework of the establishment view provided one complicates the model, and finish with observations which cannot be accommodated at all. Observations of the latter type which we shall be discussing last have been handled in the following way. First they are delayed at the refereeing stage as long as possible with the hope that the author will give up. If this does not occur and they are published, the second line of defense is to ignore them. If they do give rise to some comment, the best approach is to argue simply that they are hopelessly wrong, and then if all else fails, an observer maybe threatened with loss of telescope time unless he changes his program. As we shall show, observations which survive these procedures lead to radical theory.

Large Scale Structure of the Universe

We start with the observations of the extreme lumpiness of the universe as far as luminous objects are concerned, which must be contrasted with the extreme smoothness of the microwave background. One way to minimize this difficulty is to invoke the presence of hidden matter of baryonic or non-baryonic form which has a much smoother distribution. This is being done. However, as has been pointed out by Wilkinson, we are approaching a point of crisis. If fluctuations of the microwave background are not seen on scales of ~10 arc minutes at a level of $\sim 10^{-5}$, it will be difficult to maintain that the matter has ever been coupled to the radiation. A model in which this coupling has never occurred will be mentioned later.

Observations of Discrete Sources Which Bear On The Evolving Universe

Given that the Friedmann model of the expanding universe is correct, it should be possible to derive from the observations of galaxies and other objects various properties. For example, we should be able to determine the deceleration parameter q_o and also derive the properties of the curvature. Also, if all of the galaxies were condensed at the same epoch in the expansion, it should be possible by looking back to larger and larger redshifts to find evidence for evolution in the galaxies themselves.

Of course it should be stressed that the preferred cosmological model, the Friedmann model, has nothing to say about the existence of condensed objects. Thus in a fundamental way studies of the properties of galaxies as a function of redshift are open-ended, in the sense that we have no unique theory of the numbers of densities of luminosities as a function of redshifts, to compare observations with. Galaxy formation theories can be limited by observational constraints, but they are in no way basic, since the initial fluctuations are assumed.

Determination of q_o

Attempts to derive q_o lead to conflicting results. It is well known that a small value of $q_o \leqslant 0.1$ is obtained (with zero cosmological constant) from attempts to estimate the matter-energy density. Wampler (1987) has obtained a large value of $q_o \simeq 3$ from the Hubble diagram for radio galaxies and, separately, QSOs. Not only is it very peculiar that QSOs and galaxies give the same value for q_o, but a non-zero cosmological constant is required to reconcile the time-scale with that for stars or for the radioactive isotopes. It is also well known that attempts to obtain q_o from the Hubble diagram are complicated by the evolution of the galaxies.

The Angular Diameter Redshift Rotation

All Friedmann models with zero cosmological constant and $q_o \geqslant 0$ give rise to an angular diameter (Θ-z) redshift relation which reaches a minimum close to z = 1.25. The only accurate test of this relationship for large z is that which has been carried out for double radio sources. In this case it has been shown (cf. Kapahi 1987) that $\Theta \propto z^{-1}$ for redshifts far in excess of 1.25. Taken at its face value this is a terrible blow for the Friedmann model, since it suggests that there is no non-Euclidian effect present. In order to preserve the status quo it is commonly argued that the linear sizes of radio sources evolve in such a way that the evolution just cancel out the non-Euclidian effect. This is an extreme example of the approach taken by those who wish to maintain the conventional model.

Counts of Objects

Hubble first attempted to carry out this test more than 50 years ago. For a variety of reasons he abandoned the attempt. Much more recently, Tyson and Jarvis, and Kron and Koo and others have counted very faint galaxies. The slope of the log N-m curve is somewhat flatter than 0.6 which is the Euclidian value. It is probably closer to 0.4 in the apparent magnitude range 20-24. It has been shown that such a slope can be simulated by an empty (q_o = 0) model provided that a luminosity function for galaxies similar to the one seen locally is assumed. However, there is no need to assume any evolution in the number density or luminosity distribution.

The counting of radio sources led to a great deal of controversy in the 1950s and 1960s. It is well known that for sources randomly distributed in Euclidian space the log N- log S curve has a slope of -1.5. Alternatively, instead of plotting N against S the source count problem can be approached in the following way. If S_m is the minimum flux density that can be picked up in a given survey, and a source $S > S_m$ is found and its redshift is known, we can calculate two volumes. The volume V is the volume of the spherical region with radius d, corresponding to the redshift of the object, and V_m is the volume with radius d_m such that the source could just be detected at flux level S_m. If a set of sources is randomly distributed the average value of V/V_m will be 0.5, corresponding to log N-log S slope of 3/2. This is the luminosity volume test and it was first used for QSOs by Schmidt (1968).

The log N-log S method has been used to test surveys of radio sources at a number of frequencies. It is found that there is a significant rise in the differential count ratio $\Delta N/\Delta N_o$ as we go from the highest flux point to lower flux densities. If the sources are very powerful and distant this means that there is an evolutionary effect, requiring that the number density was much greater in the past. However, the highest flux point in the differential counts shows a deficit of high flux sources relative to the Euclidean value. If the sources are not very strong, this deficit could be interpreted as a local one, implying that we are in a region with fewer radio sources than normal.

The establishment view, first expressed by Ryle, is that the results demonstrate evolution,* thus providing strong evidence against the steady state cosmology. Hoyle has consistently taken the second position, that we lie in a "local hole" as far as strong sources are concerned. However, there is yet another approach. For the best studied sample of radio sources the 3CR catalogue, we have almost a complete set of optical identifications and redshifts. The optical objects are of two kinds, galaxies and QSOs. With all of the redshifts known (all but a few percent) we can test directly for evolution. As far as the radio galaxies are concerned several studies with continuously increasing numbers of redshifts and fewer and fewer unidentified sources showed that there was no evidence for evolution of the identified sources, so that everything depended on the dwindling number of unidentified sources. The most recent study by DasGupta, Narlikar and Burbidge (1987) using an all sky survey by Wall and Peacock suggests that it is possible to explain the results without evolution. After nearly a year, this paper is still in the hands of referees, since clearly because it concludes that there is not evidence for evolution it is being dissected in minute detail by individuals who clearly made up their minds years ago. (This subject is now old enough to have bred scientists who have been taught that evolution is observationally proved!)

The case of the QSOs is different. If the redshifts are of cosmological origin, it has been shown by Schmidt and others using the luminosity-volume test that within the framework of Friedmann cosmology extensive evolution is present. The doubt here is associated with the nature of the redshifts, and we shall come to this in the next section.

To summarize; while it is generally assumed that we live in a Friedmann universe, it has proved hard to demonstrate the non-Euclidean nature of the universe by studying the objects in it, and even harder to find clear cut evidence for evolution in the objects themselves or in numbers or luminosities. The Θ-z relation in particular is most baffling.

We end this part of the discussion by pointing out that there is at least one approach which is compatible with all of the observations but which

---

*It must always be remembered however, that there is no "theory" which predicts evolution or its form. Those who believe in it continue to deduce its nature from the observations in an ad hoc manner.

departs strongly from the hot big bang model. this is an empirical appraoch first described by Narlikar and Burbidge (1981). In this it is argued that the microwave radiation arose in a big bang some 10^{12} years ago, and that the matter density associated with this is close to zero. All of the discrete objects now observed have been generated in smaller expanding bubbles (minature big bangs) which have finite sizes. We showed that all of the observed properties described above could be explained in this way. An obvious weakness is that in such a universe there is no obvious reason why we should be moving at such a slow speed (< 1000 km^{-1}) with respect to the microwave background radiation.

This completes our discussion of evidence in the "discordant" universe which while not in contradiction to the hot big bang model requires that a rather ad hoc approach be taken.

We turn finally to evidence which is totally in conflict with the simplest model.

Redshifts

Underlying all of the discussion so far is the belief that the redshifts are due to the expansion of the universe. Hubble first demonstrated a convincing velocity-distance relation in 1929. While there was considerable discussion in the literature of the period about alternative explanations, particularly the tired light hypothesis, the view that we are seeing a Doppler effect prevailed, and as we stated earlier, the velocity-distance relation with the microwave background radiation are the two major pillars on which the big bang cosmology rests.

Since Hubble's original relation was based on very local galaxies which we know now are not partaking in the local expansion, it is remarkable that he obtained a simple linear relation (cf. Burbidge 1981). As we stated earlier, the later results obtained by Sandage and his colleagues and others show that there is an extremely tight relation between apparent magnitude and redshift out at least to redshifts of order $z = 0.5$ ($= c\Delta\lambda/\lambda$) for intrinsically bright elliptical galaxies. Provided that this can be shown to be a Doppler shift, and this has not yet been established conclusively, this is to me strong evidence for the conventional interpretation of redshifts for "normal" galaxies.

Now we know that the redshift (z_o) that we measure in an object is in principle a product of several terms, including the component of random motion in the line of sight, z_r, the cosmological redshift z_c, and a term of non-cosmological (intrinsic) origin z_i, so that

$$(1+z_o) = (1+z_r)\,(1+z_i)\,(1+z_c) \qquad (1)$$

The real question is: "Is z_i vanishing small, in all astronomical objects? We have always known that it is not zero, since all objects have a very small component of redshift of gravitational origin.

The "new" element in this discussion, which has now been with us for 15-20 years is that the observations suggest that in some classes of objects z_i is large and dominates the other terms in (1). Particularly at this meeting it is not necessary to describe this evidence in detail since many of those who have found it are present, and their papers are contained in this volume. I summarize the evidence briefly.

Galaxies.

Tifft (these Proceedings and in many papers published over more than a decade) has found that the differential redshifts of galaxies in some clusters, and in physical pairs of galaxies shows an intrinsic redshift quantization with the quantum unit $\Delta z_q = 0.0002417$ ($\Delta V_q = 72.5$ km sec^{-1}). He has summarized this work, and Napier (these Proceedings) appears to have confirmed it.

The same effect has been found by Arp and Sulentic (1985) for groups of galaxies containing a central bright galaxy and a group of satellite galaxies. Here, not only is there a strong statistical average excess Δz for the satellite galaxies relative to the central galaxy, but this excess value appears to be quantized in units of Δz_q.

Arp has produced a great deal of observational evidence which suggests that non-cosmological redshifts components are dominant in many galaxies within the local supercluster. The evidence is largely statistical in nature. Arp has also produced more direct evidence involving cases in which galaxies with very different redshifts are physically connected. The most spectacular case of this kind is NGC 7603 and its companion, but Arp has also published several other cases of a similar kind. In addition he has found one case in which it appears that a galaxy with a comparatively large redshift lies in front of a galaxy with a smaller redshift.

The results of Tifft and Arp on galaxies, taken at face value, lead to the conclusion that most of the observed redshift of nearby objects is intrinsic so that no Hubble relation can exist. This is not necessarily an objection to the existence of the Hubble law on the larger scale, since all of the most recent investigations along conventional lines have suggested that there are large departures from the Hubble flow in the region of the Virgo Supercluster, i.e. out to distances of 10-20 Mpc. It is also the case that the redshifts of dwarf galaxies which play such an important role in the Arp-Sulentic study have never been shown to obey a Hubble relation.

The greatest difficulty that one encounters in supposing that most of the redshifts of nearby galaxies are intrinsic, is that so little is left to be interpreted as due to motions. We have a truly static local universe.

Quasi-Stellar Objects

In contrast to galaxies, the very large redshifts found for the QSOs immediately led Terrell to argue that these had nothing to do with cosmology, but that the QSOs were ejected from the center of our

Galaxy and had all gone past the solar system so that only redshifts were seen, i.e. the redshifts were due to local high velocities of recession. By 1966 Hoyle and Burbidge (1966) concluded either that the objects were at cosmological distances, or that they had been ejected from nearby galaxies. A very likely candidate was NGC 5128, the powerful radio source Centaurus A. The indirect arguments in favor of the QSOs being local objects were based partly on the absence of any discernable Hubble relation, and partly on the difficulty in understanding how such large fluxes of synchrotron radiation could arise and be maintained in the very small volumes indicated from the flux variations. Contrary to what is often stated, we did not consider that the large energies were a factor in this problem, since we already knew that they were required to explain the powerful radio sources which lie in galaxies which we assumed did lie at cosmological redshifts. Up to about 1966 all of the arguments suggesting that the bulk of the observed redshift of QSOs is intrinsic were indirect. The direct evidence came when significant numbers of QSOs with large redshifts were shown to be physically associated with bright galaxies with small redshifts, or when pairs of QSOs with different redshifts were found to be associated. Physical association can be demonstrated in one of two ways - by statistical means, or by finding luminous connections. Since very few luminous connections appear to exist the statistical method has the necessity been used the most.

To summarize this evidence briefly:

(a) The luminous connection between Mk 205 and NGC 4319 originally found by Arp after the pair was discovered by Weedman is the most striking example of association. The recent observations and analysis of the connection by Sulentic put its existence beyond dispute.

(b) The first paper of Arp (1967) showing apparent pairings of powerful radio sources across peculiar galaxies provided some of the first pieces of statistical evidence relating low-redshift galaxies to high-redshift objects. In 1971, Burbidge, Burbidge, Solomon and Strittmatter (1971) used two well defined samples, the 3CR QSOs and the Shapley Ames Galaxies to show that there was a highly significant excess number of QSOs within 6′ of galaxies. This result was confirmed using Monte Carlo techniques by Kippenhahn and de Vries (1974).

Arp in a series of papers published over the last twenty years continues to find QSOs near to bright galaxies in numbers which he has always claimed were much higher than would be expected by chance. His statistics have often been criticized, generally because it has been claimed that his samples were not well defined. However, there was no doubt from the earliest times that for whatever reason he has been able to find far more QSOs near to galaxies than were indicated by any estimates of surface density based on surveys. In 1979 (Burbidge 1979) I did an analysis of all QSOs within 10′ of bright galaxies which were known at that time, and showed that within 3′ there was a large excess, I included both QSOs found by Arp and those found by others.

Arp has also found many alignments of QSOs across galaxies, and he and Hazard found two remarkable triplets, both in perfect alignment and with very different redshifts (Arp and Hazard 1980). Much of this work was summarized in a paper I gave at the Texas Conference in 1980 (Burbidge 1981).

In my view the totality of the evidence of association should leave us in no doubt that intrinsic redshift components are dominant in many QSOs.

There is one other aspect of the QSO redshifts which leads me to the conclusion that no simple cosmological model will work. This concerns the distribution of redshifts. I showed in 1967 and 1968 that there was a peak in the redshifts in both emission and absorption at $z = 1.95$. With the increase in the number of redshifts it became clear that there was more than one peak. By 1978 (Burbidge 1978), it had become clear that there were peaks in the z distribution at 0.3, 0.6, 0.96, 1.41, and 1.96, and Karlsson (1977) and others showed that these values fitted the relation

$$\log (1 + z) = k + 0.089\, n \qquad (2)$$

Attempts were made to argue that these peaks were due in part to observational selection, but this is clearly not the case as was discussed by Burbidge (1978) and Depaquit et al. (1985). In the most recent catalogue of QSOs (Hewitt and Burbidge 1987) the peaks remain, superposed on a fairly smooth underlying distribution.

As I have often stressed, to see any peaks at all is most remarkable, since any intrinsic component of redshift z_i will easily be smeared out by the other components which should be non-negligible [cf. eq.(1)]. Peaks beyond 1.96 are not seen, although there are now many QSOs with $z > 2$. Whether or not this is due to the overall drop off in redshifts beyond 2 is not yet known.

This ends our brief summary of the evidence that tells us that some redshifts of galaxies and QSOs are not of cosmological origin.

Where does that lead us as far as "new ideas" are concerned?

NEW COSMOLOGIES

The discovery of intrinsic redshifts both for galaxies and QSOs poses a huge problem for modern cosmology and for fifteen or twenty years the existence of the effect has been doubted in part for this very reason.

If we face the problem squarely we can go in one of two directions. We can argue that this result must be applied in all cases, and question the meaning of the Hubble diagram in general. This means that we ignore the fact that there is a good Hubble relation for elliptical galaxies and also some evidence that some QSOs do lie at their cosmological distances.

The alternative approach is to try to fit the new results into the existing framework or perhaps the framework of an empirical cosmology described at the end of the last section. Which I am aware that the history of science shows that, in general, scientific revolutions are more radical than even their creators had expected, I tend to favor the second approach. This implies that there is a genetic relationship between a wide class of active objects, - broadly speaking all of those objects in which the redshifts are largely intrinsic, and some kinds of "normal" galaxies. Strictly speaking, we do not know whether the active objects arise from the normal systems, or vice versa. Assuming that the former is the case, it is natural to argue that this is evidence for the ejection of massive objects, as young systems, from older systems. If this ejection process is taking place all around us, then the predominance of redshifts and the absence of blue shifts* either means that the shifts are not due to motion, or that the objects are radiating in reverse cones down their tails.

Both of these possibilities have been discussed in the literature. Hoyle and Narlikar developed a theory in which the masses of the electrons in such objects were different from those in normal galaxies giving rise to intrinsic redshifts. The idea is then that as the objects age, the masses of the electrons reach their normal values and the redshifts decay. These ideas are incomplete, since they do not explain the ejection of the anomalous objects.

Originally Strittmatter, (cf. Burbidge and Burbidge 1967) and more recently Hoyle have discussed the directed ejection hypothesis. The geometry of the cones of radiation such that only redshifts are seen has been worked out. In either interpretation, ejection from a normal system is required, and the ejection process itself is not understood. Saslaw (these Proceedings) has discussed an updated version of the sling-shot mechanism.

In work carried out about two years ago (Burbidge 1986) I attempted to reconcile these ideas (ejection of comparatively local QSOs) with the evidence that some QSOs have redshifts very close to those of the surrounding galaxies and that some BL Lac objects lie in galaxies. I have argued that in these cases (a) the QSOs are being ejected with very low velocities with respect to the parent galaxies, and (b) that BL Lac objects are actually beamed QSOs coming towards us also ejected from the parent galaxies.

---

*However it should not be forgotten that none of the modern data have been tested for blue shifts. In early times when the observers were not so sure, blueshifts were looked for. This is no longer the case, and the problem probably requires further study.

The ultimate energy sources in the nuclei of galaxies giving rise to this class of ejection phenomena are not known. We either have the possibility of grafting these ideas onto the popular view in which the energy arises in accretion onto a massive black hole, or we can argue that creation processes are actually taking place in these regions. This latter point of view dates from Ambartsumian, and in terms of a modified theory of gravitation to Ne'eman, and most extensively to Hoyle and Narlikar.

Finally, it must be conceded that as far as the quantized nature of the redshifts of galaxies and QSOs is concerned, there is as yet no understanding.

CONCLUSION

I have summarized the ideas which are most widely accepted at present in cosmology. At the same time I have tried to show that there are many observed facts which require that the simple picture be modified and some observations which suggest that a radical new approach is required. It will be interesting to review the subject again when we celebrate Chip Arp's 75th birthday just after the turn of the century.

REFERENCES

Alfven, H., These Proceedings.
Arp, H.C. (1967). Astrophys. J. 148, 321.
Arp, H.C. and Sulentic, J.W. (1985). Ap. J., 291, 88.
Burbidge, E.M., Burbidge, G., Solomon, P.M. and Strittmatter, P.A. (1971). Ap. J., 170, 233.
Burbidge, G., Burbidge, E.M. and Sandage, A.R. (1963). Rev. Mod. Phys., 35, 947.
Burbidge, G. and Burbidge, M. (1967). "Quasi-Stellar Objects" (W.H. Freeman, San Francisco).
Burbidge, G. (1978). Phys. Scripta, 17, 237.
Burbidge, G. (1979). Nature, 282, 451.
Burbidge, G. (1981). Annals N.Y. Academy of Sciences, 375, 123.
Burbidge, G. (1986). "Structure and Evolution of Active Galaxies", G. Guiricin et al. Editors, p. 47. D. Reidel, Dordrecht.
Depaquit, S., Pecker, J.-C., and Vigier, J.P. (1984). Astron. Nachrichten, 305, 339.
DasGupta, P., Narlikar, J.V., and Burbidge, G. (1987). Submitted to Astron. J.
Eggen, O.S., Lynden-Bell, D. and Sandage, A.R. (1962). Ap. J., 136, 748.
Hewitt, A. and Burbidge, G. (1987). Astrophys. J. Suppl., 63, 1.
Hoyle, F. (1953). Ap. J., 118, 513.
Hoyle, F. and Burbidge, G. (1966). Ap. J., 144, 534.
Hoyle, F. (1968). Proc. Roy. Soc., 308, 1.
Kapahi, V. (1987). "Observational Cosmology" (Ed. A. Hewitt and G. Burbidge). p. 251, D. Reidel, Dordrecht.
Karlsson, K.G. (1977). Astron. and Astrophys., 58, 237.

Kippenhahn, R. and de Vries, H.L. (1978). Astrophysics and Space Sci., 26, 131.
Napier, W. These Proceedings.
Narlikar, J.V. and Burbidge, G.R. (1981). Astrophysics & Space Sci., 74, 111.
Saslaw, W. These Proceedings.
Schmidt, M. (1968). Ap. J. 151, 393.
Tifft, W. These Proceedings.
Wampler, E.J. (1987). Preprint.

DISCUSSION

H. Arp: In response to Dr. Sciama's question to Dr. Woltjer I would say that there are gross anisotropies in quasar distribution. For example, for quasars $1.4 \leq z \leq 2.4$ there are 3.5 times more in the R.A.$=0^h$ direction than in the R.A.$\approx 12^h$ direction on the sky. There is a concentration in the general region of the sky near NGC 5128 for quasars with $z \leq 0.5$. These must be reflected in anisotropies of the X-ray background if the X-ray background has a major contribution from quasars.

G. Burbidge: The X-ray background is quite isotropic. Anisotropies of the kind that you describe would suggest that QSO's only contribute a small fraction to the X-ray background. The shape of the spectrum also demonstrates this.

N. Sharp: I wanted to mention my favorite anomaly in observational astronomy, known for over 50 years and of high statistical significance (Borchkadze & Kogoshvili, AA, **53**, 431, 1976), which is the excess of S-shaped spirals over anti-S-shaped (z-shaped) spirals, in most directions on the sky. This effect is also present in data from observers with no apparent cultural bias (e.g. an alphabet with no "s").

G. Burbidge: This anomaly requires much greater publicity than it has had so far. I have no idea of how to explain it.

NEW -- AND OLD -- IDEAS ON THE UNIVERSAL J(M) RELATIONSHIP

Virginia Trimble
Dept. of Physics, Univ. of California, Irvine CA 92717 USA
and
Astronomy Program, Univ. of Maryland, College Park MD 20742

Abstract. A plot of angular momentum vs. mass for a range of astronomical objects, from asteroids to superclusters, shows a remarkably tight clustering around a power law of index 2. The present contribution addresses both possible interpretations of this correlation and the wider question of whether any special interpretation is, in fact, needed.

THE DATA AND SUGGESTED INTERPRETATIONS

Brosche (1962,1963) seems to have been the first to plot angular momentum and mass of astronomical objects in logarithmic coordinates and show that the full range (asteroids, moons, planets, stars, galaxies, and clusters and systems of these) was well fit by log J = A + B log M, with B near 2. He concluded that this was a new, independent fact about the universe, requiring explanation, but did not attempt to provide one. Later authors have been less reticent.

The proposed explanations of which I am aware can be divided into rough categories, ordered according to their decreasing distance from conventional astronomical theory. First come higher-order or heirarchical analogues to the quantum mechanical quantization of angular momentum, with the unit of quantization being either spin times a power of the fine structure constant (Barnothy 1974) or the combination c^5/GH_0^2 (Liu et al. 1985). Next come considerations of the unification of gravitation with the other forces, linking J(M) to the Dirac large numbers (Sternglass 1983; Sisteró 1983; Wesson 1984; Oldershaw 1986). A third category ascribes the origin of angular momentum to processes occurring in the early universe, either the evaporation of PBH's (Dorrington 1982) or expansion from Ambartsumyan-style dense cores (Muradyan 1984; Abramyan and Sedrakyan 1985).

The last class of interpretations links J(M) to other known physical processes or phenomena, for instance magnetic moments (De Sabbata & Gasperini 1984) or strings (Tassie 1987). The most conventional of these are primordial turbulence (Kuiper 1955; McCrea 1960,1978; Brosche 1970, 1977; S. Shore pr. comm. 1985; Arquilla & Goldsmith 1986; Fleck 1987) and tidal torques or collisions (White 1985; Wesson 1985; Halling 1985). These last two, when applied to a single class of object at a time, are merely part of modelling the formation process and non-controversial. F.D.A. Hartwick (pr. comm. 1985) notes, for instance, that $J \propto M^2$ is equivalent to $V \propto R^{1/6}$, which is just what you get from a relaxed Zeldofich spectrum with power-law index n = $-\frac{1}{2}$.

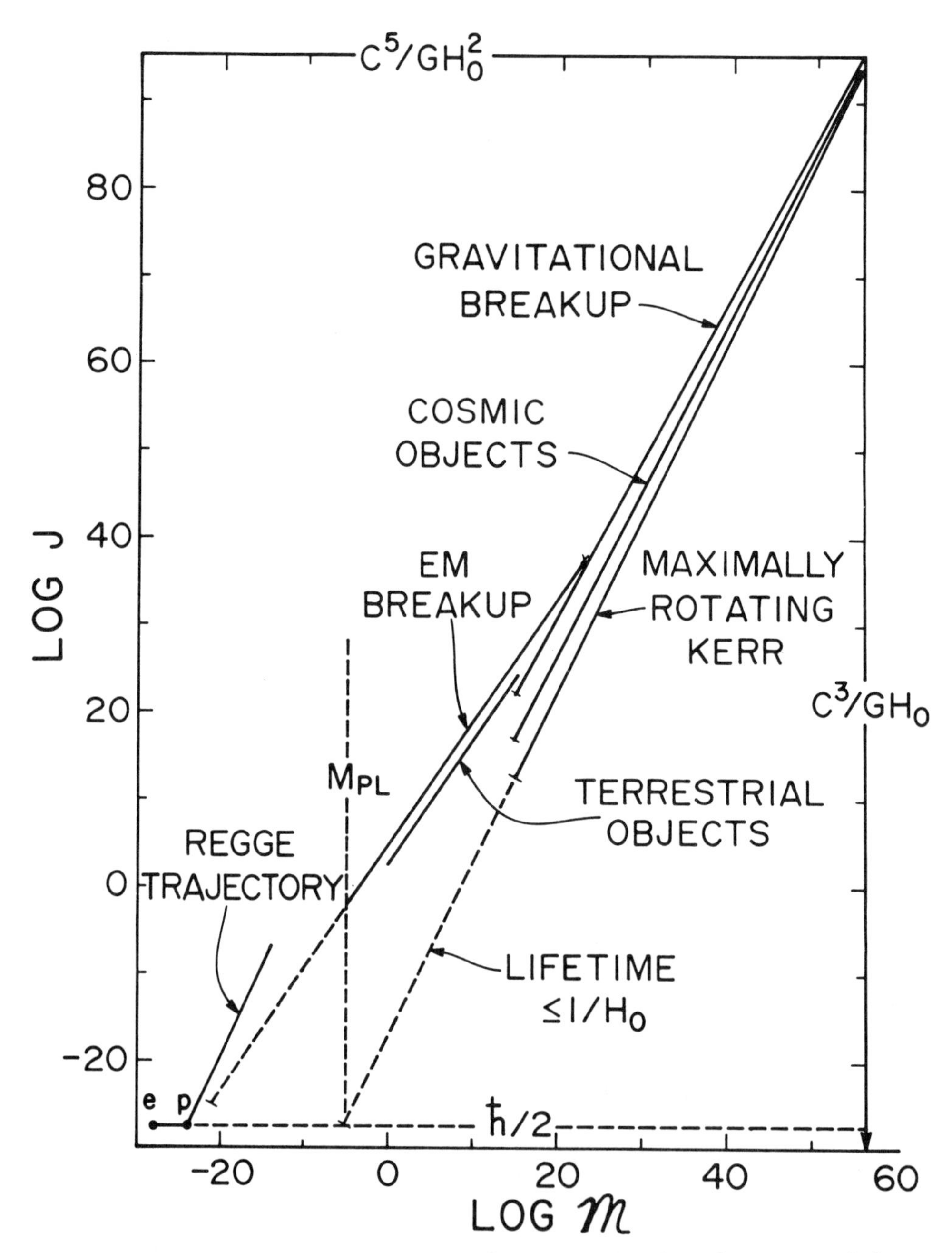

Figure: Angular momenta and masses of astronomical and terrestrial objects and elementary particles and some of the processes that constrain these.

IS THERE ANYTHING TO BE EXPLAINED?

I first encountered this problem via binary stars and concluded (Trimble 1984; Trimble & Walker 1986) that the slopes of 1.9 for eclipsing systems and 5/3 for spectroscopic and visual ones were a reflection of observational selection effects acting upon objects governed by Kepler's laws and (for contact systems) a mass-radius relation. This led to a broader consideration of other astronomical systems and the conclusion that for them, also, J(M) was not a new, independent piece of information. The evidence for this was a plot in which J for each sort of system was replaced by an error bar covering the full range over which the system could form and remain stable. A slope near two was still the best fit. A similar plot for terrestrial objects showed an equally tight relation with slope 3/2, reflecting saturation of the electromagnetic force. Brosche (1986) has recently responded that most classes of systems are grouped more tightly around the mean relation than stability argument require, so that there is still something to be explained.

The figure shows a still wider range of objects and phenomena in the J-M plane; the narrowness of the zone occupied is remarkable, though the limit to J for the universe as a whole from 3K isotropy (Barrow et al. 1985) puts it 5-6 dex below an extrapolation of the line for cosmic objects.

REFERENCES

Abramanyan, M.G. & D.M. Sedrakyan 1985. Astrofiz. 23, 396
Arquilla, R. & D.F. Goldsmith 1986. Astrophys. J. 303, 356
Barnothy, J. 1974. IAU Symp. 69, 25
Barrow, J.D. et al. 1985. Mon. Not. R. Astron. Soc. 213, 917
Brosche, P. 1962. Astron. Nach. 286, 241
..... 1963. Zs. f. Astrophys. 57, 143
..... 1970. Astron. Astrophys. 6, 240
..... 1977. Astrophys. Space Sci. 51, 401
..... 1986. Comments Astrophys. 11, 213
De Sabbata, V. & M. Gasperini 1984. Lett. Nuovo Cimento 38, 93
Dorrington, G.D. 1982. Lett. Nuovo Cimento 34, 409
Halling, R. 1984. Astrophys. Space Sci. 103, 379
Kuiper, G.P. 1955. Publ. Astron. Soc. Pac. 67, 387
Liu Y.-Z. et al. 1985. Astrophys. Space Sci. 116, 215
McCrea, W.H. 1960. Proc. Roy. Soc. A256, 245
..... 1978. in S.F. Dermott et al. (eds.) *The Origin of the Solar System* (Wiley) p. 75
Muradyan. R.M. 1984. Astrofiz. 21, 396
Oldershaw, R.L. 1986. Astrophys. Space Sci. 126, 199
Sisteró, R.F. 1983. Astrophys. Lett. 23, 235
Sternglass, E.J. 1983. Bull. Amer. Astron. Soc. 15, 937
Tassie, L.J. 1987. Nature 323, 40
Trimble, V. 1984. Comments Astrophys. 10, 127
Trimble, V. & D. Walker 1986. Astrophys. Space Sci. 126, 243
Wesson, P.S. 1984. Space Sci. Rev. 39, 153
..... 1985. Earth, Moon, Planets 30, 275
White, S.D.M. 1985. Astrophys. J. 286, 38

NONCOSMOLOGICAL REDSHIFTS : THEORETICAL ALTERNATIVES

Jayant V. Narlikar
Tata Institute of Fundamental Research
Bombay 400005, India

Abstract. Starting from the various alternative theoretical interpretations offered for redshifts, this paper describes two models in some detail. The first model belongs to conventional physics and makes use of the Doppler effect. The second model makes use of the variable mass hypothesis arising from the Hoyle-Narlikar theory of gravitation. Some observable consequences of these models and testable predictions are briefly outlined.

Introduction

In a recent survey of observational evidence for noncosmological redshifts (Narlikar, 1986, hereafter referred to as Paper I). I had mentioned the paradoxical situation that exists vis-a-vis theory and observations. On the one hand it is argued that much of the evidence is unacceptable because it cannot be interpreted, that is there is no theory available for it. At the same time unorthodox theories belonging to 'conventional' or 'new' physics are set aside because it is claimed that there is no observational evidence that calls for them.

Indeed, if history of science is any guide, theory and observations have never exactly been in phase. For example, Maxwell's electromagnetic theory predicted electromagnetic waves whose existence was experimentally demonstrated by Hertz several years later. Likewise, the existence of spectral lines was known and accepted long before quantum theory appeared on the scene.

The only difference between such examples and the claims of anomalous redshifts is that the former were found in laboratory experiments while the latter involve astronomical observations. However, it can no longer be argued that astronomy can and should use only that physics that has been tested in the laboratory. Most of the present speculations about the very early universe and dark matter are based on physical theories that have not been tested in the laboratory, nor have they yet come up with any observable consequences. Anomalous redshifts at least provide concrete observations of the present-day universe that can be checked and rechecked.

The cartoon shown in Figure 1 illustrates the frustrating situation concerning the cosmological hypothesis (CH). Awkward observations are

Figure 1 : Chip Arp and Geoffrey Burbidge trying to kill the fly CH.

either 'wished away' on the grounds of statistics not always sound, as found in the case of the Arp-Hazard triplets (Edmunds and George, 1982; Narlikar and Subramanian, 1982) or artificially reinterpreted as, for example, in the case of the quasar found near the galaxy 2237+0305 (Huchra, et al, 1985, Burbidge, 1985).

To fix ideas I will define CH as the statement that the entire redshift z of any extragalactic object arises from the Hubble's law. To allow for some flexibility one may permit Doppler redshifts of upto $\sim$500 kms^{-1} as are observed in clusters of galaxies. One may allow a gravitational redshift component of comparable magnitude if the light has to climb out of potential wells typical of galaxies or clusters.

The alternative to CH, the noncosmological hypothesis simply states that a substantial part of the redshift of at least a few extragalactic objects are of noncosmological nature. Specifically one may write

$$1 + z = \left(1 + z_c\right)\left(1 + z_{NC}\right) \tag{1}$$

where z_c is cosmological (arising from the expansion of the universe) and z_{NC} is due to some other causes. The killing of the fly in Figure 1 amounts to admitting that $z_{NC} \neq 0$. To get out of the vicious circle described in the first paragraph of the article, I will assume that the present evidence (e.g. that reviewed in Paper I) warrants this assumption. What theoretical alternatives then exist for noncosmological interpretations?

The alternatives may be divided into two classes: the first is limited to known physics while the second calls for new physics. To the first class belong the two well known alternatives of Doppler and gravitational redshifts. The second class is necessarily open but the alternatives often discussed in it are the variable mass hypothesis (VMH) of the Hoyle-Narlikar theory of gravity and the tired light theory discussed in this conference by Vigier and Pecker. In the present talk I will describe the first of these alternatives in each class.

The Doppler Hypothesis

Proposed by Terrell (1964) soon after the first two quasars were discovered, the Doppler hypothesis has undergone a few modifications. Instead of arguing that quasars were ejected from the centre of the Galaxy, as Terrell had proposed, Hoyle and Burbidge (1966) suggested that they were ejected from active galactic nuclei. Such nuclei provide manifest signs of explosions and are more plausible candidates for ejectors of quasars then the relatively sedate nucleus of our Galaxy. However, this hypothesis immediately ran into the difficulty of the apparent absence of blueshifted quasars. Indeed, as first demonstrated by Strittmatter (1967) the blueshifted quasars should far outnumber the redshifted ones (because of their brightness) in a flux limited sample. To get round the difficulty Strittmatter had suggested that quasars might be emitting light preferentially in the backward direction.

Subsequently Hoyle (1980) calculated that a quasar moving with velocity V with respect to the cosmological rest frame (CRF) and emitting radiation in a backward cone of angle

$$\theta = 2 \cos^{-1} \frac{c - \sqrt{c^2 - V^2}}{V} , \; V = |\underset{\sim}{V}| , \qquad (2)$$

in its rest frame would never be seen blueshifted by any observer stationary in the CRF. What physical process, however, would limit the quasar to radiate backwards?

Three motion-dependent alternatives suggest themselves. Hoyle's own suggestion (op.cit) was that piled up plasma in the forward direction would absorb quasar's radiation so as to make it invisible from the front. In practice this does not work because the intergalactic medium (IGM) is not dense enough to cause effective absorption. A second possibility is suggested by the adaptation of a model of Rudermann and Spiegel (1971) in which the quasar may focus incoming IGM into a dense wake which would radiate. Again, putting actual numbers into the problem shows that the mechanism does not work. The details of both these calculations are given in a paper by myself and Subramanian (1983) which is mainly concerned with exploring a third alternative that does work. Hereafter this paper will be referred to as Paper II.

The idea is based on a modification of the twin exhaust model of Blandford and Rees (1974). In the original Blandford-Rees model the source, which is a massive object (either a compact one like a black hole or a dense star cluster) surrounded by a plasma cloud develops a twin deLaval nozzle along the line of least resistence through the cloud (which happens to be its axis of rotation). Plasma is squirted out in highly collimated jets in opposite directions.

When this model is viewed in the case of a quasar ejected at high speed through the IGM a modification becomes necessary. The line of least resistence is now the backward direction, where the jet should emerge. The ram pressure of the IGM (moving relativistically relative to the quasar) is high enough to prevent a forward jet from developing. The highlights of the model are as follows.

To begin with, the model has four alternative scenarios. The central gravitating source in the ejected quasar for example could either be a uniform starcluster (Type I) or a compact massive object (Type II). For each of these two source types the IGM could interact with the gas cloud in two ways: Case A: it goes clean through it, or Case B: it produces a bow shock. Figure 2 illustrates one of these four scenarios.

Several questions arise when this model is investigated in detail. I mention some below:

a) Can the gas cloud remain intact in spite of its motion at high speed against the IGM?

The answer to this question is 'yes' provided the gravitational binding of the cloud is sufficiently strong. The inequalities expressing this condition for all the four scenarios are described in Paper II.

b) Does the IGM distort the internal composition of the gas cloud?

Again, details given in Paper II indicate that the answer is 'No', provided certain inequalities are satisfied.

Figure 2 : The components of the quasar model where the gravitating mass is a compact object. The jet is in backward direction. The dots denote emission line clouds.

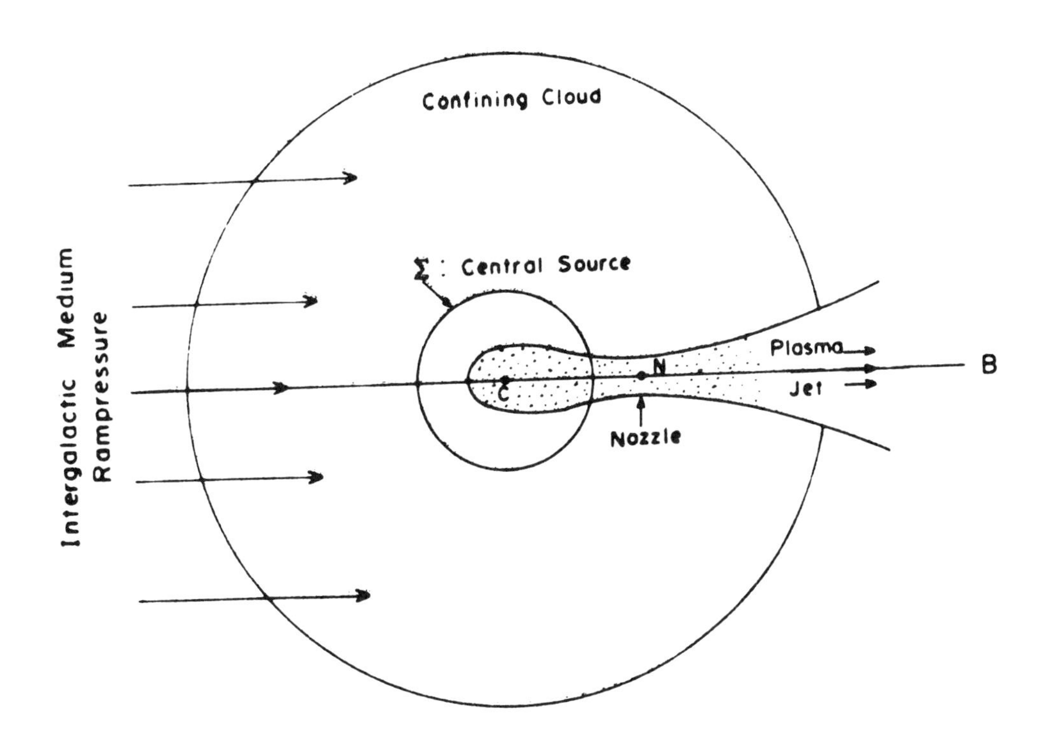

c) How is the jet formed?

Here the scenario differs between the two cases A and B. In case A, the plasma would have magnetic fields which would stop the incoming IGM through a collisionless shock. The plasma then 'feels' the IGM pressure and begins to decelerate. The resistence to the plasma progressively drops as one goes over from the forward to the backward direction. The plasma then seeks the way out along the line of least resistence and thus forms a backward jet.

In Case B, transient pressure gradients are set up impeding plasma expansion in the forward direction while accelerating it in the backward direction. When the transient effects are over (within $\sim 10^4$ yrs) the plasma keeps flowing along the backward direction along the channel scooped out by the initial jet.

d) What is the size of the jet? What are its other physical parameters?

Adaptation of the Blandford-Rees model gives the answer to this question. Details are worked out in Paper II for all four cases. The model for local quasars is, of course, a scaled down version of the Blandford-Rees model.

e) Where does the continuum emission come from?

Calculations using a synchrotron model suggest that the luminosity in the range 10^{41} - 10^{41} erg s^{-1} (for a local QSO) can be generated in a magnetic field $\sim 10^{-4}$G and electron energies $\gamma m c^2$ for $\gamma \sim 10^6$. Spectral analysis of this radiation shows that it is less powerful in the radio than in the optical. To see a 'radio' quasar therefore one needs to be more or less preferentially located in the backward direction. Then the Doppler enhancement can uplift radio emission to observable levels. For example, for a plasma flowing with a jet velocity $c/\sqrt{3}$ forward radiation along the jet is 27 times as intense as in the direction opposite the jet. This may explain why radio quasars are comparatively rare.

f) What is the source of line radiation?

To explain line radiation the model postulates the existence of clouds of characteristic mass $\sim 10^{-2}$ $M_\odot$ and radius $\sim 5 \times 10^{-4}$ pc in the jet. The clouds are driven forward by the ram pressure of the plasma jet while the gravity of the source pulls them back. These two opposing forces determine the location of a typical cloud in the jet.

Only a small proportion ($\sim 10^{-3}$) of the plasma cloud by mass in the form of dust is sufficient to absorb the emission lines. Thus in the forward direction the object would appear lineless.

These questions and their answers indicate that a viable Doppler model can be constructed in sufficient detail to make it testable as a realistic hypothesis. What tests can one use? A few are suggested below.

(i) Of course, the most direct test is the detection of proper motions. VLBI techniques capable of measuring angular proper motions of 0.5-5 m arc-s yr^{-1} against the cosmological rest frame should detect the effect especially if the quasars are closer than ∿ 100-300 Mpc.

(ii) The quasars should as a rule have only one jet. Observations to date support this conclusion. Proper motions, if detected, should be opposite to the direction of the jet.

(iii) The cloud surrounding the quasar should be detected as a fuzz. The larger the redshift, the smaller are the physical dimensions of the cloud and hence the shorter the fuzz.

(iv) The apparent rarity of radio loud quasars is understood naturally by the Doppler model.

(v) The alignments of quasars or their location near low redshift galaxies may be understood as the Doppler ejection phenomenon.

The Variable Mass Hypothesis

More than two decades ago Fred Hoyle and I (1964) proposed a theory of gravitation that explicitly incorporated Mach's Principle (1893) That is, it started with the assumption that a long range scalar interaction between any two particles in the universe generates inertia for each of them. Thus the total inertial mass m_a of any particle a can be written as the sum of inertial contributions $m_a^{(b)}$ from all other particles in the universe.

$$m_a = \sum_{b \neq a} m^{(b)} \quad (3)$$

Apart from this input which quantifies Mach's principle, this theory (hereafter called the HN theory) assumed the curved spacetime framework of general relativity, as well as that the mass-interaction (3) is conformally invariant. The following important result then emerges from the theory:

If the solution of any gravitational problem is described by inertial masses m_a,--- and the spacetime metric g_{ik} (i,k = 0,1,2,3) then the same problem can also be described by another solution in which the particle masses are $m_a \Omega^{-1}$,--- and the metric is $\Omega^2 g_{ik}$, for any well behaved function $\Omega>0$. (Hoyle and Narlikar, 1966).

In the simplest set of solutions the function Ω, called the conformal function, can be so chosen as to make all particle masses constant. The theory then becomes identical to general relativity. However, as shown by A.K. Kembhavi (1978) the 'constant mass' constraint is an artificial one. In particular, if a hypersurface in the original solution had all particle masses vanishing, then the above conformal transformation to general relativity leads to that hypersurface becoming singular. Thus the inevitability of spacetime singularity in general relativity is seen in the HN theory as being due to the existence of zero-mass hypersurfaces.

To illustrate this effect consider the standard k = 0 Friedman model, given by

$$ds_F^2 = dt^2 - \left(\frac{3t}{2}\right)^{4/3}\left[dr^2 + r^2\,(d\theta^2 + \sin^2\theta d\phi^2)\right] \tag{4}$$

where we have taken the Hubble constant at the present epoch (t = 2/3) to be unity. This spacetime manifold is singular on the hypersurface t = 0.

However, this is a conformal transform of the flat spacetime given by the Minkowski line element

$$ds_M^2 = d\tau^2 - \left[dr^2 + r^2\,(d\theta^2 + \sin^2\theta d\phi^2)\right], \tag{5}$$

where

$$\Omega = \left(\frac{3t}{2}\right)^{2/3}, \quad \tau = (12t)^{1/3}. \tag{6}$$

In the HN cosmology, a homogeneous isotropic universe is described in its simplest form by the line element (5) but with particle masses given by

$$m(\tau) = \mu\tau^2, \qquad \mu = \text{constant}. \tag{7}$$

The masses all vanish on the hypersurface $\tau = 0$, which is the same as the singular hypersurface $t = 0$ of the standard model. The insistance that all masses be constant forces the relativistic solution to have an unphysical Ω that vanishes on $t = 0$ leading to a spacetime singularity. The big bang of standard cosmology thus corresponds to the zero mass epoch in the HN cosmology.

In general the HN cosmology has variable mass solutions with $m = 0$ hypersurfaces occurring now and then. All such hypersurfaces lead to singular geometries if one forces on the spacetime a conformal function that makes masses constant.

Returning to the Minkowski solution, it is easy to see how redshifts arise in the HN theory. A galaxy G_0 at $r = 0$, $\tau = \tau_0$ views another G, at $r > 0$, when its epoch was $\tau = \tau_0 - r$ (we have taken $c = 1$ here). The particle masses at this epoch were systematically smaller than those in G_0 at τ_0. Hence the spectral wavelengths of lines in G's radiation will be systematically longer than those measured in a locally generated radiation by an observer in G_o. Using (7) we get the observed redshift as

$$z_G = \left(\frac{\tau_o}{\tau_o - r}\right)^2 - 1. \tag{8}$$

Thus redshifts can be produced without cosmological expansion, from variable particle masses. This is the Variable Mass Hypothesis (VMH).

Notice that unlike the Friedman solution the $m = 0$ hypersurfaces allow the spacetime to be extended to $\tau < 0$. There is no corresponding pre-big bang era ($t < 0$) in standard cosmology. Later work (Narlikar 1977) showed that redshift anomalies can occur when such hypersurfaces develop kinks.

The effect of a kink is illustrated in Figure 3. Imagine the worldlines of the observer galaxy G_o and the source galaxy G crossing the zero mass hypersurface at $\tau=0$. The galaxy G has a neighbour quasar Q whose worldline crosses the $m = 0$ hypersurface at $\tau = \tau_Q > 0$. Particle masses in Q change according to the rule

$$m = \mu\ (\tau - \tau_Q)^2. \tag{9}$$

Hence the quasar redshift as measured at G_o is

$$z_Q = \left(\frac{\tau_o}{\tau_o - \tau_Q - r}\right)^2 -1, \tag{10}$$

with the result that $z_Q > z_G$. Thus Q has an anomalous part $z_{NC} > 0$.

In physical terms thus result has the following interpretation. We may in loose terms consider the epoch $m = 0$ as one of creation although, as shown in Figure 3, the worldlines of particles existed prior to it. Thus bulk of the matter in the universe had $m = 0$ at $\tau = 0$. The 'later creation' in the kink may be thought of as occurring in a minor explosion at certain isolated spots in the universe. The quasar Q in the above example could be considered as fired out of the nucleus of the neighbouring galaxy G.

The dynamical details of this phenomenon have been analysed by Narlikar and Das (1980, hereafter referred to as Paper III). The salient points are summarised below.

a) An upward kink ($\tau > 0$) like the one in Figure 3 would inevitably lead to $z_{NC} > 0$, i.e., to anomalous redshifts. For anomalous blue-shifts ($z_{NC} < 0$) one needs $\tau_Q < 0$, i.e., galactic explosions to occur before the big bang. The ejecta of such explosions are unlikely to survive the big bang. So one should see anomalous redshifts only.

b) If the galactic explosions occurred soon after the big bang, i.e. if $\tau_Q << 1$, then we expect the anomalous component of the redshift to be small. We may consider the anomalous redshifts of companion galaxies observed by Arp and others to arise in this way. The companion galaxies are thus the evolved states of quasars. Present day quasars may likewise acquire galactic forms at later epochs.

c) Since the particles at the creation epoch have zero rest mass, they initially travel with the speed of light. Can such particles remain trapped in the gravitational field of the parent galaxy? Narlikar and Das (op. cit) find that since the mass of the ejected object rapidly

Figure 3 : The 'upward' kink in the (r, τ) plane in the zero mass hypersurface has a quasar worldline (Q) passing through it at $\tau = \tau_Q$. The neighbouring galaxy (G) has its worldline passing through the hypersurface at $\tau = 0$.

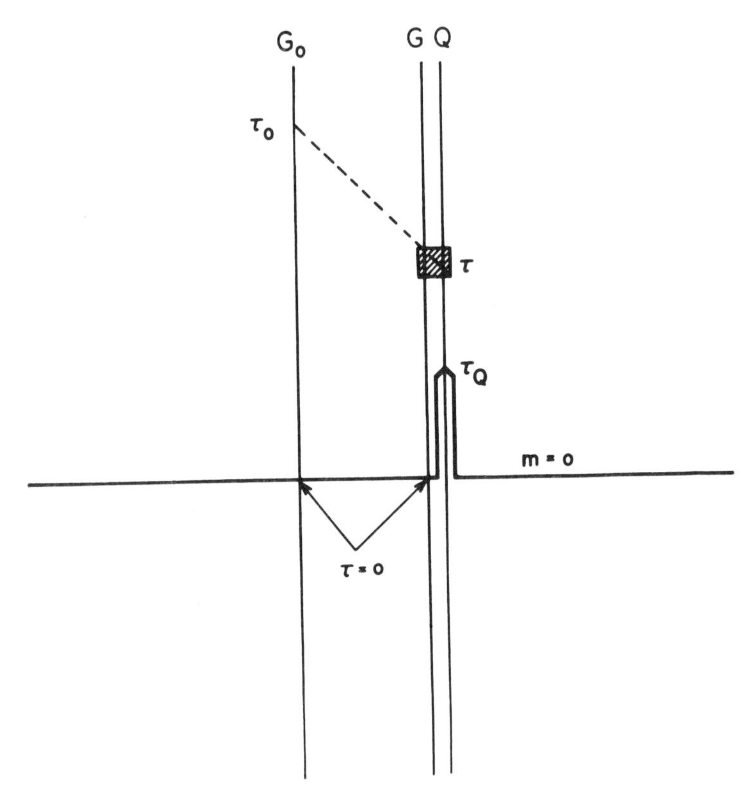

grows, it slows down. Denoting by η the time elapsed since the explosion, it can be shown that the relativistic time dilatation factor γ for the ejected mass declines in the early stages as

$$\gamma \sim \left(\frac{\eta_o}{\eta}\right)^2 \left(1 + \frac{\eta^4}{\eta_o^4}\right)^{1/2}, \tag{11}$$

where η_o is a constant of the motion.

It is clear that whether the quasar escapes the galaxy or remains trapped in it depends on the magnitude of η_o, and on the mass M of the parent galaxy. For each M one finds that there is a critical η_c such that for $\eta_o < \eta_c$ the quasar has already got into a bound orbit around the galaxy at the time of its observation (τ). For $\eta_o > \eta_c$ the quasar has either escaped altogether, or its 'turn round time' is later than the epoch τ at which it was observed.

For a bound highly eccentric orbit one may calculate the maximum separation R_{max} between G and Q. R_{max} depends on z_Q and M. For example, for $M = 10^{11} M_{\odot}$, $z_Q = 2$ we have $\eta_c = 1.03 \times 10^{-5}$ and $R_{max} = 398$. For the same mass $z_Q = 0.5$ leads to $R_{max} = 449$. Thus R_{max} decreases slowly as z_Q increases. It is also found that $R_{max} \propto M^{1/3}$ and that it decreases slowly with z_G.

d) It could be argued on the basis of the VMH that all quasars were so created through delayed galactic explosions and that the bound quasars are the ones showing anomalous redshifts. Quasars which have escaped are seen as field quasars.

The VMH can be tested observationally in a number of ways. Some are indicated below.

(i) The typical Q - G separation will be less than R_{max} for two reasons. At the time of observation the quasar need not be at maximum separation. Moreover, for a highly eccentric orbit the projected separation perpendicular to the line of sight will be less than R_{max}. Averaging with respect to position on the orbit and projection angles suggests that typical separation may be $\sim$ 10% of the R_{max} value. Since R_{max} slowly decreases with z_G we expect θ to drop as $z_G^{-1-\delta}$ where δ is small and positive. Likewise we expect θ to decrease slowly with z_Q. In Paper III it is shown how these expectations are borne out by observations.

(ii) If a number of quasars are ejected in a single explosion we expect them to have the same redshift. Also two quasars so ejected would be aligned across the galaxy with equal redshifts. Observations of this type have been reported from time to time (see for example, Arp et. al 1979, Arp 1980, also, Paper I).

(iii) The Arp-Hazard triplets (Arp and Hazard, 1980) can be given a novel interpretation in VMH. The well aligned triplets have redshifts

(2.15, 0.51, 1.72) and (2.12, 0.54, 1.61). The corresponding members of the two triplets have nearly the same redshifts! By joining their positions we find that they could have been ejected in a triple explosion from a compact region. This region should be searched for post-explosion remnants.

(iv) Evidence for smaller particle masses in higher redshift quasars could also come from the astrophysics of these objects. For example, in a given magnetic field a less massive electron would generate greater synchrotron luminosity than a more massive one.

Concluding Remarks

These two models illustrate what may be possible by way of theoretical interpretation of noncosmological redshifts. Further work needs to be done in both the models so far described. For example, one needs to know more about the ejection process in the Doppler model. The VMH might throw some light on the periodicities observed in redshift distributions (Hewitt and Burbidge 1986, Depaquit et al 1985, Tifft, 1987), through the phenomenon of quantization of particle masses.

Whatever be the ultimate fate fate of such theories they serve a useful purpose in breaking the vicious circle mentioned in the beginning of this contribution.

References

Arp, H. (1980). Ap. J. 236, 63.
Arp, H. and Hazard, C. (1980). Ap.J. 240, 726.
Arp, H., Sulentic, J.W. and di Tullio, G. (1979). Ap. J. 229, 489.
Blandford, R.D. and Rees, M.J. (1974). M.N.R.A.S. 164, 395.
Burbidge, G.R. (1985), A.J. 90, 1399.
Depaquit, S., Pecker, J.-C. and Vigier, J.-P. (1985). Astronom. Nach., 306, I, 7.
Edmunds, M.G. and George, G.H. (1981). Nature. 290, 481.
Hewitt, A. and Burbidge, G.R. (1986). A new catalogue of quasistellar objects. In Quasars, ed. G. Swarup and V.K. Kapahi, pp.51-52 Dordrecht : D.Reidel Publishing Company.
Hoyle, F. (1980). Astrophysics and Relativity, Preprint 63, Cardiff: University College.
Hoyle, F. and Burbidge, G.R. (1966), Ap. J. 144, 534.
Hoyle, F. and Narlikar, J.V. (1964). Proc. Roy. Soc. A 282, 191.
Hoyle, F. and Narlikar, J.V. (1966). Proc. Roy. Soc. A 294, 138.
Huchra, J., Gorenstein, M., Kent, S., Shapiro, I., Smith, G., Horine, E. and Perley, R. (1985). A.J. 90, 691.
Kembhavi, A.K. (1978). M.N.R.A.S. 185, 807.
Mach, E. (1893). The Science of Mechanics. London : Watts & Co.
Narlikar, J.V. (1977). Ann. Phys. 107, 325.
Narlikar, J.V. (1986). Noncosmological redshifts. In Quasars, ed. G. Swarup and V.K. Kapahi, pp. 463-473. Dordrecht: D.Reidel Publishing Company. (Paper I).

Narlikar, J.V. and Das, P.K. (1980). Ap. J. 240, 401 (Paper III).
Narlikar, J.V. and Subramanian, K. (1982). Ap. J. 260, 469.
Narlikar, J.V. and Subramanian, K. (1983). Ap. J. 273, 44 (Paper II).
Rudermann, M.A. and Spiegel, E.A. (1971). Ap. J. 165, 1.
Strittmatter, P.S. (1967). Work cited in Quasi-stellar Objects by G.R. Burbidge and E.M. Burbidge, pp. 164-174. San Francisco: W.H. Freeman and Company.
Terrell, J. (1964). Science 145, 918.
Tifft, W.J. (1987). Quantization in redshifts. In New Ideas in Astronomy (Chip Arp's Sixtieth Birthday Symposium Proceedings) ed. F. Bertola, J. Sulentic and B. Madore. Cambridge: Cambridge University Press.

DISCUSSION

W. Tifft: Could you comment on what would be required to introduce quantization into the redshift? Could you, for example, view the increase in mass to occur in quantum steps? Can one estimate expected rates or time scales for observable changes?

J. Narlikar: It is possible to look at the VMH in a quantum way. For example, if one quantises the scale factor in Einstein's equations one gets stationary states of the universe. In the HN theory this corresponds to stationary states of mass. I believe this should lead to quantised redshifts. But this theory has not yet been worked out in detail to give quantitative answers to your questions.

J.-P. Vigier: Can you explain with your mechanism the discrepant redshifts of companion galaxies?

J. Narlikar: Yes. As I explained, the redshift of matter decreases steadily as it gets older. In this evolutionary sequence, the companion galaxies occur at a much later stage than QSO's. Thus one can imagine an object fired from a galaxy first appearing as a quasar and then eventually it accretes matter and becomes a companion galaxy.

T. Jaakkola: I wonder why you and Fred Hoyle have delivered (deserted? ed) the perfect cosmological principal, which has a long and honorous tradition and which is in fact in a very good accordance with the empirical cosmological data?

J. Narlikar: The perfect Cosmological Principle (PCP) was used by Bondi&Gold to describe the large scale structure of the universe. In the formulation given by Hoyle and me (using the C-field) the steady state solution obeying the PCP follows from the field equations. However, there are other solutions also. I feel one should not at this stage be to restrictive in our choice of solutions.

ALTERNATIVE INTERPRETATION OF THE COSMOLOGICAL REDSHIFT IN TERMS OF VACUUM GRAVITATIONAL DRAG

J. P. VIGIER
Institut Henri Poincaré
75231 PARIS CEDEX 05

SUMMARY

The causal stochastic interpretation of quantum microobjects (including photons) as collective excitations on the top of a real stochastic covariant zero point "vacuum" or "aether" suggested by Dirac(1951) (see also Cufaro-Petroni & Vigier, 1982) implies the introduction of small dissipative forces corresponding to the vacuum's gravitational drag. It is shown here that a justification of such quantum vacuum damping terms in Dirac's superfluid aether model (Sinha et al., 1986; hereafter SSV) implies a stochastic derivation of known quantum equations completed by very small dissipative irreversible corrective Langevin type forces. Such forces could appear at astrophysical level in the form of non velocity redshift mechanisms.

I. INTRODUCTION

The idea that the cosmological redshift is not due to an expansion of the Universe but to some "tired light" mechanism is as old as its discovery by Hubble. Its possible observational consequences were later confronted to those of possible precise cosmological tests in a famous paper by Hubble and Tolman(1935) at a time when observations were too scarce to conclude. Despite its revival at various intervals in the literature (i.e. by Zwicky and Pecker et al.(1976,77), "tired light" models have intrinsic theoretical difficulties and lack supporting laboratory evidence.

This situation is now rapidly changing for the following reasons:

A) In the field of astronomical observations, it has been recently established by Arp(1987) (utilizing the Tully-Fisher distance indicator) that if Sb galaxies of the same luminosity class as M31 and M81 define a narrow Hubble relation with $H_o = 65^{+15}_{-8}$ km/sec/Mpc the Sc galaxies (specially ScI) deviate strongly towards higher redshifts from a linear log redshift apparent magnitude relation.

- This non linearity of the "Hubble flow" has been confirmed independently by Segal and Giraud(1984) who have discovered (also using the Tully-Fisher independent distance indicators) a significant increase of H_o from the center towards the periphery of the Virgo supercluster. Moreover, one has never explained in conventional terms the increase of the Hubble constant of galactic sources seen through intervening clusters of galaxies (i.e. the so-called Karoji-Nottale (1976a,b) effect).

- One has not interpreted the significant excess of spiral redshifts in spiral-elliptical galaxy pairs (Giraud et al., 1982; Sulentic, 1982), or well known anomalies in the QSO-Galaxy associations.

- More important still due to a recent analysis of La Violette(1986), it now appears that four known cosmological tests fit "tired-light" models better than expansion models. To interpret within big-bang models, the observed data one would need ad-hoc galaxy evolution models which have (at least) not (yet) been explained or justified.

B) In the field of the Quantum Mechanical Physical Theory of Light, redshifts can only result (in our present state of knowledge) from known properties of light i.e.

a) from direct source motions (Doppler effect)

b) from light motion within varying gravitational fields (Einstein effect)

c) from recently discovered coherence properties of extended light sources (Wolf effect)

- from "tired-light" mechanisms

Any "tired-light" model should satisfy two strong constraints, i.e. a) yield redshifts as functions of distance but independent of frequencies and b) prevent line and source image broadening- both satisfied by the Doppler effect due to receding sources.

such as

d) from Compton scattering or possible other types of elastic photon scattering with very light particles (axions, etc.).

e) from interactions of light with vacuum now associated with realistic causal stochastic interpretations of Quantum Mechanics (Einstein, de Broglie, Bohm, Nelson, Guerra etc).

Laboratory experiments have shown that a) certainly exists (has been tested),

b) certainly exists (has been tested),

c) certainly exists (has been tested) but is a small effect,

d) has not been observed in the laboratory but presents difficulties (such as image broadening) to interpret possible non-velocity redshift observations,

e) has not yet been directly tested but is now suggested as a viable alternative by the development of the Bohr-Einstein controversy on the interpretation of Quantum Mechanics. It provides a new possible basis for a gravitational drag model of the cosmological redshift along lines originally explored by Zwicky and Ehrenfest et al.(1931; hereafter EPT). Indeed very recent neutron interferometry experiments (Rauch et al., 1986) suggest that quantum microobjects are waves and particles simultaneously– so that the real de Broglie waves (i.e. Maxwellian waves surrounding particle-like photons in the case of light) must propagate on a real covariant subquantal "aether" of the type, first suggested by Dirac(1951). (Very recent laser experiments directly confirm the existence of real zero point electromagnetic "vacuum" fields) with possible new vacuum friction phenomena.

- A theoretical model of Dirac's "aether" in the form of a covariant superfluid medium carrying real Einstein-de Broglie waves (and particle aspects) as excited collective motions has recently been proposed by Sudarshan et al. (briefly discussed in SSV). Built with fermion and antifermion non zero mass ($m_v \neq 0$) elements and fields, this superfluid carries not only fermionic and bosonic excitations (particles) but also contains a normal fluid component which evidently implies a small gravitational friction (drag), if the photon mass m_γ satisfies $m_\gamma \gg m_v$. In this model where a single "photon" is now described (SSV) by a particle-like photon moving (with its group velocity) within a wave packet (pulse) of light practically represented by Maxwell's equations (which corresponds to the limit case when $m_\gamma \to 0$) of Proca's equations (see Gueret & Vigier, 1982). The pulse represents, for example, a photon emitted by a distant source in a given spectral line. Such vacuum models imply a frictional drag if we remark:

- that in the ground state of Dirac's aether, the non-zero mass vacuum elements have a constant stochastic surface density on the mass shell $p_\mu p^\mu = -m^2c^4$, the vacuum distribution (including the normal fluid mode) is invariant for all observers so that Dirac's aether is the only known covariant distribution in physics.

- that this vacuum being a superfluid offers no first order resistance to particle and associated wave motions but that, since the excited fermions and boson (photon) waves (built with excited fermion pairs) have non zero masses (i.e. $m > 0$) and a velocity $v < c$, Dirac's vacuum necessarily corresponds to a dispersive medium. Indeed, all non zero mass particles satisfy the Einstein-de Broglie relation

$$E = h\nu = \frac{mc^2}{\sqrt{1 - v^2/c^2}}$$

which connects their frequency and wave-group velocity v.

- that since particles correspond to stochastic Brown-Markov processes one should expect second order dissipative effects on their energy momenta according to Einstein's Brownian motion theory.

As we will show, light pulses travelling in such a vacuum exert a gravitational pull on the normal mode vacuum elements which can be compared to test particles which interact gravitationally with light: an interaction which should be analysed within the frame of Einstein's theory of general relativity. The aim of this work is to describe this particular drag model and to discuss some of its physical and cosmological implications. In section II, we briefly recall the physical properties of recent stochastic "vacuum models" discussed in the literature. In section III, we recall the theory of the reciprocal gravitational interaction of photons with test particles (or "vacuum" elements) first presented by EPT and its consequences on the propagation of light (or scalar particles) in a covariant stochastic medium of the Dirac type. In sections IV and V, we analyse (in classical and quantum terms) the "tired-light effect" which results from this model. In conclusion, we present some of its implications and cosmological consequences.

II. PHYSICAL PROPERTIES OF DIRAC'S "AETHER"

Recent advances in the stochastic interpretation of Quantum Mechanics consider quantum "particles" as relatively stable and conserved excitations (pilot-waves plus solitons) on the top of a real covariant vacuum (random fields) of the Dirac type. Such "particles" will only be registered on the large scale level since all known apparatus are only sensitive to those features of the field that will last for some time but not to those features that fluctuate rapidly. Thus the "vacuum" will produce no visible effects at the large scale level since its fields will cancel themselves out on the average and space will be effectively "empty" for every large-scale process: exactly like a perfect crystal lattice or a superfluid is effectively empty for an electron in the lowest band; even though the space is full of atoms.

Along these lines of research, it has been shown (by Nelson, Bohm et al., Guerra-Ruggiero, Vigier, Cufaro-Petroni et al., (see Vigier, 1982 and SSV for references), and EPT) that it was possible to describe Dirac's covariant aether in terms of a covariant superfluid offering no laboratory evidence (except perhaps in the $K^\circ \rightarrow 3\Pi$ decay) of friction and irreversibility to the propagation of quantum waves considered as collective excitations on the top of such a material "aether". Moreover, SSV have assumed that such a vacuum corresponds to a covariant chaotic distribution of pairs of fermionic and antifermionic fields whose energy-momentum density is defined by Einstein's vacuum term $(T_{\mu\nu})_{vac} = \rho_{vac} g_{\mu\nu}$ where ρ_{vac} is the vacuum energy density. Einstein's field equations with the vacuum term, thus take the form:

$$G^{\mu\nu} = 8\pi G(T^{\mu\nu} + T^{\mu\nu}_{vac}) = 8\pi G(T^{\mu\nu} + g^{\mu\nu}\rho_{vac}) \tag{2.1}$$

which contain the usual cosmological term $\Lambda g_{\mu\nu}$ connected to ρ_{vac} by

$$\Lambda = \frac{8\pi G}{c^4}\rho_{vac} = \kappa \cdot \rho_{vac} \tag{2.2}$$

Λ being practically (at our level) a scalar constant whose variations can generally be neglected except in special cosmological situations such as black hole vicinity etc. Of course, the energy of the vacuum (i.e. zero point energy) due to subquantum fluctuations would also (following Sakharov and other authors) contribute to all local gravitational fields.

If one denotes by $m_v \ll m_\gamma \ll m_e$ the bare rest mass of the vacuum elements the vacuum's temperature $T = m_v c^2 / K$ (where K represents Boltzmann's constant) is practically equal to zero, so that it presents practically no observable friction to the usual massive quantum particles. Any small viscosity related to the presence of the normal fluid component would only show up on the cosmological scale.

Of course, the introduction of Dirac's covariant non-zero mass stochastic chaotic fermion distribution implies the existence of an associated chaotic field behaviour i.e. of a permanent fluctuating contribution to the $g_{\mu\nu}$ term which should always be written $g_{\mu\nu} = \bar{g}_{\mu\nu} + \delta g_{\mu\nu}$ where $\bar{g}_{\mu\nu}$ denotes a local average value and $< \delta g_{\mu\nu} >= 0$. In the Stochastic Interpretation of Quantum Mechanics (Dirac,1951) (SIQM), the existence of such fluctuations is indispensible to interpret/justify (if one accepts the existence of real time-like trajectories of all particle aspects of individual microobjects) the statistical predictions of quantum mechanics. In other terms in SIQM one starts from Smolin's (1986) basic assumption that quantum fluctuations are, in fact, real statistical fluctuations which reflect the basic stochastic character of Dirac's aether. Such real fluctuations behave as ordinary statistical fluctuations in the physical states of physical systems. However, since they result from a process which is superimposed on the evolution of the local dynamical variables and introduce an element of randomness into the evolution of the dynamical variables, their action is in general dissipative.

Since experimental evidence supporting general relativity and quantum mechanics suggests that such (sub)-quantum fluctuations also have properties which are not shared by any other known dissipative phenomena in Nature, this implies (Smolin, 1986):

1) "that in the absence of gravitational fields there exist preferred states of motion for a particle moving through empty space (the vacuum) such that the action of quantum fluctuations on bodies undergoing any of these preferred motions is (practically) non dissipative".

The term (practically) introduced by us means, as we shall see later, that any vacuum energy dissipation is (at our earthly experimental level but perhaps not over cosmological distances) unobservable–a property suggested by the reversible character of the quantum mechanical formalism, even in its present derivation-justification by a stochastic formalism. This implies that any realistic vacuum model should indeed (practically) behave like a superfluid. Moreover, no energy dissipation (except as we shall see in possible "tired-light effects") is observed in the motion of massive particles along geodetic trajectories–so that in the absence of gravitational fields, these preferred states of motion coincide precisely with the inertial motions, i.e. the states of preferred motion in which the laws of mechanics take on a simple form.

2) That if the dissipative effects of quantum fluctuations are absent for bodies undergoing inertial motion, the simplest hypothesis is that the dissipative effects of the field on a body are for the simple case of uniform acceleration proportional to the acceleration of the body. This is exactly what is expressed in the Langevin equation. It can be shown moreover that in the absence of gravitational fields, the effects of quantum fluctuations of a field on a body undergoing uniform accelerations g, through a region of space in which the field is its ground state (like Dirac's aether), are exactly equivalent to the effects of immersing the particle at rest in a thermal bath at temperature

$$T_g = \hbar g / 2\pi c \tag{2.3}$$

where $\hbar$ sets the scale of quantum effects (Smolin, 1986). This illustrates in a striking way de Broglie's assimilation of the subquantum stochastic aether to a "hidden thermostat". As also remarked by Smolin(1986), if the acceleration is not uniform but is slowly changing on the time scale $\hbar/T_g$, then this equivalence is true to the first order in $(\hbar/T_g)(\dot{g}/g) = 2\pi c\dot{g}/g^2$ which is exactly what one finds when one calculates explicitly the effect of quantum fluctuations in the Minkowski vacuum state on particles undergoing various motions.

As one knows, in the SIQM the effects of (sub)quantum fluctuations on the particle (soliton) trajectories implies that they undergo an intrinsic Brownian-Markov motion in configuration and phase space from which one has shown:

- that the corresponding associated positive probabilities satisfy an H-theorem both in the non-relativistic and relativistic case (Bohm & Vigier, 1954).

- that one can justify the Heisenberg uncertainty $< \Delta p_\mu, \Delta x_\mu >= D\delta_{\mu\nu}\Delta\tau$ with $D = \hbar/2m$ if one accepts de Broglie's phase coherence (continuity) principle... so that $m_{quantum} \cong m_{inertial}$.

Before passing to the next point, we want to emphasize the physical importance of Smolin's (1986) principle. In SIQM, it implies that if one associates (sub)quantum fluctuations with real fluctuations and if they have the property that their action is (practically) non dissipative only on particles undergoing certain preferred classes of motions (such as the de Broglie-Bohm trajectories) and if these motions are, in the absence of gravitational fields, inertial motions, then it follows that in the general case in the presence of gravitational fields there may be no states of motion on which the quantum fluctuations do not act dissipatively.

As a consequence of this brief analysis, the vacuum appears (exactly like all real stochastic media generating Brownian motion) as both a (very weakly) dissipative and dispersive covariant medium. Indeed, as remarked long ago by de Broglie himself, in the absence of matter all non zero-mass particles, the relation $E = h\nu = mc^2 \cdot (1 - v^2/c^2)^{-\frac{1}{2}}$ for free particles implies the relation

$$n = \frac{c}{v} = (1 - \frac{m^2c^4}{h^2\nu^2})^{-\frac{1}{2}} \tag{2.4}$$

which introduces a frequency dependent vacuum index of refraction. Moreover, the vacuum's critical velocity has been shown (SSV) to be c, so that no particle (wave) excitation can supersede the velocity of light.

Clearly for the propagation of electromagnetic waves the $g_\mu \nu$ field of general relativity behaves as a real physical medium (see de Felice, **General Relativity and Gravitation**, Vol.2, (1971) 347). *Indeed the gravitational field acts in this case as an optical medium.* Given a curved space-time with a metric tensor $g_{\mu\nu}$, Maxwell's equations may be written as if they were valid in a flat space time in which there is an optical medium with a constitutive equation. When optical phenomena are considered, this medium turns out to be equivalent to the gravitational field. Following our program, we shall now discuss possible gravitational dissipative interactions between Dirac's non zero-mass vacuum elements and particles moving in such a stochastic medium.

III. DIRAC'S AETHER AS A DISSIPATIVE MEDIUM

We now discuss possible gravitational dissipative interactions between Dirac's non zero-mass vacuum elements and heavier particles (waves) moving in such a stochastic medium.

Clearly Einstein's general relativity implies action and reaction between the $g_{\mu\nu}$ field and any moving object characterized by an energy momentum distribution $T_{\mu\nu}$ since

$$G_{\mu\nu} = R_{\mu\nu} - \frac{1}{2}g_{\mu\nu}R + \Lambda g_{\mu\nu} = 8\pi G(T_{\mu\nu} + g_{\mu\nu}\rho_{vac})$$

In other terms, the motion of light is modified by the surrounding $g_{\mu\nu}$ field but vice-versa the passage of a light pulse modifies the surrounding $g_{\mu\nu}$ field. This effect is, of course, very small and has been generally neglected in the literature.

We shall limit here this discussion to the case of very light particles (i.e. photons and utilize Einstein's approximate solution of the equations of general relativity (valid in weak fields) to describe

- the effect of passing pulses of light on the $g_{\mu\nu}$ fields in their neighborhood

- the gravitational red-shift induced by this disturbance on the passing photon motion.

In other terms, we shall discuss the action (local vacuum disturbance) induced on the vacuum distribution by a particle passing through it and the dissipative reaction on the particle's motion induced by this action. As we shall see, this action and reaction:

1) satisfies (as it should) the principle of four momentum conservation

2) implies the existence of a local gravitational drag (equivalent to a local dissipative Doppler shift

3) does not blur the image of distant sources

4) explains the photon's (particle's) red-shift energy losses in terms of increased vacuum temperature fluctuations

The distortion of the $g_{\mu\nu}$ field by a passing pulse of light has already been analysed in the literature in a very important paper (EPT) which we can briefly summarize as follows.

- In accordance with Einstein's equation for weak fields, EPT have analysed the effects of steady pencils and passing pulses of light on the line elements (corresponding for example to isolated spectral lines) in their neighborhood and deduced the corresponding acceleration of test particles (here vacuum elements with $m_\gamma \gg m_v$) in such fields.

EPT have shown:

A) that test rays moving parallel to the pencil or pulse do so with uniform unit velocity the same as that in the pencil or pulse itself but

B) that test rays moving in other directions experience a gravitational action towards the photon's propagation path (equivalent to a pinch effect) since a test particle placed at points equally distant from the two ends of the track of a pulse experiences no net integrated acceleration parallel to the track but experiences a net acceleration towards the track by twice the amount which would be calculated by Newton's theory.

Having analysed the nature of the motions of the vacuum elements implied by the passage of a photon or pinch of light, one can analyse the reaction of the perturbed local vacuum distribution on the photon's motion, i.e. the consequence of the vacuum presence on the "free" particle behavior.

Four consequences appear immediately:

A) Since Dirac's aether implies a covariant isotropic distribution of vacuum elements at each point, the local vacuum disturbance has cylindrical symmetry around the world line followed by the passing (disturbing) particle at each point.

B) Since the undisturbed vacuum distribution seen from any inertial frame is identical (covariant) the corresponding gravitational disturbance is

$$\frac{\delta\nu}{\nu} = (\phi' - \phi)/c^2$$

where ϕ' and ϕ represent the gravitational potential behind and in front of the moving pulse.

C) In other terms, the vacuum's gravitational drag is equivalent to a local gravitational redshift: equivalent to a local Doppler effect. If the density of the vacuum's normal fluid component is big enough (i.e. $\rho_v(m_v) \gg \rho_\gamma(m_\gamma)$) there is practically no photon (particle) deflection... exactly like there is no observable deflection of a heavy cannon ball passing through a gas of very light particles. Indeed, if we are dealing with N vacuum elements (which represents the excess of the elements behind the moving particle with respect to the distribution in front) which each induce a loss of energy and a slight angular deflection one knows that the former contribution is proportional to N and the latter to $\sqrt{N}$ ... so that (with N big enough) one can reduce the deflection to insignificant contributions w.r.t. the red-shift. This shows that this model bypasses objections formerly raised against scattering mechanisms.

D) Due to the covariant character of Dirac's aether distribution, the pinch effect is practically independent of the light pinch's (wave packet) velocity, and the total redshift $z = \delta\nu/\nu$ only depends on the total length of the distance travelled within the medium so that we have

$$z = \frac{\delta\nu}{\nu} = -H\delta r \qquad \text{or} \qquad \frac{\delta\lambda}{\lambda} = e^{Hr} - 1$$

where H is a term depending on the Dirac aether density, their vacuum masses m_v and m_γ.

As one knows, Doppler type effects do not broaden the lines nor the source's dimensions so that this local gravitational drag type (tired- light) model first suggested (without Dirac's aether) by Zwicky(1929,35) escapes all objections presented against elastic scattering models previously proposed and discussed in the literature.

Of course, any such form of vacuum drag implies a modification of both classical and quantum equations of motion even for so called "free" particles moving in the so called "empty" space. We shall thus conclude this section with a brief discussion of this problem in the spin zero case since the introduction of spin does not alter the physical picture significantly.

IV. CLASSICAL RELATIVISTIC MOTIONS IN A DISSIPATIVE VACUUM

In the case of a classical spinless particle moving along a time-like world-line defined by coordinates $x_\mu = x_\mu(\tau)$ (τ denoting the proper time) within Dirac's aether (with rest mass elements m_v) considered as a real dissipative stochastic medium, we accept as starting point the now classical damping model of Greenberger(1979a,b) which compares this problem to the passage of a drop of water through a chaotic fog of droplets or of a cannon ball through a chaos of particles. In such a situation, the passing heavy object can be expected

- to have (even at rest) a variable mass associated to a variable surrounding atmosphere of vacuum elements and self energy contributions.

- to increase its effective mass along its motion by building up a surrounding atmosphere (or absorbing fog elements) with comoving chaotic elements.

- to attract at least temporarily (as shown in Section III) elements towards its track surrounding vacuum (an effect comparable to an electromagnetic pinch effect behind or passing heavily charged wave packet) which would not be balanced by the vacuum elements in front.

The passing (relatively) heavy particle will accordingly be progressively slowed down and lose energy (i.e. frequency since $E = h\nu$) in favour of the vacuum's energy, and a particle of initial constant rest mass m_o would thus (very slowly and very slightly) reduce its velocity and increase its mass during its motion. In other terms, the existence of our covariant "aether" implies, in general, the introduction of a real variable mass m which we shall consider as a dynamical variable canonically conjugated to the proper time τ of the particle's trajectory. Vacuum damping can be described as a collective phenomenon produced by the interaction of a passing microobject within Dirac's aether. It is the dissipation of its energy to this background that causes the damping and at the same time the particle receives some energy from the background's fluctuations.

Let us now analyse in some detail the classical relativistic problem of a free particle in a real chaotic surrounding medium of the Dirac aether type

---

The physics behind this model is clear. In classical language, the photon's behaviour can (qualitatively) be compared to the slowing down of a positively charged heavy ion (of charge +Ne) penetrating into an isotropic cloud (gas) of electrons of charge -e. Such a particle would attract the surrounding electrons towards its track leaving behind it a trail of enhanced density. Evidently the electromagnetic drag due to this trail is comparable to a temporary "pinch effect" of electrons which would not be balanced by the electrons in front, so that the heavy ion would progressively be slowed down, i.e. loose energy and frequency since E =hν in the EPT model. The mechanism described above is similar, with gravitational replacing electromagnetic interactions.

The excess gravitational attraction generated by the enhanced density trail of Dirac's superfluid "aether" elements implies that the photon moves in a decreasing gravitational potential practically independent of its velocity (very close to $c = 1$): a situation comparable to that of photons leaving the earth vertically as in the Pound-Rebka experiments. .

In general, its motion can be represented as usual by the canonical coordinates $x_\mu(\tau)$ and $p_\mu(\tau)$ (which depend on the proper time τ of the particle's world line), but one must also assume following Schild(1953) and Greenberger(1979a,b) that the passing of such a heavy particle of mass m through a chaos of lighter particles ($m \gg m_v$) implies as stated above:

1) that its rest mass m always fluctuates as a consequence of a variable surrounding "atmosphere" of vacuum elements around a mean value $\bar{m}$ even when the particle is at rest so that (bar denoting a mean value in any interval $d\tau$) we have $m = m(\tau)$ with $m(\tau) = \bar{m} + \delta m(\tau)$ and $\overline{\delta m} = 0$. This yields $\bar{m} = \bar{m}_o$ and $\overline{m^2} = \overline{m}_o^2 + (\delta m)^2$.

2) that the particle's average rest mass $\bar{m}$ should increase with τ in general by building an increasing "atmosphere" of vacuum particles with comoving chaotic elements. If this growth is proportional to distance, one has a variable average mass term $\bar{m} = \bar{m}_o \exp(\gamma\tau)$.

In other terms, this amounts to considering τ and $\bar{m}c^2$ as canonical variables (an assumption which fits with Bohr's uncertainty relation $\Delta\tau\Delta\bar{m}c^2 \geq \hbar$) and has been justified by Schild(1953) with quantum self energy arguments. Indeed, if we recall the relativistic equation of a particle being $dp_\mu/d\tau = F_\mu$ with $p_\mu p^\mu = -m^2c^2$ and $p_\mu = \partial L/\partial x^\mu$ we have $L = m(dx^\mu dx_\mu)^{\frac{1}{2}} + eA_\mu$ and $md^2x_\mu/d\tau^2 = e(F_{\mu\nu})\dot{x}^\mu$ and see that the requirement that the rest mass is constant is the constraint $F_\mu \cdot p^\mu = 0$ which is not satisfied in general by all scalar potentials V since $\partial V/\partial x^\mu = F_\mu$ can present components parallel to p_μ. Introducing, for example, the general scalar parameter u with $du = d\tau/\bar{m}$ and the symbol $\mathring{A} = dA/du$ and $\dot{A} = dA/d\tau$ (so that $\dot{A} = \mathring{A}m^{-1}$) the charged particle motion in this "aether" is now defined by the action principle

$$\delta \int_{u_1}^{u_2} L(x_\mu, p_\mu)du = 0 \tag{4.1}$$

with the new evolution parameter, called canonical proper time by Caldirola (see P. Caldirola 1983, *Il. Nuov. Cim.*, **77B**,241) with

$$L = m\frac{\mathring{x}_\mu \mathring{x}^\mu}{2} - V \tag{4.2}$$

This yields

$$p_\mu^* = \frac{\partial L}{\partial \mathring{x}_\mu} = m\mathring{X}_\mu$$

where p^*_μ represents the new canonical conjugate momenta. From this one, deduces the new Hamiltonian

$$H = \frac{p^*_\mu p^{\mu *}}{2m} + V \tag{4.3}$$

and the equation of motion

$$\mathring{P}^*_\mu = -\partial_\mu V$$

which yields, when going back to the ordinary variables, τ, x_μ, p_μ, the relation

$$\frac{d}{d\tau} p_\mu + \gamma p_\mu = -\partial_\mu V \tag{4.4}$$

which coincides with the classical relations for dissipative systems: first proposed long ago by Levi Civita (1896, *Att. R. Ist. Veneto Sci.*, **53**, 1004).

The preceding calculations also imply that the canonical momenta $p_\mu = \bar{m}\dot{p}_\mu$ are no longer equal to the usual kinetic momenta $p^{cin}_\mu = m\dot{q}_\mu$ and that the corresponding equations of motion with $U = V\exp(\tau\gamma)$ i.e.

$$\ddot{q}_\mu + 2\gamma\dot{q}_\mu + m^{-1}\partial V/\partial q_\mu = 0 \tag{4.13}$$

add Langevin's damping force $F^\mu_d = -\gamma m V^\mu$ to the problem of a particle moving in a potential V, if we assume that "white noise" is negligible in Dirac's aether model.

In other terms, the gravitational vacuum drag can be represented by a continuous loss of energy of the particle to the vacuum (with total energy momentum conservation) since we have

$$\dot{p}_\mu = -\dot{\gamma} p_\mu \qquad \text{i.e.} \qquad p_\mu = \mathring{p}_\mu \exp(-\gamma\tau)$$

which corresponds to a continuous decrease of its velocity. Writing as usual the relation $d\tau = dt\sqrt{1 - v^2/c^2}$ with $v = d\tau/dt$ we have

$$dE/E = -\gamma d\tau = dt\sqrt{1 - v^2/c^2} = -\gamma dt\sqrt{1 - v^2/c^2} = -(\gamma/v)dr$$

if dr and v represent the distance elements and velocities seen in the laboratory frames. In the limit where $v \sim c$ this relation becomes (with $E = h\nu$).

$$d\nu/\nu = z = -(\gamma/c)dr \simeq -H dr$$

so that $Z = \exp(-Hr)$ which is exactly the well known "tired-light" form of Hubble's law. One sees also the canonical momenta $p_\mu = \bar{m}\dot{x}_\mu$ are no longer equal to the kinetic momenta $m\dot{x}_\mu$ and that the vacuum introduces a damping force $F_\mu = -\gamma\bar{m}v_\mu$ which has exactly the form proposed by Langevin and Nosé in relativistic damping processes. Relativistic particle motions in vacuum now evidently correspond to a time irreversible process.

V. GRAVITATIONAL DRAG IN THE EINSTEIN-DE BROGLIE THEORY OF LIGHT

As one knows the Einstein-de Broglie Theory of Light yields practically equivalent results to the usual classical and quantum theories of light but starts from different premises i.e. generalizing the stochastic derivation of the Klein-Gordon equation of Cufaro-Petroni et al.(1985)

- Light is assumed to correspond to non zero mass particles so that $m_\gamma \gg m_v \neq 0$

- It is built with real Proca-Maxwell waves (Einstein, "gespenster wellen" or de Broglie's "pilot waves" which carry particle like photons which travel along average drift lines and beat in phase (in their rest frames one has $E = h\nu = m_\gamma c^2$) with their surrounding wave-fields.

-In such a model, the field corresponds to J = 1 so that there are three spin states

$J_3 = \pm 1$ the usual em fields (transverse photons)

$J_3 = 0$ the Coulomb field (longitudinal photons)

$J_3 = \pm 1$ and $J_3 = 0$ being practically decoupled ($J_3 = 0$ behaves like a scalar when $m_\gamma \to 0$)

- The field quantities are defined by a vector field A_μ (see the Aharonov-Bohm effect) which satisfies

$$\partial^\mu F_{\mu\nu} = \frac{m^2 \gamma c^2}{h^2} A_\mu \qquad \longleftrightarrow \qquad \begin{array}{c} \Box A_\mu = \frac{m^2 c^2}{h^2} A_\mu \\ \partial^\mu A_\mu = 0 \end{array}$$

The passage from the J = 0 (scalar) to the J = 1 (vector) case is straightforward. One substitutes to the scalar Lagrangian of the SIQM

$$\begin{aligned} \mathcal{L} &= \hbar^2 c^2 \partial_\mu \psi^\star \partial^\mu \psi + (mc^2 \hbar / 2i) \left[\psi^\star \frac{\partial \psi}{\partial \tau} - \psi \frac{\partial \psi^\star}{\partial \tau} \right] \\ &= mc^2 e^{2p} \frac{\partial S}{\partial \tau} + e^{2p} \left[\partial_\mu S \partial^\mu + h \partial_\mu P \partial^\mu P \right] \end{aligned}$$

The Lagrangian

$$\begin{aligned} \mathcal{L} = {} & mc^2 e^{2p} \frac{\partial S}{\partial \tau} + e^{2p} \left[(\partial_\mu S + r_\mu)(\partial^\mu S + r^\mu) + \hbar \partial_\mu P \partial^\mu P \right] \\ & + \frac{e^{2p}}{4} (\partial_\mu r_\nu - \partial_\nu r_\mu)(\partial^\mu r^\nu - \partial^\nu r^\mu) + \alpha \partial_\mu S \cdot r^\mu \\ & + \beta (r_\mu r^\mu - 1) + \gamma \partial_\mu P r^\mu + \alpha \dot{\alpha} + \beta \dot{\beta} + \gamma \dot{\gamma} + \delta \dot{\delta} \end{aligned}$$

This can be justified:

- by dropping the assumption that the drift velocity $v_\mu = 1/2\,(b_\mu^+ - b_\mu^-) = \partial_\mu S$ is irrotational and replacing it by the general form

$$v_\mu = \frac{1}{2} b_\mu^+ - b_\mu^- = \partial_\mu S + r_\mu$$

so that curl $v_\mu = \partial_\mu r_\nu - \partial_\nu r_\mu = f_{\mu\nu}$

- by adding to $\mathcal{L}$ the contribution of this vortex motion $\sim (f_{\mu\nu} f^{\mu\nu}) e^{2p}$

- by imposing the constant length and transverse character (space like nature) of this rotational contribution v_μ imposed by the Lagrange multipliers α, β, γ and δ i.e.

$$\partial_\mu S \cdot r^\mu = 0$$

$$\partial_\mu P \cdot r^\mu = 0$$

$$r_\mu r^\mu = 1$$

The field equations $\delta\mathcal{L}/\delta P = 0$, $\delta\mathcal{L}/\delta S = 0$ $\delta\mathcal{L}/\delta r_\mu = 0$ along with the preceding constraints yield:

- for $\delta\mathcal{L}/\delta P = 0$ the Hamilton-Jacobi equation

$$\partial_\mu S \partial^\mu S - \hbar^2 (\Box P + \partial_\mu P \partial^\mu P) - f_{\mu\nu} f^{\mu\nu} - m^2 c^2 = 0$$

- for $\delta\mathcal{L}/\delta S$ conservation equations $\partial_\mu (e^{2p} \partial^\mu S) = 0$ and $\partial_\mu \{e^{2p} (\partial^\mu S + r^\mu) = 0$

- for the transverse real velocity $\partial_\mu (e^{2p} f^{\mu\nu}) = r^\mu$, $\partial_\mu r^\mu = 0$ and $r_\mu r^\mu = 1$ which can be combined into:

$$\phi(x_\mu, \tau) = \frac{mc^2}{\hbar} \frac{\partial \phi_\mu}{\partial \tau} (x_\mu, \tau)$$

with

$$\begin{aligned} \phi_\mu(x_\mu, \tau) &= \exp\left(\frac{im\gamma c^2}{\hbar} \tau\right) A_\mu(x_\mu) \\ &= \exp\left(\frac{im\gamma c^2}{\hbar} \tau\right) e^{p+iS} \cdot r_\mu \\ &= \exp\left(\frac{im\gamma c^2}{\hbar} \tau\right) A_\mu(x_\mu) \end{aligned}$$

This yields the usual Proca equation on the vector potential $A_\mu = e^{\rho + is} \cdot r_\mu$ i.e.

$$\Box A_\mu = \frac{m^2 c^2}{\hbar^2} A_\mu$$

To express the vacuum gravitational damping of this Proca-Maxwell field, one utilizes exactly the same procedure as in the scalar Klein-Gordon case[25]

- we replace m by $\bar{m} = m\exp(\gamma\tau)$
- we add the interaction $\frac{\lambda}{2}\bar{m}c^2 \cdot \rho$ and obtain the Lagrangian

$$\mathcal{L} = \bar{m}c^2 e^{2p}\frac{\partial S}{\partial \tau} + e^{2p}\Big[(\partial_\mu S + r_\mu)(\partial^\mu S + r^\mu)$$

$$+ \hbar\partial_\mu P\partial^\mu P + \frac{1}{4}\partial_\mu r_\nu \partial_\nu r_\mu)(\partial^\mu r^\nu \partial^\nu r^\mu)\Big]$$

$$+ \alpha\partial_\mu S \cdot r^\mu + \mu(r^\mu r_\mu - 1) + \gamma\partial_\mu P r^\mu + \delta\partial^\mu r_\mu - \lambda e^{2p}\bar{m}^2 \cdot S$$

with $S = s(\tau) + S(x_\mu)\rho = \rho(x_\mu) r = r(x_\mu)$ and $\dot{S} = \dot{s} = -mc^2$

An immediate calculation shows that one obtains similar results to the scalar case i.e. the Hubble law.

Indeed, as in the scalar case, one derives from $\delta\mathcal{L}/\delta\rho$ the Hamilton-Jacobi equation: $\partial_\mu S\partial^\mu S + \bar{M}^2c^2 = 0$ with

$$\bar{M}^2c^2 = \hbar^2(\Box P + \partial_\mu P\partial^\mu P) + f_{\mu\nu}f^{\mu\nu} - \bar{m}^2c^2 - \lambda\bar{m}^2\hbar^2\delta(X_\mu)$$

where the photon's four-momenta $p_\mu = \partial_\mu S$ is no longer parallel to their four velocity $\dot{x}_\mu$ in general. From this, one deduces (as in the scalar case) that

$$\partial_\nu(\partial_\mu S\partial^\mu S) = 2\partial^\mu S\partial_\mu\partial_\nu S = -\partial_\nu \bar{M}^2c^2$$

so that denoting by $\mathring{S}$ (and $\mathring{A} = dA/d\tau'$) the time-like world lines derivative along followed by the photon's energy-impulsion (tangent to unitary four vectors w_ν) we get

$$w^\mu\partial_\mu\partial_\nu P = \mathring{P}_\nu = -\partial_\nu\bar{M}c^2 = -(\frac{1}{2}\bar{M})\partial_\nu\bar{M}^2c^2$$

i.e.

$$\frac{dP_\nu}{d?\prime} = \mathring{P}_\nu = -\frac{1}{2\bar{M}}Q - {}_\mu P_\nu$$

if we write $\mu = (1/2)\bar{M}^{-1}.\lambda\hbar^2mc^2$ as factor which characterizes the vacuum's friction coefficient. Since for plane or spherical waves (with distant centers) $\bar{Q} = 0$ we recover in that case the tired-light Hubble type effect

$$\dot{P}_\mu = -\mu P_\nu$$

which corresponds to Hubble's and Tolman's initial assumption.

CONCLUSION

We conclude thus with two remarks:

The first is that the preceding interpretation of the cosmological redshift can be extended to interpret the missing mass problem. Indeed, if the vacuum contains an invisible covariant mass distribution, one can assume that huge mass concentrations in galaxies (and galactic clusters) would provoke a concentration of the vacuum's hidden superfluid elements and of its normal fluid component, thus increasing gravitational attraction (i.e. non velocity redshifts) around them. One could thus interpret the missing mass factor(La Violette, 1986) necessary to interpret:

- the behaviour of galactic clusters

- the variations of Hubble's "constant" mentioned above.

- the strange velocity behaviour of neutral hydrogen in the halos of spiral galaxies (which widely differ from Keplerian rotation laws (the vacuum mass concentrations would exceed the halos' dimensions).

- the fact that the Universe appears (up to $z \sim 0.5$) as nearly Euclidian in La Violette's (1986) analysis.

The second remark is that contrary to one of La Violette's assumptions, the tired-shift mechanism preserves energy-momentum. Indeed, the energy lost by light increases in principle the vacuum's fluctuations and can reappear in terms of pair creation within the vacuum itself (more predominantly) in the neighborhood of huge mass concentrations–a mechanism which fits very well with some assumptions of Hoyle and Narlikar.

Of course, the proposed vacuum gravitational drag mechanism presented here is not incompatible with a weaker form of expansion (which would push back into time the supposed origin of the universe... if it exists) but it justifies a re-examination of Einstein's static model or steady state theories in the light of La Violette's results.

Another exciting aspect of the present situation is the forthcoming use of the Hubble space telescope which will carry galactic observations from $z \sim 0.5$ to 1.5 and allow observations of one supplementary magnitude. This will offer the possibility to extend the range of four tests discussed by La Violette and to see if very distant galaxies and clusters show any sign of evolution. One could check, for example, Zwicky's(1929,35) suggestion that the type composition of clusters of galaxies, their ratio with field galaxies and the rate of creation of supernovae do not change with increasing distance.

REFERENCES

Arp, H. 1987, "Quasars, Redshifts and Controversies", (Interstellar Media, Berkeley).

Bohm, D. & Vigier, J. P. 1954, *Phys. Rev.*, **96**, 208.

Cufaro-Petroni, N. & Vigier J. P. 1982, *Astr. Nach.*,, **303**, 55.

Cufaro-Petroni, N., et al. 1985, *Phys. Rev. D*, **32**, 1375.

Dirac, P. A. M. 1951, *Nature*, **168**, 906. (1951), 906. See also

Ehrenfest, P., Podolski R. and Tolman, R. C. 1931, *Phys. Rev.*, **37**, 602. (EPT)

Giraud, E. 1984, "Galaxies Normales autour du Flot de Hubble" Thesis, Universite des Sciences et Techniques du Languedoc.

Giraud E., Moles, M. and Vigier, J. P. 1982, *C.R. Acad. Sc. Paris*, **294**, Ser. 2, p. 195.

Greenberger, D. 1979a, *J. Math. Phys.*, **20**, 762.

Greenberger, D. 1979b, *J. Math. Phys.*, **20**, 771.

Gueret P. & Vigier, J. P. 1982, *Il Nuovo Cim. Lett.*, **35**, 260.

Hubble E. & Tolman, R. C. 1935, *Ap.J.*, **82**, 302.

Karoji, H. & Nottale, L. 1976, *Nature*, **259**, 259.

Karoji, H., Nottale L. and Vigier, J. P. 1976, *C.R. Acad. Sc. Paris*, **282B**, 103.

LaViolette, P. A. 1986, *Ap.J.*, **301**, 544.

Pecker, J. C. 1976, *Coll. C.N.R.S.*, **263**.

Pecker, J. C. 1977, *C.N.R.S. Ed.*, **451**.

Rauch, H., et al. 1986, *Phys. Rev.*, **34**, 2600.

Schild, A.1953, *Phys. Rev.*, **92**, 1009.

Sinha, K. P., Sudarshan, E. C. G., Vigier, J. P. 1986, *Phys. Lett.*, **114A**, 291. (SSV)

Smolin, l. 1986, *Class. Quantum Grav.*, **3**, 347.

Sulentic, J. W. 1982, *Ap.J.*, **252**, 439.

Zwicky, F. 1929, *Proc. Nat. Acad. Sci.*, **15**, 773.

Zwicky, F. 1935, *Phys. Rev.*, **48**, 802.

DISCUSSION

E. Giraud: Could you repeat why there is no scattering in your new model of tired-light?

J.-P. Vigier: The vacuum's gravitational drag is not a scattering process; 1) If the vacuum element's mass is $m_V \ll m_\gamma$ then their density is huge and their density fluctuations with respect to the cylindrical symmetry of the motion can be made as small as one needs, 2) These fluctuations decrease relatively along the path since the effective mass (very slowly) increases. So there is no cumulative effect.

J.-C. Pecker: How do you reply to Borner's argument, given some minutes ago, according which the Λ value of the vacuum would be much too large?

J.-P. Vigier: The words "too large" apply to the usual view but certainly not to a superfluid "model". Now $\Lambda \geq 0$ implies that gravitational forces become repulsive at a distance (a nice interpretation of the lumping of matter in the universe). If it describes superfluid density it can be quite large and should be evaluated by this lumping.

THE PRESENT STATUS OF THE INFLATIONARY UNIVERSE

G. Boerner
Max-Planck-Institut für Physik und Astrophysik
Institut für Astrophysik
Karl-Schwarzschild-Str. 1
8046 Garching, FRG

1. CLASSICAL COSMOLOGY AND INFLATION

The standard big-bang model has various successes with the interpreation of observations, but it fails to explain several interesting cosmological puzzles: The density parameter Ω in an expanding Friedmann-Lemaitre (FL) model is always rather close to 1

$$\left|\frac{1-\Omega}{\Omega}\right| \sim S^{-2/3} \left(\frac{M_{Pl}}{T}\right)^2 , \qquad (1)$$

since the constant entropy S in a comoving value is

$$S = s_o R_o^3 = 10^{87} . \qquad (2)$$

(s_o is the present entropy density of photons and neutrinos; M_{Pl} is the Planck Mass $\sim 10^{19}$ GeV; T the cosmic temperature). Therefore even for the present epoch ($T \simeq 3K$) Ω is of order 1. At earlier epochs $\Omega \simeq 1$ with high precision, e.g.

$$|1-\Omega^{-1}| \sim 10^{-15} \quad \text{at} \quad T \simeq 1 \text{ MeV} .$$

Within the standard model this property is determined by the constant value for S, which is given as an initial condition.

Taken together with the horizon structure of a FL-model the value of S implies that opposite regions in the sky were causally unconnected until the time of the decoupling of matter and radiation. Then the isotropy of the microwave background is a puzzle if we want to explain it from physical processes - it is a trivial consequence of the initial conditions in the standard model.

The horizon structure also prevents a straightforward amalgamation of particle theories and cosmology: The symmetry breaking of a GUT theory leads to the production of at least one monopole per horizon volume. The monopole density would be so large that the universe would recollapse immediately.

This "monopole catastrophe" was the original motivation for the "inflationary universe" (Guth 1981; Sato 1981). This model appears in many different shapes (Linde 1982; for a review Brandenberger 1985) but the basic feature is a rather simple change in the thermal history of the standard model: One assumes that for a certain time interval the energy density remains large and almost constant, leading to an exponential deSitter expansion. A thermalization stage follows in which the energy density is converted to radiation. The expansion factor increases by a factor Z during this "inflationary expansion", and correspondingly after thermalization the entropy in a comoving volume has increased by a factor Z^3. Thus $Z \sim 10^{29}$ can produce the correct numerical value of eq. (2), and solve the problems of classical cosmology: The whole observable universe is contained within one horizon dimension, and Ω = 1 is a prediction of the model.

The simplicity of this concept is deceptive, however, since it touches on deep and difficult problems of particle physics and general relativity.

2. INPUT FROM PARTICLE PHYSICS

There is no generally accepted model for a unified theory of elementary particles, but quite generally the concept of a spontaneous symmetry breaking of a large gauge symmetry leads to the introduction of self-interacting scalar fields into GUT theories. These "Higgs" fields ϕ have a free energy density (or an effective potential V_{eff}) which has a nonzero "vacuum energy" $V_{eff}(0)-V_{eff}(\sigma)$ which can appear like a constant energy density in the Friedmann equations. $\phi = 0$ corresponds to the symmetric and $\phi = \sigma$ to the "broken symmetry" state of the model.

This effective potential must then have all the properties necessary for the occurrence of an inflationary phase. All the constructions investigated so far have failed to demonstrate that this can be achieved.

i) There are strict limits on any cosmological constant Λ at the present epoch. One must have

$$\frac{\Lambda}{8\pi G} \leqslant \rho_c \sim 10^{-46}\ (\mathrm{GeV})^4\ , \tag{3}$$

and this is much smaller than the vacuum energies connected with electroweak ($\sim 10^8$ (GeV)^4) or QCD ($\sim 10^{-1}$ (GeV)^4) phase transitions. It seems difficult to understand the cosmological constant as a vacuum energy density now, and it is therefore naive to exploit it this way during the first 10^{-35} seconds of the universe.

ii) The model involves the concept of spontaneous symmetry breaking which is one of the difficult questions of particle physics. Numerical studies of Abelian Higgs models on a fixed lattice have yielded mixed results: For compact models the expectation value $\langle\phi\rangle = 0$ everywhere, whereas a transition to $\langle\phi\rangle \neq 0$ was found in the "Landau-Gauge" for non-compact models (e.g. Borgs & Nill 1986).

iii) Several technical problems concerning the transition from $<\phi> = 0$ to $<\phi> = \sigma$ have not been solved. Thus ϕ must be treated as a quantum field, and then neither the equation of motion nor the estimates of the tunneling probability (Callan & Coleman 1977) inspire much confidence.

iv) To predict the amplitude of density fluctuations correctly, the self-interaction must be very small, but then the coupling to massless particles also becomes negligible, and the thermalization phase at the end of the deSitter expansion does not occur (Linde 1985).

3. INPUT FROM GENERAL RELATIVITY

Most models start already in a homogeous and isotropic FL universe at $t < 10^{-35}$ sec. But the inflationary concept is of value only, if it works in more general initial conditions. A few more general cases have been investigated.

i) It is found that in anisotropic and homogeneous models the anisotropy is strongly reduced by an inflationary phase (Rothman & Ellis 1986). Inhomogeneous and anisotropic cosmologies give rise to a stable state $\phi = 0$ if the initial anisotropy is too large, only for reasonably small values does the universe reenter a FL-like stage (Barrow & Turner 1982; Börner & Götz 1987).

ii) There are many choices in deSitter space for a time direction. How then can the choice of a spatially homogeneous time direction be guaranteed during the transition from a vacuum-energy dominated deSitter space to a radiation-dominated FL universe?

References

Barrow, J.D., Turner, M.S. (1982), Nature 292, 35.
Börner, G., Götz, G. (1984), in preparation.
Borgs, C., Nill, F. (1986), Commun. Math. Phys. 104, 349.
Brandenberger, R.H. (1985), Rev. Mod. Phys. 57, 1.
Callan, C., Coleman, S. (1977), Phys. Rev. D16, 1762.
Guth, A.H. (1981), Phys. Rev. D23, 347.
Linde, A.D. (1982), Phys. Lett. 108B, 389.
Linde, A.D. (1985), Comments Astrophys. 16, 229.
Rothman, T., Ellis, G.F.R. (1986), Univ. of Capetown, preprint.
Sato, K. (1981a), M.N.R.A.S. 195, 487; (1981b), Phys. Lett. 99B, 66.

DISCUSSION

S. Bonometto: What is the present status of ideas allowing both inflation and a substantial amount of strings in the present horizon?

G. Börner: Any topological defect (monopolar strings) appearing before inflation is removed from the present observable universe by the exponential expansion phase. To have strings in the present universe one resorts to an additional symmetry breaking at an energy scale $\sim 10^{12}$ GeV (Turok 1986) which is below the scale $\geq 10^{14}$ GeV at which inflation occurs.

J. C. Pecker: What price have we to pay for the reconciliation of inflation with R. G.? —-Incidentally, "bubbles" have been computed in R.G. Friedmann-like models, with $\Lambda > 0$, by Souriau (1975?) in order to represent the Peebles large cell structures; this is completely independent of the "early" universe history.

G. Börner: The present "inflationary model" is just a modification of the simplest solutions of the Einstein field equations. Thus there is no price to pay; except, perhaps, that the initial conditions of the Friedmann-Lemaitre universe models have to be replaced by some complicated particle physics properties.

ALTERNATIVES TO MISSING MASS

Robert H. Sanders
Kapteyn Astronomical Institute, Groningen, The Netherlands

In any system such as a galaxy or cluster of galaxies we can estimate a luminous mass by adding up the total light and assuming a reasonable mass-to-light ratio. There may be some argument about what is a reasonable mass-to-light ratio but, I would say that for a spiral galaxy that would be something like the M/L of the stellar population in our Galaxy in the neighborhood of the sun-- 2 to 4 in solar units. In an elliptical galaxy, which contains a generally older population of stars-- that is with no massive bright young blue stars-- the reasonable mass-to-light ratio might be something more like ten. So while there is clearly a dependence of M/L on stellar population (ie., the distribution of stars by mass) the variation in true M/L for luminous matter (stars with a thermonuclear energy source) might well be no more than a factor of 3 or 4. So having estimated the luminous mass let's call this M_l. Now in some systems (such as spiral galaxies with extended rotation curves) it is also possible to estimate the dynamical mass (M_d) within some characteristic volume. This dynamical estimate employs, of course, the usual laws of Newtonian gravity and dynamics. If $M_d/M_l = 1$ then I would say there is no mass discrepancy. If $M_d/M_l > 1$ then there is a mass discrepancy but given the uncertainties in the estimates, this ratio should be considerably greater than one before we can be convinced that it is real.

Now having made this definition now let me state simply that there is no evidence for significant mass discrepancies within the bright optical disks of spiral galaxies. This assertion, first made convincingly by Kalnajs (1983) has been strongly supported by several recent studies in which the rotation curve calculated from the radial light distribution is compared with the observed rotation curve. For example, Rubin and her collaborators (Rubin 1986) have measured optical rotation curves (that is in visual emission lines) for about 50 spiral galaxies covering a factor of 100 in luminosity and a factor of 10 in size. Kent (1986) obtained visual surface photometry for about 40 of these galaxies and, following Kalnajs, calculated the expected shape of rotation curves assuming that the mass traces the light. Some of his results are shown in Fig 1. We see that the predicted rotation curve agrees often in detail with the observed rotation curves. The mass-to-light ratios required for these fits are again reasonable for spiral galaxies. So from optical rotation curves and optical surface photometry alone there is no evidence for significant mass discrepancies in spiral galaxies.

Fig. 1: Four galaxies from the study by Kent (1986). The points show the rotation curve derived by optical emission line observations and the solid lines are the curves calculated from the mean radial light profile.

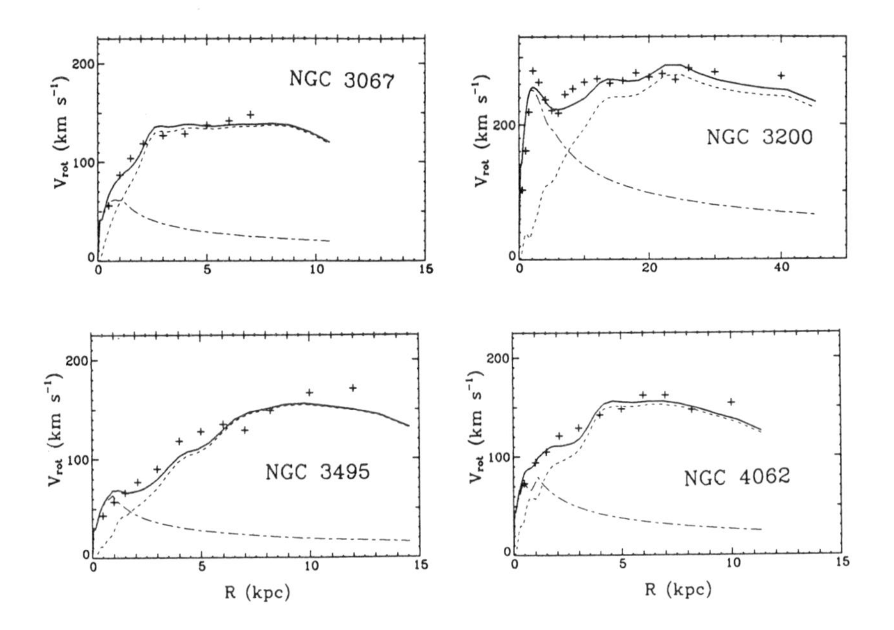

The only evidence for mass discrepancies in spiral galaxies comes from the extended 21 cm line rotation curves of spiral galaxies which extend well beyond the optical disk. There was a strong indication of this already in the early observations of M31 by Roberts and Whitehurst (1975) and in the extensive Westerbork observations of Bosma (1978). But more recent high sensitivity 21 cm line observations have made the case for significant mass discrepancies in spiral galaxies indisputable as we see in Fig. 2. These are two galaxies with a radial light profile determined by Wevers (1984) and with neutral hydrogen rotation curves determined by Begeman (in preparation, 1987).

These rotation curves extend far beyond the optical disk at the brightness level of 25 mag/(arc sec)^2. The solid curves show the rotation curves expected from the radial distribution of light. The indicated mass-to-light ratios (1.9 and 4.0) are chosen so that the calculated curve matches the amplitude of the observed curve in the bright inner regions. Here it is obvious that there is a significant discrepancy. The total mass discrepancy as I defined it above is about four or five (ie. four or five times more dynamical mass than visible mass out to the last measured point). If one chooses the mass distribution in the disk to fit the entire rotation curve then one finds a local mass discrepancy in the outer most measured regions in excess of 1000.

Fig. 2: 21 cm line rotation curves (points) for NGC 2403 and NGC 3198. The solid lines are again the rotation curves predicted from the radial light distribution.

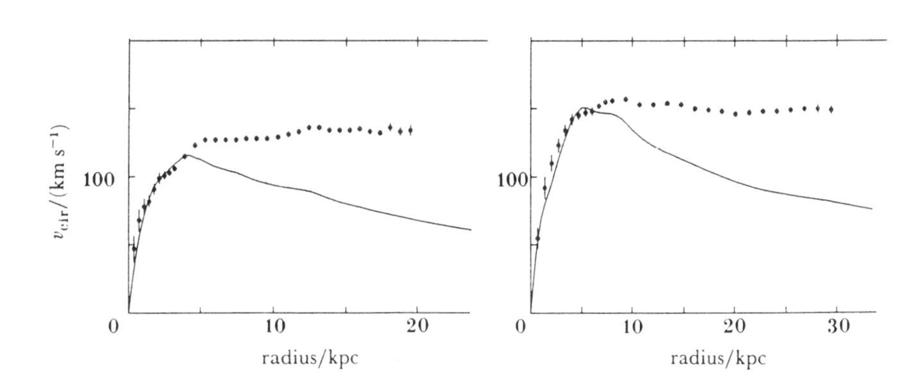

I want to emphasize that there is no way around this-- the determination of the rotation curve and interpretation of the rotation curve as tracer of the radial force is quite unambiguous. The entire two dimensional velocity field of the galaxy is used to derive the rotation curve. Moreover the neutral hydrogen distribution and the velocity field show no asymmetries or distortions which one might associate with an unrelaxed distribution of gas or warps of the gas layer. That is to say, the extended hydrogen distribution is coplanar with the optical disk and there is no ambiguity in correcting for inclination of this layer with respect to our line-of-sight.

These observations of spiral galaxies constitute the most direct and detailed quantitative evidence for mass discrepancies in any gravitationally bound system in the Universe. I don't mean to say that there is not substantial evidence for mass discrepancies in groups and clusters of galaxies. I am just saying that 21 cm line observations of these spiral galaxies is telling us something very precise about the magnitude of the mass discrepancy and its relationship to visible matter.

Another way of saying it is that the motion of the neutral hydrogen in the outer parts of spiral galaxies is now a quite accurate probe of the gravitational potential there. This is quite analagous to the motion of the planets around the sun. It has been known for some time that the planetary motion is a probe of the radial dependence of the gravitational force in the solar system and that turns out to be $1/r^2$ on the scale of 10^{14} cm. And similarly the motion of the gas beyond the optical disk of spiral galaxies is a probe of the gravitational force law on the scale of 10^{22} cm and that turns out to be $1/r$. Now there may be an important difference between these two cases and that is: in the solar system we know that the probe is very extended with respect to the mass distribution-- the source of the gravitational field-- which is the sun. Therefore, the $1/r^2$ dependence is clearly the law of gravity on these scales. However, in spiral galaxies the probe may not be extended with respect to the mass distribution so the $1/r$ dependence could be telling us more about the mass distribution in the outer regions of spiral galaxies and that would be $\rho \sim 1/r^2$. **But this is an assumption.** If we are very conservative and we only believe our eyes and our radio telescopes then the neutral hydrogen-- the probe on scale of galaxies-- is also very extended with respect to the mass distribution which is the visible galaxy. In this case, the $1/r$ dependence would be the law of gravity on the scale of 10^{22} cm. Therefore there are two extreme, and in my opinion, equally reasonable points of view: The first is that extended 21 cm rotation curves are probing the mass distribution in the outer regions of spiral galaxies. Since there is no light from these regions the mass is dark. The second is that the 21 cm rotation curves are probing the law of gravity on the scales and that is more like $1/r$ rather than $1/r^2$. It is the first alternative that has-- for understandable reasons-- received most attention, but I think that we should not ignore the second alternative and its consequence that our present theoretical description of gravity-- general relativity-- is incomplete.

But let us consider for a moment the first alternative, and the implications of the observations of spiral galaxies for dark matter hypotheses. The conventional view is that spiral galaxies lie at the center of an extended halo which is more or less spheroidal and essentially non-luminous. One can then fit the total rotation curve by adding the halo rotation curve (quadratically of course) to the disk rotation curve as is done here by van Albada and Sancisi (1986) for NGC 3198 (Fig.3). Of course there is no unique way in which an observed rotation curve may be decomposed into disk and halo contributions as is shown by these several model fits in which the contribution of the disk becomes progressively smaller, but there are several arguments (given by van Albada and Sancisi) which support the suggestion that the maximum disk solution is most nearly correct.

Fig. 3: Fits of exponential disk and halo to the observed rotation curve for NGC 3198 (van Albada and Sancisi 1986). The upper left figure is the maximum disk model and the subsequent figures show models with the disk contribution reduced to 0.75, 0.50, and 0.25 of the the maximum disk.

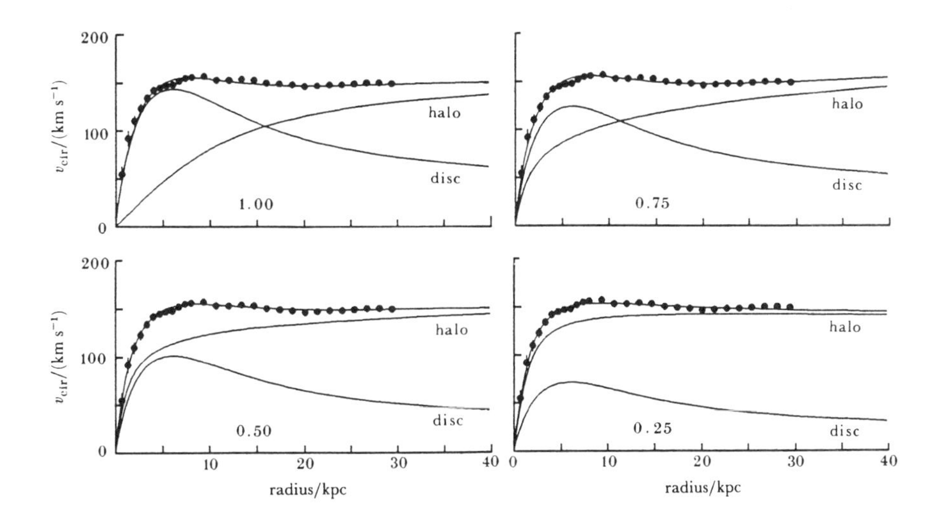

That is to say, the rotation curve in the very inner region is almost entirely dominated by the visible disk and that the ratio of unseen to visible matter within the optical disk is quite small. But the most compelling argument I have already given. That is-- if one had only optical rotation curves and optical surface photometry there would be no reason to believe that there are mass discrepancies in spiral galaxies; in other words, there is no substantial halo contribution to the total mass within the bright optical regions of spiral galaxies; in other words; the decomposition of the rotation curve into disk and halo components is unique in spiral galaxies-- it is the maximum disk solution.

If the maximum disk solution is generally true then it has far reaching consequences with respect to dark matter hypotheses:

1. The almost flat rotation curve extending from the bright inner regions to the outer most measured point is due to a very careful matching of falling disk rotation curve and a

rising halo rotation curve. This is what has been called the "conspiracy" between dark and luminous matter in spiral galaxies. Obviously, if the maximum disk solution did not apply-- if the halo dominated everywhere-- there would be no conspiracy. The flat rotation curve would result from one component only; it takes two to "conspire".

2. The existence of a well-defined luminosity-maximum rotation velocity relationship (Tully-Fisher law) implies that it is the amount or mass of luminous matter that sets the maximum rotation velocity. Combined with the fact that the rotation curve is flat beyond the optical disk, it would appear then that the asymptotic circular velocity in the halo is also set by the amount of luminous matter. In other words a little bit of visible matter down in the middle of this giant extended dark thing somehow determines the radial force law in the dark thing. It would seem as though the tail is wagging the dog.

Now the real meaning of the conspiracy is that in every spiral galaxy the halo parameters must be somehow fine tuned to the particular disk-- and this fine tuning must occur in galaxies differing by factors of 10 in size and 100 in luminosity. Of course we don't want to fine tune the halo in every single galaxy separately-- so the solution of the conspiracy problem in the context of dark matter must be found in some very general aspect of galaxy formation. My only point here is that these detailed observations of the extended 21 cm line rotation curves severely constrain dark matter theories or perhaps theories of galaxy formation in the context of dark matter.

A more natural and economical solution to these problems of mass discrepancies in spiral galaxies and in the Universe in general might be found in an enlargement of the theory of gravity. In the recent literature there have been several papers suggesting naive or ad hoc modifications of Newton's laws of gravity or dynamics (Milgrom 1983, Tohine 1983, Sanders 1984, Kuhn and Krugylac 1987). Those suggestions in which the departure from the $1/r^2$ law occurs beyond a critical length scale are already ruled out by the observations. Comparable mass discrepancies appear in galaxies ranging in size from 8 to 80 kpc, whereas this sort of modification would predict that larger galaxies should exhibit larger discrepancies. Moreover, these suggestions unavoidably lead to a Tully-Fisher law of the form $L \propto v^2$, whereas the observed relationship is $L \propto v^4$.

More successful is the suggestion by Milgrom where the departure from the inverse square law occurs below a critical acceleration. In original form this was a modification of Newton's second law of dynamics which was written by Milgrom as

$$F = ma\mu(a/a_o) \tag{1}$$

where

$$\mu(x) = 1, \; x \gg 1$$
$$(2)$$
$$\mu(x) = x, \; x \ll 1$$

and a_o is a new constant of nature with units of acceleration and a magnitude comparable to the gravitational acceleration in the outer parts of spiral galaxies. It is obvious that the rotation curve of a galaxy should be asymptotically flat :

$$V_a{}^4 = Gma_o \quad (3)$$

and that the Tully-Fisher law is $L \propto v_a{}^4$, where v_a is the asypmtotic velocity. This modification however does not in an obvious way solve the conspiracy problem; ie., it does not explain why the rotation curves of galaxies are flat even into the inner regions where the appropriate law of gravity is Newton's. This requires that average surface density of galaxies should be constant and about equal to a_o/G. For example, if the mean surface density is five times larger than this value the rotation curve decreases from its peak value in the disk by almost 50% to the asymptotic value at large radii. It is an observational fact that the mean surface density in spiral (and elliptical) galaxies is about constant (Freeman 1970) but a modification which attempts to explain flat rotation curves should account for that fact and not rely upon it.

A more useful approach might be to consider at the outset modifications not of Newton's law (which we know is only a descriptive empirical relationship) but of the current theory of gravity which is general relativity (Bekenstein and Milgrom 1984, Sanders 1986). There have been many attempts to modify general relativity but none of these attempts have been experimentally motivated. That is because there are no local experimental contradictions to general relativity. But what is being suggested here is that contradictions do arise in the limit of of large distances from mass concentrations or very weak fields where accurate astronomical observations now probe for the first time.

I would like to sketch for you a kind of toy relativistic theory which might work. It relies upon two concepts which have been around for some time: a cosmological constant λ and an additional long range force which is associated with a scalar field ϕ . It is well-known that a cosmological constant can solve the cosmic missing mass problem. Natural physical processes in the very early evolution of the universe might force the universe to be flat or very nearly flat now, which implies that, in the absence of a cosmological constant, Ω_o =1. But the observed density of luminous matter would imply something more like $\Omega_o \sim 0.003$ and the standard model for nucleosynthesis of the light elements implies that $\Omega_o \sim 0.1$ at least for baryonic matter. Thus the need arises for ten times more non-baryonic matter. But, of course if there is a cosmological constant then, in a flat universe

$$\Omega_o = 1 - \lambda/3H_o{}^2 \tag{4}$$

and can be much less than one if $\lambda \sim 3H_o^2$. The point I want to make now is that a cosmological constant can also account for mass discrepancies in galaxies if it arises as the potential energy associated with a scalar field.

The idea of a scalar field is not new and was the essential element of the modified theory of gravity suggested more than twenty years ago by Brans and Dicke (1961). In such a theory the field action is written

$$S_f = -\int (-g)^{\frac{1}{2}}[R - \phi_{,\alpha}\phi^{,\alpha}]d^4x \tag{5}$$

where where the symbols and conventions have their usual meaning. The particle action is

$$S_p = -mc\int (-\Psi^2(\phi)\frac{dx^\mu}{d\rho}\ \frac{dx^\nu}{d\rho}\ g_{\mu\nu})^{1/2}d\rho \tag{6}$$

where Ψ is an arbitrary function of ϕ and expresses the joint coupling of the scalar field with gravity to matter. In Brans-Dicke theory this function has the form

$$\Psi \ \propto \ \exp(\frac{-\sqrt{\alpha}}{2}\ \phi) \tag{7}$$

(see Wagoner 1970) and the field equation for the scalar field becomes

$$\Box\phi = \ 4\pi\sqrt{\alpha}GTc^{-4} \tag{8}$$

The parameter α might be referred to as the "scalar coupling constant. In the formalism of Brans-Dicke this scalar coupling is expressed in terms a parameter ω which is related to α as

$$\omega = \ \frac{1}{\alpha} - \frac{3}{2} \tag{9}$$

When the theory is written in this way the scalar field appears explicitly in the equation of motion for a particle

$$\frac{dP_i}{d\tau} = \ 1/2\Psi^2u^ju^hg_{jk,i} + c^2\frac{\sqrt{\alpha}}{2}\phi_{,i} \tag{10}$$

where P_i is the 4-momentum of a particle.

Therefore in the weak field limit this can be viewed as a two-field theory of gravity where the usual gravitational potential is

$$\Phi = \Phi_1 + \Phi_2 \tag{11}$$

and

$$\Phi_1 = 1/2g_{00}$$
$$\Phi_2 = \frac{\sqrt{\alpha}}{2}\phi \tag{12}$$

Brans-Dicke theory has fallen out of favor in recent years because solar system experiments and the observed decay rate of the binary pulsar restrict ω to be greater than 100. Thus, the scalar force, if it exists is 100 times weaker than gravity and not of much dynamical importance.

But recently both the cosmological constant and the scalar field have re-emerged united as an aspect of quantum field theory in the early universe-- in particular, in inflationary scenarios where φ is a Higgs field associated with the breaking of some symmetry (Brandenberger 1985). In this case the action is written

$$S_f = -\int(-g)^{\frac{1}{2}}[R - \frac{1}{2}\phi_{,\alpha}\phi^{,\alpha} + V(\phi)]d^4x \tag{13}$$

where V(φ) is the energy density of the vacuum corresponding to some value of φ . Such potentials are characterized by a non-vanishing value of V at φ =0 and a minimum value at some $\varphi \neq 0$.

Fig. 4: The assumed potential and derivative (expanded scale) as a function of

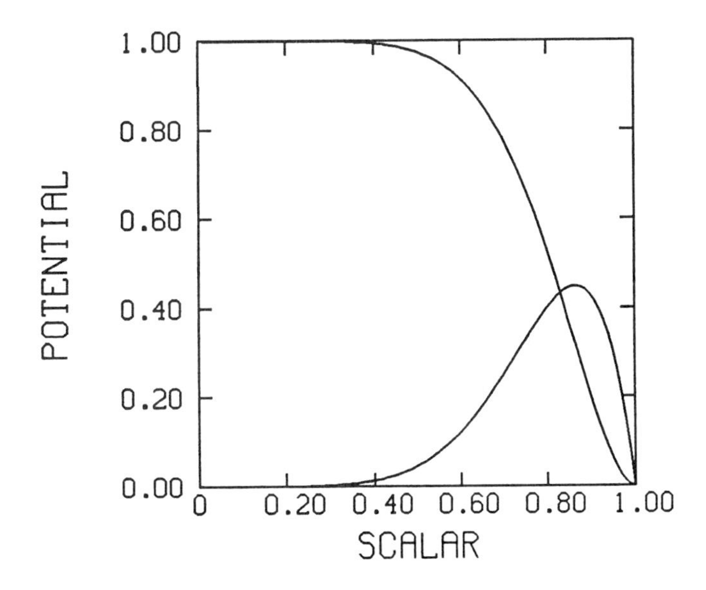

Now for the sake of our toy let us assume that there exists such a scalar field in the present universe. For simplicity let us also assume that the potential has the form

$$V(\phi) = n\lambda\left(\frac{\phi}{\sigma}\right)^n\left(\frac{ln\phi}{\sigma} - \frac{1}{n}\right) + \lambda\,, \quad \phi < \sigma$$

$$V(\phi) = 0\,, \quad \phi > \sigma \tag{14}$$

$$V(\phi) = \lambda\,, \quad \phi < 0$$

so it is a two energy level universe. This is shown in Fig. 4 for the case where n=8.

And finally we assume that the scalar field couples to matter exactly as in Brans-Dicke theory. In the static limit the the field equation for the scalar is

$$\nabla^2\phi = \frac{-4\pi\sqrt{\alpha}G\rho}{c^2} + V'(\phi) \tag{15}$$

Written in this way a positive mass density contributes to a positive scalar field.

Now consider a mass concentration or "galaxy" in an empty static universe. We expect $\varphi \to 0$ at $r \to \infty$ but we may choose σ such that $\varphi \to \sigma$ in the interior of the mass concentration. So empty space has a positive energy density, but inside a substantial mass concentration the energy density of space is lower. The mass concentration polarizes the vacuum. Look again at the field equation for φ . The potential gradient is negative and thus enters with the same sign as the density term; that is, it acts as a source for the scalar field. V' is an effective "halo" for the mass concentration which means that the scalar force can fall off less rapidly than $1/r^2$ and eventually dominate over gravity. I show you one example where the mass concentration is a sphere with a mass of 10^{11} $M_\odot$ and a density distribution like that of a galaxy. The value of V(0) is chosen to be $3H_0^2$ and α =0.004 which corresponds to ω =250 so the presence of the scalar field does not conflict with any local test of gravity. I have also chosen σ such that the bottom of the well is reached near the center of the "galaxy".

The rotation curve of the galaxy is shown in Fig. 5 along with pure gravity rotation curve. We see that such a scalar force, even though it is much weaker than gravity locally can dominate on large scales. This is because of the large V' source for the scalar field-- the polarization of the vacuum.

With respect to this idea, it is interesting to look at a rather striking numerical coincidence. If there were a cosmological constant with a value of roughly $3H_0^2$ then the energy density of space would be about 10^{-9} ergs/cm^3. I mentioned before that the mean surface density of galaxies is roughly constant.

$$\sum \simeq M/R^2 \simeq const.$$

This also means that the gravitational potential energy density in galaxies

$$GM^2/R^4 \simeq const.$$

is roughly constant and equal to about 2 x 10^{-10} ergs/cm^3. It is perhaps dangerous to make too much of such a coincidence but it does seem in a vague way to be consistent with the kind of picture I have sketched above: the energy density of empty space is higher than the energy density in mass concentrations; part of the difference has reappeared as gravitational energy density of the mass concentration.

Fig. 5: The rotation curve of a spherical galaxy in which a scalar field has been added. The field is characterized by a Brans-Dicke = 250 and a potential of the form shown in Fig. 4. The dashed curve shows the rotation curve resulting from gravity alone.

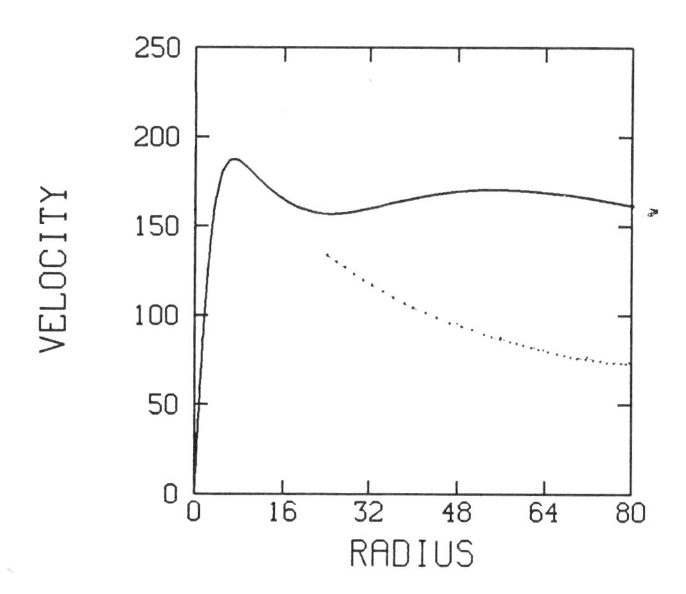

I stress that this is a toy theory. I simply use it to demonstrate that the concept of a cosmological constant arising as the vacuum energy density associated with a scalar field can not only solve the cosmic missing mass problem, but also the galaxy missing mass problem. The theory as it stands cannot account for the systematics of mass discrepancies in galaxies: it produces roughly the same effective "halo" for every galaxy independent of the galaxy mass. although a potential which is also a function of the curvature scalar might solve that problem.

Bekenstein and Milgrom (1984) have previously proposed a scalar-tensor theory with a potential energy term, but in their case the potential is a function of a second dimensionless variable which is

$$\chi = c^4 \frac{\phi_{,\alpha}\phi^{,\alpha}}{a_o{}^2} \tag{16}$$

where a_o is the Milgrom acceleration parameter. Choosing properly the form of the potential in the limits where x is much greater than or much less than 1 it is possible to reproduce in the weak field limit the phenomenology of Milgrom's modified dynamics. But, as Bekenstein and Milgrom point out, this kind of non-quadratic Lagrangian theory inevitably leads to acausal propagation (faster than light) propagation of scalar waves and should probably be disregarded.

It is fair to say that we still don't have an acceptable relativistic theory of stronger gravity on extragalactic scales but any such theory will probably involve a field in addition to the usual tensor field of general relativity where the effects of this additional field are somehow locally supressed.

Let me just conclude by making a few general remarks. The advantage of such ideas is that they carry considerable predictive power and are eminently falsifyable. If there is no dark matter then we must be able to explain the rotation curve of every galaxy (without obvious distortions in its velocity field) by the observed distribution of visible mass. If one is proposing that the extended flat rotation curves are a result of physical law then one well-established counter-example is sufficient to destroy the hypothesis. For example a galaxy with a decreasing rotation curve beyond the optical disk (decreasing according to the old Kepler law) would rule out these suggestions. The dark matter hypothesis is considerably more flexible. Some may consider this flexibility an advantage-- I do not.

When it comes down to dark matter vs. new physics I think that it is again instructive to look at the solar system. The outstanding success of the dark matter hypothesis in terms of Newtonian gravity was the discovery of the planet Neptune in the last century. Very precise astronomical observations revealed irregularities in the motion of Uranus. This lead Adams in England and LeVerrier in France to independently predict the existence of an outer planet perturbing the motion of Uranus. They not only predicted its existence but also rather precisely described its orbit-- and lo and behold it was found. The case of Mercury was quite different however. LeVerrier thought that the anomalous precession of the orbit of Mercury could be explained by small planets lying between Mercury and the sun-- again dark matter-- but none were found. Newcomb proposed that small particles -- the same that cause the zodiacal light-- might be responsible and Seeliger produced what was regarded as an acceptable astrophysical model. But after 1916 this model seemed terribly contrived and artificial-- because as we now know-- in the limit of very strong gravitational fields new effects appear which must be described by a more complete theory of gravity. But there is another limit and that is the limit of very weak fields where accurate astronomical observations now explore for the first time. So perhaps we should not be too surprised to find new physical effects in this limit as well-- new effects which might again require an enlargement of our theory of gravity.

References.

Albada, T.S. van & Sancisi, R. 1986, **Phil. Trans. R. Soc. Lond.**, 320, 447.
Bosma, A. 1978, Ph.D. thesis, Univ. of Groningen.
Begeman, K.G. 1987, Ph.D. thesis Univ. of Groningen.
Bekenstein J.D. & Milgrom, M. 1984, **Astrphys. J.**, 286, 7.
Brandenberger, R.H. 1985, **Rev. Mod. Phys.**, 57, 1.
Brans C. & Dicke, R.H. 1961, **Phys. Rev.**, 124, 925.
Freeman, K.C. 1970, **Astrophys. J.**, 160, 811.
Kalnajs, A.J. 1983, in **Internal Kinematics and Dynamics of Galaxies IAU Symp.No.100**, ed. E. Athanassoula, Reidel, Dordrecht.
Kent, S.M. 1986, **Astron. J.**, 91, 1301.
Kuhn, J.R. & Krugylac, L. 1987, **Astrophys. J.**, 313, 1.
Milgrom, M. 1983, **Astrophys. J.**, 270, 365.
Roberts, M.S. & Whitehurst, R.N. 1975, **Astrophys. J.**, 201, 327.
Rubin, V.C. 1986, in **Dark Matter in the Universe, IAU Symp.No.117**, eds. G. Knapp & J. Kormendy, p.51, Reidel, Dordrecht.
Sanders, R.H. 1984, **Astron. Astrophys.**, 136, L21.
Sanders, R.H. 1986, **Mon. Not. R.A.S.**, 223, 539.
Tohline, J.E. 1983, in **Internal Kinematics and Dynamics of Galaxies, IAU Symp.No.100**, ed. E. Athanassoula, Reidel, Dordrecht.
Wagoner, R.V. 1970, **Phys. Rev. D.**, 1, 12.
Wevers, B.M.H.R. 1984, Ph.D. thesis, Univ. of Groningen.

DISCUSSION

P. Shaver: The maximum disk solution is important in your argument, but you haven't fully justified it in your talk. Would you care to elaborate on it?

R. Sanders: There are essentially four arguments in favor of the maximum disk: 1) As is seen in the work of Kent, small details in the observed rotation curve are reproduced by the light/mass distribution in the disk. We would not expect this to be the case if the gravitational field were "diluted" by a substantial halo contribution. 2) The mass-to-light ratios for the maximum disk solution are not unreasonable (2-6 in solar units). 3) The existence of a well defined luminosity-velocity relation (Tully-Fisher) suggests that at least there is not a large variation in the ratio of dark-to-visible matter in spiral galaxies covering a large range of luminosity. 4) The presence of well developed two-arm spiral structure seems to require that a fairly large fraction of mass inside the visible galaxy should be in the active disk (Athanassoula *et al*, 1985, IAU Symp. 117).

R. Kraft: There is one case in which, over a very small distance scale (~100 pc), there is a discrepancy between mass determined from M/L-ratio and mass determined from velocity dispersion, and that is in the Draco Dwarf Galaxy (according to Aaronson). How would you deal with this, if it is indeed true?

R. Sanders: If this is true, then it provides very strong support for the kind of modification suggested by Milgrom where the appearance of mass discrepancies does not occur beyond some critical length scale but in the limit of low accelerations (or low field gradient). In general, this kind of modification predicts large mass discrepancies in objects with very low surface density of light.

J.-C. Pecker: What reply do you give to the people (such as Rees and others), who interpret the steep velocity gradient at the center of galaxies (including our own), as requiring the existence of a massive ($\sim 10^7 M_\odot$) black hole at the center of the galaxy?

R. Sanders: The rapid rise in the rotation curve in the central regions of many spiral galaxies can be easily accounted for by the presence of a centrally condensed bulge component. This is true in our own galaxy where the mass distribution estimated from the near-infrared photometry is consistent with the observed rotation curve. There may or may not be massive black holes in the centers of spiral galaxies but there is certainly no indication of a significant mass discrepancy there.

W. Tifft: Rotation curves extend into the radial domain where pairs of galaxies yield quantized (constant) differentials. I think it would be inconsistent if curves were *not* flat and suggest that the same quantum phenomenon for pairs contributes to extended rotation dynamics. I agree that a modification of gravity on the large scale is required. The *continuity* from spiral arms → companions on arms → companions is an important connection. Over this range we pass from continuity to quantization.

W. Saslaw: How would these models modify the two-point correlation function of galaxies?

R. Sanders: I expect that any modification of Newton's law which does not return to $1/r^2$ beyond some distance (or acceleration) would not be consistent with the observed spatial correlation of galaxies if the correlation develops from hierarchical clustering. This is one reason why I prefer a modification which does return to $1/r^2$ on the largest scales (but with a large effective constant of gravity). Of course, it is not easy to answer this question because cosmic N-body simulations are problematic if the force law is something like $1/r$ where mass outside the grid has an effect.

J.-P. Vigier: Can you account for your curves with possible very light (non-zero mass) particles (neutrinos photons etc.)? What would be the constraints on their distribution?

R. Sanders: Of course most people are trying to explain flat rotation curves by halos consisting (usually) of non-baryonic particles, and at every meeting on this subject we hear the usual recitation of the list of particle dark matter candidates– all, as yet, undiscovered. The point I have tried to make here is that such dark matter hypotheses must, in a natural way, account for the observed "conspiracy" and the implied coupling between dark and visible matter. In my opinion the dark matter scenario is becoming increasingly an intricate system of "crystal spheres" which will some day collapse under its own weight.

DIFFICULTIES OF STANDARD COSMOLOGIES

J.-C. Pecker
Collège de France, Paris

Abstract

The standard cosmologies do account well for some classical observations; but it is clear that other cosmologies satisfy them as well, or better, at the expense of some ad-hoc constructions. On the other side, ad-hoc constructions must indeed be also introduced to improve the standard cosmologies, in such a way that they satisfy some tests, for which non-standard cosmologies are better suited to the representation of observations.

The statistical studies of quasars lead to another class of difficulties, which may lead to new physics as a need to explain non-cosmological redshifts. How to solve them within the reference frame of the standard cosmologies ? One hardly conceives an ad-hoc construction in this case; the only way to save the standard cosmology is to deny any value to these observations.

Other basic difficulties are permanent : the inhomogeneity of the universe, its hierarchized structure, the principles of thermodynamics, the unknown physics of the first 10^{-43} s - all these problems are so far from being solved and cast a doubt on every cosmological attitude. For the time being, one should be very cautious, and not exclude any new way of thinking, providing that he is alert and critical.

Introduction

Present day cosmologies have many necessary ingredients.

First, they must satisfy the law of physics, i.e. account for all the facts observed in the terrestrial laboratory, or even in the solar system. Hence, they must satisfy the equations of General Relativity, the modern quantum physics, and all the by-products of these basic theories. But of course, they may very well by-pass these laws, in the way that the classical physics must be verified locally, but may be inadequate to describe some properties of the universe at large. More general laws might be needed, of which the laws of physics (some of them at least) need only to be a first approximation, in the same sense that both laws of newtonian gravitation and special relativity are only the first approximations of General Relativity.

After this first requirement, cosmological constructions must account in principle for all astrophysical observations. Most of them are indeed accountable for by classical physics, and there is no need to worry about them if the first requirement above is fulfilled. But some astrophysical observations or facts are well-observed facts and cannot be explained by other considerations; these facts will be called "cosmological facts", or "facts of cosmological importance". And, of course, all facts are not given the same status in this respect by all observers. I shall treat this problem - the subjectivity of the choice of the cosmological facts - at the meeting organized this week by the Istituto Gramsci.

I would like now to limit myself to the question: how well do the existing cosmologies, assumed to satisfy the physical laws, satisfy the astrophysical observations ?

The cosmological facts are indeed of a different nature.

One group of them is linked with the very nature of the redshift, whatever object is redshifted - galaxy, or quasars (we shall use, for simplicity, the word quasar to designate all kinds of active galaxies). The possibility of "no expansion" has to be considered in their light.

A second group is associated, assuming or not the cosmological redshift of galaxies to be due to expansion, with the possibility for quasars to have a redshift of some intrinsic nature ("abnormal" or "intrinsic" redshifts).

Assuming finally that both quasars and galaxies are accounted for by an expanding universe - that of the standard cosmologies - how well do the various cosmologies, standard or not, respond to the observed cosmological facts?

Taking these different groups of facts, we shall have to choose between three different attitudes, as follows:

a) All redshifts - those of galaxies, and those of quasars - are "normal"; the Big Bang universe is quite valid.

b) Galaxies are described by the classical Big Bang universe; but there are abnormal redshifts, displaying a new interesting physics at work in active galaxies and quasars.

c) No redshift is "normal" at 100%; at least part of any redshift is due to other causes than the recession one. Hence, all redshifts <u>may be</u> completely accounted for, without any expansion, in a static universe.

1 <u>The classical tests</u>

The standard cosmologies are quite happy with the classical tests. But all of these tests are just as satisfactory for the non-standard cosmologies.

<u>a) The Hubble's m-z diagramm for galaxies</u>

One can easily see the appearance of this much used diagramm for each of the suggested (standard or not) models of the Universe. To make only a small selection (figure 1):

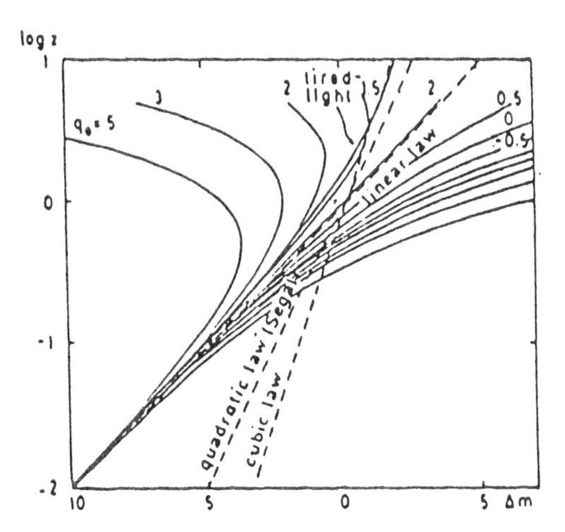

The log z-m diagramm. Thin lines: Friedmann's models for different values of q_o. Thick lines: tired-light models. Dotted lines: power laws.

FIGURE 1

The Friedmann's models are described in these diagramms by the relation :

(1) $\underline{m}_1 = \underline{K}_1 + 5 \log \underline{z} + 1.086\ (1-\underline{q}_o)\ \underline{z}$

The tired-light mechanisms lead to :

(2) $\underline{m}_2 = \underline{K}_2 + 5 \log (\ln(1+\underline{z}))$

The discrimination is possible only for $\underline{z} > 1$.

The Segal's chronogeometry (Segal, 1975) corresponds to :

(3) $\underline{m}_3 = \underline{K}_3 + 2.5 \log (2\underline{z}/(1+\underline{z})) - 2.5(1-\underline{a}) \log ((1+\underline{z})/2)$

Of course, various conceptual models such as steady-state continuous creation models of Bondi, Gold, Hoyle, or the Hoyle-Narlikar models, or the Canuto's model, or some others - should also be compared with data. Our simplified discussion should by no mean, be taken as an ostracism !

The comparison of data with theoretical relations imposes to correct for the $\underline{K}$-term either the observations or the theory. But the correction is important when $\underline{z}$ is large (figure 2).

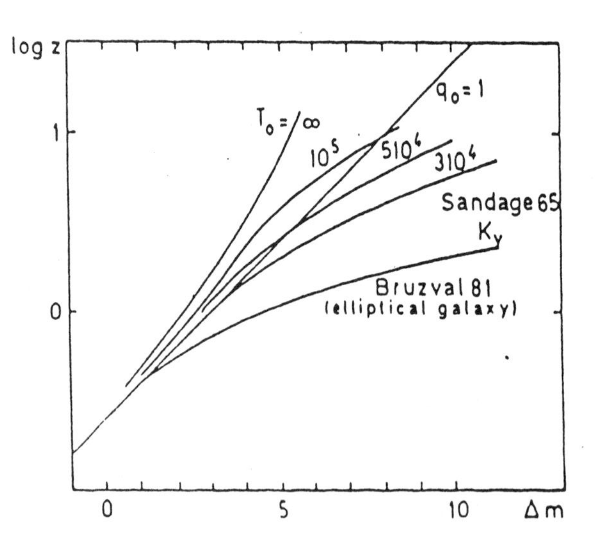

The K-correction according Sandage (1965)(for blackbodies of different temperatures T_o) and Bruzual (1981) (for an elliptical galaxy), as applied to Friedmann's model for $q_o = 1$.

FIGURE 2

A recent comparison between data and models has been done by LaViolette (1986). The result of this comparison pleads against $q_o = 0$, $\Lambda = 0$ Friedmann-type models, and for the tired-light models. But it seems to us not advisable to derive such an extreme conclusion. On one side, one should not combine different sets of data, as does LaViolette. On the other side, it is clear that one could push a little bit the comparison by stating that data might very well fit quite well Friedmann-type models built with $\Lambda = 0$, and with a value of q_o adequately selected. Truly enough, this might lead to an exaggerate value for the actual density ρ_o; but one could also play on the value of Λ in an almost free way. It should be quoted also that similar comparisons, made by Segal & Nicoll, plead, in their opinion, for chronogeometry.

Actually, it is difficult to apply properly that test: the samples should be very homogeneous; the small z objects do not allow much discrimination; the high z objects may be of a different nature, hence not suitable for inclusion in an homogeneous sampling.

That such tests generally lead to the apparent need for evolution is indeed a proof that, by themselves, the tests are inoperant.

b) The angular size-redshift relation

Tests had been performed previously using the extension of double radiosources, but they are strongly affected by the inclination of the axis. In spite of this, we had earlier concluded (1976) that the comparison was in favour of tired-light mechanisms; still the dynamical evolution of the double radiosources was a supplementary source of ambiguity; and the comparison was thus not really convincing.

LaViolette's analysis uses instead the large sample of 94 points (according Hickson, 1977) describing the size of galaxy clusters. The test then favours clearly the tired-light model. The Hickson's discussion shows that values of $q_o \neq 0$ or $\Lambda \neq 0$ do not fit better the data, unless one allows ρ_o to take negative values. Hickson's conclusion requests for evolution (collapse of clusters).

Again, the need of introducing evolution, at relatively small z, emphasizes the fact that the tests are not positive - except for the tired-light mechanisms!

c) The counts of galaxies

Once again this well-known test shows that tired-light models are to be preferred to the classical Friedmann-type models. It shows also that an euclidian static model, without shifts, cannot work at all.

Many other attempts have done, since the early days of extragalactic radioastronomy; they led to historical disputes; one aspect of the debate is that the counts, at larger flux densities (nearby sources) are affected by a strange depletion. Could this effect be associated with a certain type of evolution, a smaller number of sources being formed now than at earlier times?

2 The 2.7 K background radiation

One shall not here remind the reader of the story of the background radiation: predicted by Gamow, Alpher and Herman, discovered 10 years later by Penzias & Wilson, the discovery led then to a triumphal adoption of the Big Bang (the Big Bang "mythology", according Alfvén !) by all those who completely forgot that the same prediction was made, at the same time as Gamow's papers, by Finlay-Freundlich, with the approval of Max Born, on the basis of tired-light theory. For sure, Finlay-Freundlich did perform an incorrect computation. But more recently, it has been shown that the 2.7 K background can be made compatible with several of the many cosmologies on the market (see for example, Pecker, 1976, Canuto, Hsieh, 1977, Canuto 1978). It has to be said that this fact has been criticized on the basis that the blackbody distribution would result from the result of a good chance (Puget, Schatzman, 1974). But this argument is not really valid, as no complete calculations have been performed.

3 The abundances of light elements

In our opinion, this is one of the best arguments ever given as favouring the standard cosmologies.

Truly enough, the Big Bang cosmologies explain, at least within an order of magnitude, the mutual abundances of ^1H, ^2D, ^3He, ^4He, ^7Li. One can find, in other terms, a set of values of the parameters q_o, ρ_o, and H_o which accounts for the observed abundances. This set of parameters has been somewhat criticized; but we shall not take it too seriously, taking into account the fact that no other cosmology accounts for the data, except through some ad-hoc invention, generally quite independant from the choice of the cosmological model.

One such ad hoc construction is to assume that He has not been built in galaxies, but that, at the beginning of their life, their center was a strong source of radiation, able to expell, by the effect of radiation pressure, the hydrogen atoms - but not the helium atoms, as the lines at 304 and 584 A would be used, not to push away helium, but to ionize hydrogen. Another idea is to give to the galactic medium a much longer life time than the so-called Hubble time. These mechanisms are obviously ad hoc; and they have not been quantitatively described; they should at least deserve some attention; after all, they do not request any new physics.

4 The age of globular clusters

It is often argued, in favour of the classical models, that the age of globular clusters is of the same order as the so-called age of the universe, implying that quickly after the decoupling of matter and radiation, clusters are formed, even before the galaxies take their flattened shape. Actually, this is quite questionable.

Even without subtle evolution theories, the simple reasoning reaches an age of 18 to 20 10^9 years (G. Cayrel), furthermore the "age of the Universe" especially when one admits a value larger than 50 for the Hubble constant (in km s^{-1} Mpc^{-1}). It means that, only for that, we need $\Lambda \neq 0$.

But, this age is possibly still an underestimation: the turbulent diffusion, postulated by Schatzman and Maeder, implies a slower evolution rate during the main-sequence phase of the life of any stars; it enlarges by an unknown, but possibly non trivial, quantity the age of clusters as determined from the H.R. diagramm.

We must also face the abundances anomalies presented by various stars in some globular clusters, notably ω Cen. It may show that clusters have formed in an already inhomogeneous medium, strongly suggesting that first generation stars have occured before the formation of globular clusters, and that the galaxy itself might have taken some time to evolve before it. This seems to us difficult to reconcile with the classical universe without some new "epicycles" !

5 Abnormal redshifts of all types

It is quite clear that, at a symposium organized at the occasion of Chip Arp's birthday, the cosmological importance of the so-called "abnormal redshifts" would take an important part in this discussion.

But let us here be very clear. Once abnormal redshifts are accepted as real, in any given situation, they bring the problem of their origin, not linked with expansion; but in no way do they necessarily cast doubt on any cosmological description. If they do so, it may be through an abuse of language; abnormal redshifts possibly contribute to reinforce the idea that expansion does not explain everything, but they do not rule out expansion. The discussion of the second part of this paper does not affect therefore the standard cosmology. But it gives to the theoreticians of non-standard cosmologies a new physical effect not taken into account by standard cosmologies, hence some new strength; the prospect of this strengthening seems to be the real reason for the tenants of standard cosmologies to be more than reluctant to the very idea of abnormal redshifts.

For that indirect reason, of a purely psychological nature, the abnormal redshifts need to be taken very seriously. As they have been, or will be often mentionned during these few days, I shall limit myself to some reminders.

a) Abnormal effects in the solar system

Relativistic effects were not known at the beginning of this century; some new ideas allowed their prediction; the actual measurements have been used as a strong proof for the new ideas - those of General Relativity. Whenever GR found in the laboratory many verifications, the relativistic effects in the solar system have been measured and measured again. But they seem to disagree slightly with the GR predictions (Mérat et al., Depaquit et al. 1974, 1975). It is the case for the redshifts of radiosources during eclipses by the Sun; for the limb-shift near the solar limb; for the angular displacement of objects situated behind the Sun, outside its limb, and close to it. The implication that a new theory is needed has been criticized on the basis that these effects are explained quite naturally or even that they are mere artefacts; but the actual computations showing that it could be indeed the case, has not been actually performed. On the contrary, the tired-light mechanism seems to account for the three effects, provided the solar environment is the location of some additional cause of the "tiring" of the light. It is difficult to conclude; but some attention and much care should still be given to the experimental aspects of such suggestions.

b) Effects of intervening clusters of galaxies

It has been shown, notably by Karoji and Nottale (1976), that the galaxies located behind a cluster of galaxies were affected by an additional redshift of about 2000 km s^{-1} in the more definite cases (see Jaakkola et al. 1976).

Is this redshift really "abnormal"? Nottale has thought for a while that it could be explained in a purely relativistic way; but, as far as I know, it failed to be quantitatively convincing.

c) Type-effect in clusters

This effect is well-known by now; and, so far as I am concerned, no one has been able to interpret it in a "normal" way. Let us quote the work by Molès and Nottale (1981), by Giraud (1984) (figure 3), and of course by Arp (1987).

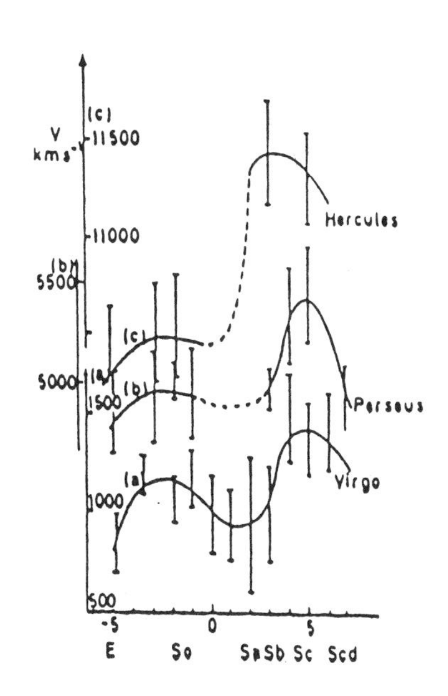

In ordinates, the velocity measured from redshift, for galaxies of different types (in abscissae) belonging to three different clusters (after Giraud, 1984).

FIGURE 3

d) Type-effects in small groups

In groups where a bright elliptical galaxy is accompanied by a number of small objects, the systematic difference in redshifts between the bright object and its companions has been noted several times (Arp, 1976; Bottinelli & Gouguenheim, 1973; Collin & al., 1974). The recent results by Arp (1987) are in this respect quite convincing.

The case of binaries is particularly clear, as shown by Arp et al. (1983), and Giraud and Vigier (1983) in the cases of Karatchensev's pairs.

In all cases, the observations are conform to the data of figure 3.

e) The quantification of redshifts

The discovery by Tifft (1976), its confirmation, by Arp and others, of a quantum redshift of 72 km s^{-1} is obviously a very disturbing fact (see earlier sessions of this meeting).

f) Are the quasars abnormally redshifted?

The question is of an enormous importance. First one, it would give to the theoreticians a strong case of abnormality. Second one, quasars have the largest z of all observed objects; hence, they have been widely used in order to test the various cosmologies. Arp has described many cases in which high z quasars are associated in space with galaxies of a much lower redshift. It has been often argued that these are due to artefacts: artefacts, in the way of searching for the interesting cases, where some biases of subjective origin may appear; artefacts also in that gravitational focusing might enhance the apparent number of distant faint objects in the vicinity, on the sky, of a close-by massive galaxy. However, precise calculations of this effect, advocated in particular by Nottale (1984), seem to be still insufficient, to the best of my knowledge, to account for the Arp's results.

Another way to look at the association of objects of different redshifts has been the statistical approach performed by Chu and associates (1985); their conclusion is that quasars are statistically associated with close-by field galaxies; this argumentation has of course been submitted to criticism, but in a very inconclusive way yet, so it seems.

6 The classical cosmological tests and the quasars

We have just questioned the normal character of the redshifts of the quasars. Still, they are often used to complete, at large z, the analysis of the classical tests described above in the case of the normal galaxies. What can be deduced from such comparisons, admitting that the z-value is a *bona fide* measurement of distance?

a) The Hubble's m-z diagramm for quasars and active galaxies

The location of quasars in that kind of diagramm has always appeared strange. The scattering of points in apparent magnitude is not as large as the dispersion in z. The obvious reason for such a behaviour is a priori that the so-called quasars are indeed objects belonging to a variety of categories, differing from each other in particular in their absolute brightness. But this may not be the only way out. Let us have a look at the most recent form of this diagramm (Hewitt & Burbidge, 1987), although, clearly, the diagramm is highly composite.

The diagramm is at first characterized by the abrupt drop of the number of objects for z larger than 2.5 or 3 - a point emphasized strongly by Véron (1986). This is not due to a limitation in apparent magnitude.

The diagramm is also characterized by the fact that the slope, either of the brightest objects (by bins, of, say, 10), or the average of the 10 brightest (by bins of 100); or the average brightness of all objects (in bins of 100 objects) - that slope is always much higher than unity.

One may look at this structure in different ways. Assuming the Hubble's law to be valid, the observations imply two evolutionary facts: (i) "before" $z = 4$, no quasar, no galaxy was formed; (ii) after $z = 4$, bright objects were formed, but they ceased to appear at z smaller than a critical value z_i which is larger for bright quasars than for not so bright objects.

This need for evolutionary processes, we have already met it, when examining the classical tests of standard cosmologies. Should one be satisfied with it? If we admit that evolution has, as its first effect, to modify the brightness of newly-born quasars, should we keep the peace of mind when considering the various standard candles used in determining the rate of expansion? Once evolution is felt as needed, one should unambiguously prove that standard candles are not affected by this evolution.

Let us assume, alternatively, that Segal's cosmology is valid : then only one evolutionary phenomenon needs to be present: the non-existence of quasars formation for $z > 4$. A tired-light mechanism leads to uninteresting conclusions, as z would not be a measure either of time, or distance of formation.

A valid hypothesis is still that quasars are affected by a different redshifting mechanism than are the normal galaxies. Then, we have to face the fact that intrinsic redshifts cannot exceed 4; there is no reason whatsoever for the z -value to unable us to determine the absolute magnitude. Only the study of specific associations will allow to say more; and it is of course a provisional weakness of the intrinsic redshift interpretation.

b) The z-distribution of quasars

This study has been done many times. Burbidge (1967), Barnothy (1976), Karlsson (1971), had found that the distribution is marked by several peaks, at regular intervals, well above the level of statistical significance. More recently, Depaquit et al. (1985) have taken again that question, and have attempted a serious discussion of the arguments explaining the effect by observational biases. They concluded that observational biases cannot explain the observed effect (figure 4).

Unless the quasar formation has appeared in successive waves, it is very hard to explain this effect in any natural way, in the frame work of the standard cosmologies.

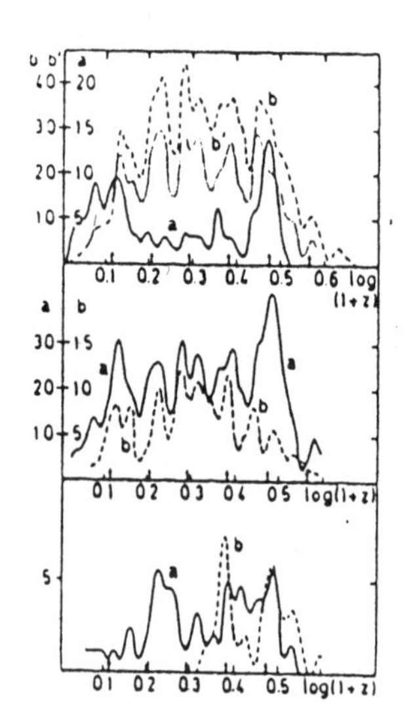

In ordinates, the number of galaxies having a given z-value, grouped by bins of equal intervals in log(1+z). Different samples give place to different curves; arrows indicate the regular intervals referred to in the text (after Depaquit et al., 1985).

FIGURE 4

c) The Souriau's equator

Another feature of the z-distribution of quasars has been discovered by Souriau and associates (Fliche et al. 1982). These authors were looking for possible evidence in favour of a symmetrical universe divided in two subuniverses, one of matter, and one of antimatter. They tried, for that purpose, to project the quasars on a 2-dimensional sphere, image, more or less of the Universe. By properly selecting the plane of projection, it is possible to find an area void of quasars (figure 5). Irrespectively of the way of discovery of this effect, it is a very convincing one.

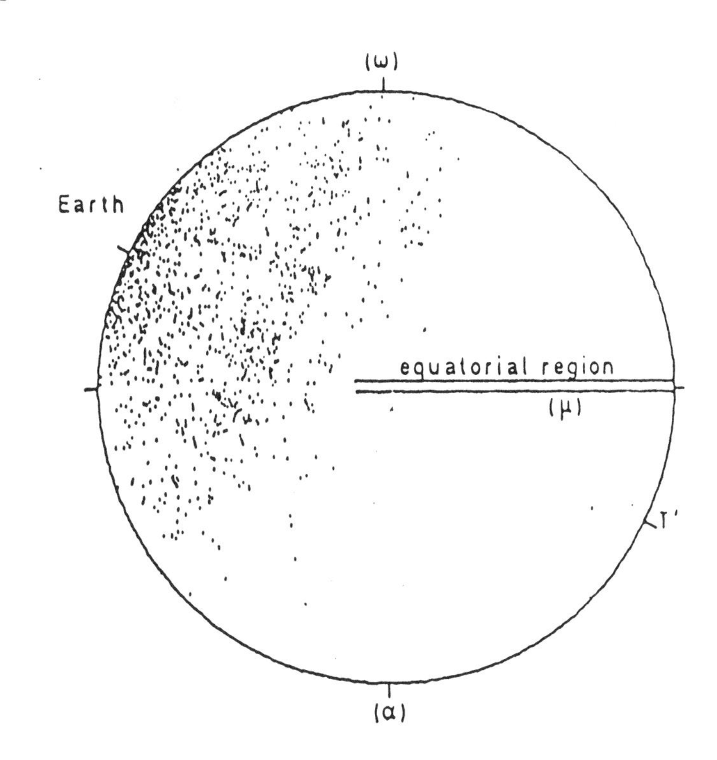

Projection of all quasars on a selected surface. In the "equatorial region", one does find a clear depletion of quasars (after Fliche et al., 1982).

FIGURE 5

Does it prove what Souriau was looking for, namely the existence of a symmetric universe? We should not rule out that this privilegied plane describes a local property of the close-by universe, whatever it means.

7 What about the hierarchical Universe?

The mass distribution of the universe is far from being homogeneous. At large, it is not even sure that the concept of an average density is well founded. If one recalls the important suggestion by De Vaucouleurs (1970), and later by B. Mandelbrot, that the mass distribution follows at large scale a fractal law of index 1.3 (which allows the Olbers-Seeliger integrals to converge), we have no way to extrapolate that relation beyond a size of the order of 10^{27} cm (i.e. 10^{9} light-years - or z = 0.1), a very small distance indeed. Shall we have to admit an abrupt stabilization of the average density at about 2 10^{-31} g cm^{-3}? Or should we instead admit that the average density of the universe as a whole tends to be equal to zero when the radius of curvature increases towards infinity? It is quite difficult to reply to these questions. It is still more difficult to imagine how to properly solve the Einstein's equations when the average density is taken as a function of the radius of curvature of the volume over which the average is taken. Intuitive (but false) reasoning would tend to the idea that, if the universe is expanding, it must have different singular points according the scale of the measurements: small bangs would affect small regions; moderately-sized bangs, moderate-sized regions; and Big Bangs, huge regions; not excluding, at still a larger scale, larger time-scale bangs. This is of course highly speculative, but it illustrates the difficulty that the hierarchical distribution implies for the standard cosmologies.

8 Other grand problems

Proponents of the standard cosmologies have indeed put a great deal of thought in the questions of the arrow of time and the second principle. Truly enough, if the Universe is isolated, then its entropy must increase; in other terms, the hydrogen must regularly disappear, till some "death" of the universe. However, the classical second principle applies well to systems in which the gravitation forces do not play a role; how do we know that, at the scale of the universe, inside which we can assume there is not wall against gravitational effects, a generalized form of entropy could not be just constant? It the universe collapses after having reached a maximum radius, which is not excluded in principle by the standard cosmologies, then hydrogen would be regenerated; is that against the second principle? What will keep, in the forthcoming universe, the trace of the increase of entropy in our own?

Another aspect of the same: let us treat the differential Einstein's equations, taking as initial condition the observed irregular distribution of matter; let them be integrated backwards (as mechanical equations can be): shall we find a highly concentrated state, analogous to the Big Bang? Instead, it is quite possible that we shall find a strange attractor, a "strange bang" indeed, not to be assimilated with a singular point of a model!

Conclusions

My task was to review the difficulties encountered by the standard cosmologies. They are of different nature. The need to satisfy to the classical tests has led to invoke an evolutionary process affecting both quasars and galaxies for z .5 or so. This is really embarrassing, in the sense that it destroys the great simplicity often advocated in favour of the standard models. But we shall not stop at this argument, as there is no reason for the universe to be described by simple laws.

The standard models fail to account for the abnormal redshifts: if these are real, then, they impose new laws of physics; and these new laws will impose a general revision of concepts.

The standards models are now unable to take into account some obvious facts; the hierarchical and irregular distribution of matter in the universe is one important difficulty.

I have not doubt, however, that, with a sufficient skill, it will be possible to reconcile standard models with all these facts. Already the inflationary models have explained the isotropy of the background radiation, allowed for a non-zero cosmological constant. Some have been asking then: "are we not drawing too many epicycles"? Indeed, it may be very well the case. I would not certainly say: "do accept the standard cosmologies, as long as you cannot definitely prove they are wrong". I would like to be more open then that, and let the door wide open to controversy, to imagination, and to the possibility of new physical laws, applicable to the universe as a whole.

Bibliography
(limited to some papers used directly in the preparation of this talk, or little known)

Alfvén, H., 1976, La Recherche, n°68, juillet-août

Arp, H., 1976, Décalages vers le rouge et expansion de l'univers, IAU Coll. n°37, CNRS, ed. Balkowski, Westerlund, p. 377

Arp, H., 1987, The Hubble relation : differences... (preprint)

Arp, H., Giraud, E., Sulentic, J.W., Vigier, J.P., 1983, Astron. Astrophys., 121, 26-28

Barnothy, J.M., Barnothy, M.F., 1976, Publ. Astron. Soc. Pacific., 88, 837 ; Bondi, H., Gold, T., 1948, Mon. Not. R. astron. Soc., 108, 252

Bottinelli, L. Gouguenheim, L. 1973, Astron. Astrophys., 26, 85

Bruzual, 1981, Ph.D. Thesis, Univ. of California, Berkeley, quoted in LaViolette, 1986

Burbidge, G.R., Burbidge, E.M., 1967, Astroph. J. Lett., 148, L107

Burbidge, G.R., O'Dell, S.L., Srittmatter, P.A., 1972, Astrophys. J., 175, 601-611

Canuto, V., 1978, Rivista del Nuovo Cimento, 1, n°21, 1-42

Canuto, V., Hsieh, S.H., 1977, Astron. Astrophys., 61, 15

Chastel, A., Heyvaerts, J., 1974, Nature, 249, 21-2 ; 1976, Astron. Astrophys., 51, 171-183

Chu, Y., Zhu, X., Burbidge, G., Hewitt, A., 1984, Astron. Astrophys., 138, 408-414.

Collin-Souffrin, S., Pecker, J.-C., Tovmassian, H., 1974, Astron. Astrophys., 30, 351

Depaquit, S., Vigier, J.-P., Pecker, J.-C., 1974, C.R. Acad. Sci., 279, 559-563

Depaquit, S., Vigier, J.-P., Pecker, J.-C., 1975, C.R. Acad. Sci., 280, 113-114

Depaquit, S. Pecker, J.-C., Vigier, J.-P. 1985, Astron. Nachr., 306, I, 7-15

Fliche, H., Souriau, J.-M., Triay, R., 1982, Astron. Astrophys., 108, 256

Giraud, E., 1981, C.R. Acad. Sci., 293, 195

Giraud, E., 1982, C.R. Acad. Sci., 294, 442

Giraud, E., 1984, Thèse, Montpellier Univ.

Hewitt, A., Burbidge, G., 1987, Astrophys. J. Suppl. ser., 63, 1-246

Hickson 1977, Astrophys. J., 217, 964

Hoyle, F., Narlikar, J., 1966, Proc. Roy. Soc. London A., 290, 143
Jaakkola, T., 1971, Nature, 234, 534
Jaakkola, T., Molès, M. 1976, Astron. Astrophys., 53, 389
Jaakkola, T., Karoji, H., Le Denmat, G., Molès, M., Nottale, L., Vigier, J.-P., Pecker, J.-C., 1976, Mon. Not. R. astron. Soc., 177, 191-213
Karlsonn, K.-G., 1971, Astron. Astrophys., 13, 33 ; 1977, 58, 237
Karoji, H., Nottale, L., 1976, Nature, 259, 31
LaViolette, P., 1986, Astrophys. J., 301, 544-553
Molès, M., Nottale, L., 1981, Astron. Astrophys., 100, 258
Mérat, P., Pecker, J.-C., Vigier, J.-P., Yourgrau, W., Astron. Astrophys., 1974, 32, 471-475
Nottale, L., Hammer, F., 1984, Astron. Astrophys., 141, 144-150
Pecker, J.-C., 1976, Décalages vers le rouge et expansion de l'univers, IAU Coll. n°37, CNRS, ed. Balkowski, Westerlund, p. 451-479
Pecker, J.-C., Tait, W., Vigier, J.-P., 1973, Nature, 241, 338
Puget, J.-L., Schatzman, E., 1974, Astron. Astrophys., 32, 477-8
Sandage, A., 1965, Astrophys. J., 141, 1560-1578
Segal, I.E., 1975, Proc. Nat. Acad. Sci., 72, 2473
Segal, I.E., 1986, March (preprint) (direct comparison of observed magnitude-redshift relations...)
Segal, I.E., Nicoll, J.F., 1986, Astrophys. J., 300, 224-241
Sulentic, J.W., 1982, Astrophys. J., 252, 439
Tifft, 1976, Astrophys. J., 206, 38
De Vaucouleurs, G., 1970, Science, 167, 1203-1213
Véron, P., 1986, Astron. Astrophys., 70, 37-42

DISCUSSION

T. Kiang: 1) The dispersion in the m-z diagram for quasars is greatly reduced by using Baldwin's relation between quasar luminosity and the equivalent width of some emission lines. Applying this relation to a set of flat-spectrum radio quasars, I and Cheng obtained an m-z diagram with a dispersion of 0.6 mag, then, on assuming no luminosity evolution, obtained a sharp determination of

$$q_0 = +1.8 \pm 0.15 dex(1\sigma)$$

(Kiang and Cheng, AA, **3**, 1983). 2) With a hierarchical universe, not only is the mean density a decreasing function of the sampling volume, the degree of clumpiness (as measured by $\sigma^2/\bar{n}$) is also an *increasing* function of the volume. But this feature should be taken as an empirical fact, more fundamental than any model, hierarchical or otherwise.

T. Kiang: Correlated with the clumpiness increasing with volume is the peculiar velocity increasing with volume–and this might explain why Hubble got his law from the galaxies of the Local Group. The peculiar velocities in the Local Group (a relative small sampling volume) are not large enough to completely wash out the Hubble expansion.

G. Burbidge: Hubble's original list of galaxies contained many that were very local indeed.

J.-C. Pecker: We should remember earlier phases of the Hubble's linear law! First, Slipher did find a qualitative relation between apparent brightness and red-shift; from that Hubble has chosen the linear law as the simplest form of a possible z-m positive correlation; but at the same time Lundmark (like Segal more recently) proposed a square-law, on the basis of essentially the same data as Hubble's!! So–preconceived ideas are permeating the whole field...

ALTERNATIVES TO THE PRESENT-DAY COSMOLOGICAL PRINCIPLES

Konrad Rudnicki
Jagellonian University Observatory
Cracow, Poland

ABSTRACT

In a most general sense by a cosmological principle one can understand a set of general convictions concerning the structure of the universe, which reach beyond observational conclusions. The present-day cosmological principles have superseded some older ones. One can expect a further evolution of the philosophical and methodological assumptions of science leading to formulation of more cosmological principles. New trends in today's science which may result in the formulation of such principles are described here, and on that base new possibilities to solve problems are shown.

THE NOTION OF A COSMOLOGICAL PRINCIPAL

Nowadays when we talk of "the cosmological principle" we usually mean the Generalized Copernican Cosmological Principle, called also the Ordinary, or Weak, Cosmological Principle, which reads that the Universe observed from every point and in every direction looks much the same. Some cosmologists prefer the Strong, or Perfect, Cosmological Principle, which additionally states that the Universe looks very much alike in all times.

The Ordinary Cosmological Principle enables us to think of the all Universe, even of those parts which are not accessible to any possible observation, i.e. situated beyond the cosmological horizon. Analogically, the Perfect Cosmological Principle can give us information about unobservable epochs of the Universe.

Cosmological principles are not laws of nature, but are a kind of higher-order rules concerning the physical world, which allow us to fill the blanks within our cosmological knowledge. This aim can be fulfilled only as far as the general beliefs (of scientific or pre-scientific character) on which a particular cosmological principle is based correspond to reality. In every historical epoch there were some characteristic dominant beliefs of fundamental character. On the other hand, we cannot consider cosmological principles as something extraneous to cosmology, pertaining rather to philosophical interpretation of scientific results. On the contrary, one must agree that without any cosmological principle no statement about the Universe as a whole would be possible unless the Universe were finite and accessible for direct or indirect observations in all its parts and all epochs. That could hardly be the case.

THE HISTORICAL COSMOLOGICAL PRINCIPLES

Though the very notion of a cosmological principle is of a rather modern provenience, some older rules of this kind can be reconstructed from ancient writings and expressed in contemporary terms. The oldest one known so far is the

Pre-Hindu Cosmological Principle(Rudnicki 1982). This principle, implicated in the highly spiritual outlook of the ancient Hindus, says that the Universe is infinite in space and time and infinitely heterogeneous. The place of ourselves in the Universe (the Earth) is in respect of its location in space and time as well as in respect of all its other properties neither extreme nor average, since an infinity knows no extreme properties and no mean ones. No mathematical cosmological model basing on the Pre-Hindu Prin- ciple can be built up, since there are no means in mathematics (at least at present) to describe infinite heterogeneity.

Another ancient cosmological principle is the Antique Cosmological Principle, reconstructed by Heller(Heller and Rudnicki 1972). It is connected with the ancient Greek view that the only real world is the physical one, and the physical is what can be touched, tasted, or smelled. The heavenly bodies were thus not physical. Thus only the Earth could be accepted as a natural centre of the Universe.

The assumption that the Earth is the centre of the Universe proved to be very prolific, and many mathematical models of the Universe were based on it. The most renowned were the systems of Hipparchus, Ptolemy, as well as that of Tycho Brahe.

Copernicus, who had the courage to consider the planets as physical bodies, used in his construction the principle that the Universe observed from each planet looks qualitatively alike. That was the original Copernican Cosmological Principle. The Ordinary Cosmological Principle mentioned above is an easy generalization of it.

COMPARISON OF THE VARIOUS COSMOLOGICAL PRINCIPLES

We can suppose that in the history of mankind there have been even more cosmological principles. But let us confine ourselves in these considerations to the five principles described in the literature in a more detailed manner : the Pre-Hindu, the Antique, the Original Copernican, the Generalized Copernican, and the Perfect, cosmological principles. We can notice some sharp contradictions between them. For example, the Pre-Hindu Principle considers the Universe to be different in every place and in every epoch, while the Perfect Principle - just the opposite. The Antique Principle would have the Earth as a natural centre of the Universe, while the Pre-Hindu, The Generalized Copernican, and the Perfect, principles do not presuppose an existence of any centre at all.

But there are also similarities. For example, all the principles, except for the Antique one, ascribe to our Earth no special position or meaning in the Universe. All those five principles do not necessarily demand spatial finiteness of the Universe (though some of them put it as one of the possibilities).

If we accept the idea that the human thinking, however it could roam out of its way, in general brings our knowledge all the closer to the real world, we can hope that subsequent cosmological principles form a sequence that eventually approaches the truth. Furthermore, if we accept that the human thinking is capable of reaching the very fundaments of existence (Steiner 1961), the attempt to formulate new, better, cosmological principles is by no means a vain play.

CONTEMPORARY TRENDS

In already classic papers as that of Ellis (1980) it has been shown that the Generalized Copernican Cosmological Principle can be considered as a hypothesis to be tested observationally. The limits of possibilities of such a testing also have been described in many papers. Some new principles of the kind have been put forth too. The Modified Copernican Cosmological Principle (Karachentsev 1974), and, even more, the Anthropic Principle (e.g. Carr 1982) take into account not only the geometry of space-time and the distribution of objects but also the observer, i.e. an element of consciousness in the world. I do not think the Anthropic Principle could possibly substitute any of the cosmological principles of the present: the concept of an intelligent being presupposed in it is much too "earthly". With a little of creative imagination one can readily conceive e.g. an intelligent human-like being consisting of a great number of very tiny parts forming something like a beehive - and with the same refute a large part of the Anthropic Principle considerations. But the underlying idea of including the observer seems a very right one. Also the idea of a law of nature as something which must have its causes (e.g. Reeves 1986) shows us, in my opinion, a way to formulate future cosmological principles, which could fill more effectively the blanks within our knowledge of the Universe.

PERSPECTIVES

When investigating the conditions in the very remote cosmological epochs with a use of cosmological data and physical laws only, we always come to certain moments of time before which no observations can bring us any information and none of the known natural laws is valid anymore. Only a suitable cosmological principle can be of help in penetrating those epochs and substituting pure fantasy with real knowledge. Contemporary science when interpreting the existing reality does not like to ask the question "why", and it has completely dropped the question "what for" from its vocabulary. Today the dominant question in physical sciences is "how", or, when probability theory is considered, "how much". If we are dealing with particular celestial bodies or restricted agglomerations of such bodies, for instance, planets, planetary systems, star clusters and galaxy clusters, this historical trend in the questions asked is quite acceptable. However, when the object of inquiry is the Universe as a whole, one must transcend that trend. In the early twentieth century, before Einstein's General Relativity, cosmogony was much a domain of philosophers. It was only with the development of the physical concepts necessary in discussing singularities in cosmological models that the balance shifted from philosophy to theoretical physics. Along with that shift there was a reinforcement of the question "why" instead of the question "how". Why, we ask, the fundamental physical constants and the laws of physics are such as they are and no other. The Anthropic Principle in cosmology has again focused our attention on the question "what for" (Rudnicki 1985). I think we should not be ashamed of the return to that old question, which seemed so outfashioned for a long time. What was once pure philosophy has now become science. This is a normal course of things. Some considerations about the entire range of possible properties of the Universe (e.g. Woszczyna 1983, 1986) indicate a wide area of cosmology where some burning questions would be solved provided a suitable cosmological principle could be a guide in elimination of non-realistic solutions. As long as a better, deeper cosmological principle is not yet formulated, some cosmological problems have to wait still for their solution. But it seems good when at least we are aware that a cosmological principle is not something fixed once for all, but has its history and its future.

REFERENCES

Carr, B.J. (1982), *Acta Cosmol.*, **11**, 143.

Ellis, G.F.R. (1980), *Annals New York Acad.Sci.*, **336**, 130.
Heller, M.& Rudnicki, K. (1972), *Analecta Cracoviensa*, **4**, 47.
Karachentsev, I.D. (1974), *Acta Cosmol.*, **2**, 43.
Reeves, H. (1986), *Proc.Inter.School Physics "Enrico Fermi"*, **86**, 522.
Rudnicki, K. (1982), *"Die Sekunde der Kosmologen" Frankfurt a/M - the chapter "Der Kosmologische Prinzip der Ur-Inder"*, pp. 12-18.
Rudnicki, K.. (1985), *"The Galileo Affair"*, Eds. Coyne,Heller, Zycinski, Citta' del Vaticano.
Steiner, R. (1961), *"Grundlinien einer Erkenntnisstheorie der Goetheschen Weltanschauung"*, Stuttgart.
Woszczyna, A. (1983), *Acta Cosmol.*, **12**, 27.
Woszczyna, A. (1986), *Acta Cosmol.*,**14**, 79.

COSMOLOGICAL MODELS WITH NON-ZERO LAMBDA

E. Joseph Wampler
European Southern Observatory

W.L. Burke
Lick Observatory

INTRODUCTION

There is strong evidence that quasar spectra contain luminosity sensitive features (Baldwin et al. 1987; Kinney et al. 1987; Wampler et al. 1984; and Baldwin 1977). When such features are used to correct the observed magnitudes of quasars the resulting Hubble diagram shows either that high redshift quasars are more luminous than their low redshift counterparts with very similar spectral properties or that the geometrical structure of the Universe causes an apparent brightening at high redshift. Furthermore, the quasar Hubble diagram seems to be a simple extension of the infrared Hubble diagram for a complete sample of 3CR radio galaxies (Wampler 1987). It was decided, therefore, to investigate the consequences of the hypothesis that the apparent brightening of distant "standard candles" can be explained by the geometrical structure of the Universe.

Any serious model of the Universe must not only be compatible with the galaxy-quasar Hubble diagram but must satisfy all other relevant observational constraints. These include the observationally determinable parameter H_0T_0, the radio source counts (log N - log S), the existence of the 3°K background radiation, the angular diameter redshift relationship and the observed law of galaxy number density as a function of redshift (Loh 1986). Perhaps surprisingly, the introduction of a non-zero cosmological constant into the Lemaitre equations gives model universes that appear capable of explaining many of the observational facts as they stand today without invoking _ad hoc_ evolution laws which must be individually adjusted in order to fit existing observations to the currently popular Friedmann universes with zero cosmological constant and $\Omega_0 \lesssim 1$.

After briefly describing the cosmological models we will discuss the observational constraints with the assumption of no evolution.

MATHEMATICAL FRAMEWORK

We start with the Lemaitre equations:

$$2\,\frac{R''}{R} + \left(\frac{R'}{R}\right)^2 + \frac{K}{R^2} = \Lambda \tag{1}$$

and

$$\left(\frac{R'}{R}\right)^2 + \frac{K}{R^2} = \frac{8\pi\rho}{3} + \frac{\Lambda}{3} \tag{2}$$

Here ρ is the matter density, K is the curvature parameter and is +1, 0, -1 for closed, critical or open universes, respectively, and Λ is the cosmological constant. If $\Lambda > 0$ the vacuum energy is positive and the pressure is negative. Each prime denotes a derivative with respect to time. The requirement of energy conservation gives:

$$\frac{d}{dt}\left[\frac{8\pi\rho R^3}{3}\right] = 0\ . \tag{3}$$

We have numerically integrated the Lemaitre equations, 1) and 2), for appropriate sets of values for Λ. We started each integration with a critical value for ρ and continued until we reached a predetermined value for H_oT_o. For each solution we then calculate the parameters

$$\Omega_o \equiv \frac{8\pi\rho_o}{3H_o}\ ; \qquad \lambda \equiv \frac{\Lambda}{3H_o^2} \tag{4}$$

and plot a Hubble diagram which can be compared with observations. Model universes that were judged acceptable were located on an Ω-λ plot.

In universes that obey the most general Einstein field equations and conserve energy the angular sizes of distant objects can be determined directly from the observed Hubble diagram. Choosing the notation used in Baldwin et al. (1978) we have:

$$\frac{d}{\theta} = \frac{\text{proper size}}{\text{angular size}} = R\ S(\chi) = \frac{R_o}{(1+z)}\ S(\chi) \tag{5}$$

but

$$\phi = \log_{10}\left[\frac{z}{(1+z)\ S(\chi)}\right] + \text{constant} \tag{6}$$

so

$$\log_{10}\theta = \phi + 2\log_{10}(1+z) - \log_{10}z + \text{cst} \tag{7}$$

In a static Euclidean universe $\theta \propto 1/z$ so we have simply

$$\log_{10}\theta_{GR} - \log_{10}\theta_E = \phi - 2\log_{10}(1+z) \tag{8}$$

Here θ_{GR} is the angular size of an object in a General Relativistic Universe and θ_E is the angular size in a Euclidean universe. Because ϕ is simply the departure of the observational points from a $z^2/(1+z)$ law, the predicted angular size law in these universes can be determined directly from the observed Hubble diagram without having to determine the details of a particular model universe. This should not be surprising. The curvature of the Universe magnifies the image of a distant source but, like a lens, it does not change the surface brightness of the image. Thus the departure of the observed magnitudes from the $z^2/(1+z)$ law in a Hubble diagram is a direct measure of the "magnification" of the image.

The density of sources in co-moving coordinates was taken to be constant. In this study we have concentrated on closed universes ($k = +1$) and for each model we have calculated the percentage volume out to a given redshift. From the tabulations of percentage volume and relative flux as a function of redshift it is a simple task to determine the log n - log S relationships appropriate for a particular model and luminosity function.

In Table 1 we give details of two of the model universes: one with $H_oT_o = 0.65$ and $\Omega_o = 1.78$; the other with $H_oT_o = 0.8$ and $\Omega_o = 1.07$. For each of these closed universes we give the arc distance, χ, corresponding to a redshift z, the volume at redshift z relative to the total volume and the magnitude at redshift z of an object of zero magnitude at redshift z = 0.1.

Table 1. Model Universes

z	$H_oT_o = 0.65$; $\Omega = 1.78$			$H_oT_o = 0.80$; $\Omega = 1.07$		
	Arc distance	Rel. volume	Rel. Mag.	Arc distance	Rel. volume	Rel. Mag.
0.1	0.1231	0.000	0.00	0.1109	0.000	0.00
0.2	0.2332	0.003	1.47	0.2164	0.002	1.49
0.3	0.3311	0.008	2.29	0.3152	0.007	2.42
0.5	0.4954	0.025	3.28	0.4913	0.024	3.49
0.7	0.6269	0.048	3.89	0.6393	0.051	4.13
1.0	0.7806	0.089	4.44	0.8176	0.101	4.74
1.5	0.9638	0.158	5.02	1.0332	0.189	5.34
2.0	1.0935	0.218	5.39	1.1857	0.267	5.70
3.0	1.2677	0.313	5.85	1.3896	0.386	6.15
4.0	1.3820	0.381	6.16	1.5224	0.469	6.42

COMPARISON WITH OBSERVATIONS

Fig. 1 shows a combined Hubble diagram for galaxies and quasars taken from Wampler (1987). We have supplemented the data given by Wampler (1987) with additional IUE data that has recently become available (Kinney et al. 1987). Table 2 gives the relevant new data. Following Wampler (1987) the observed magnitudes were corrected using the observed CIVλ1550 equivalent widths and a (1+z) band stretching term. Thus, these magnitudes are not flux densities. Because it is common to include the band stretching term in galaxy K-corrections it is included here for the quasar data. Using this definition for the magnitudes, a straight line in the magnitude - log z Hubble diagram corresponds to $q_0 = 1$ when $\Lambda = 0$.

Table 2. Quasar Data*

Name	z	W_0(CIV) (Å)	AB_ν(1550)	Corr. mag.	Name	z	W_0(CIV) (Å)	AB_ν(1550)	Corr. mag.
0405-123	0.574	41	15.50	15.23	1512+370	0.371	65	17.13	15.77
0414-060	0.781	51	16.28	15.70	1912+550	0.398	86	17.45	15.55
0454-220	0.534	26	15.95	16.57	2128+123	0.501	103	17.03	14.84
1100+772	0.311	46	16.38	15.68	2251+113	0.323	29	17.08	17.32
1137+660	0.652	66	16.35	15.17	2308+098	0.432	78	16.28	14.60
1318+290	0.549	133	17.60	14.93	2344+092	0.677	39	16.95	16.85
1425+267	0.366	59	17.63	16.47					

* Kinney et al. (1987)

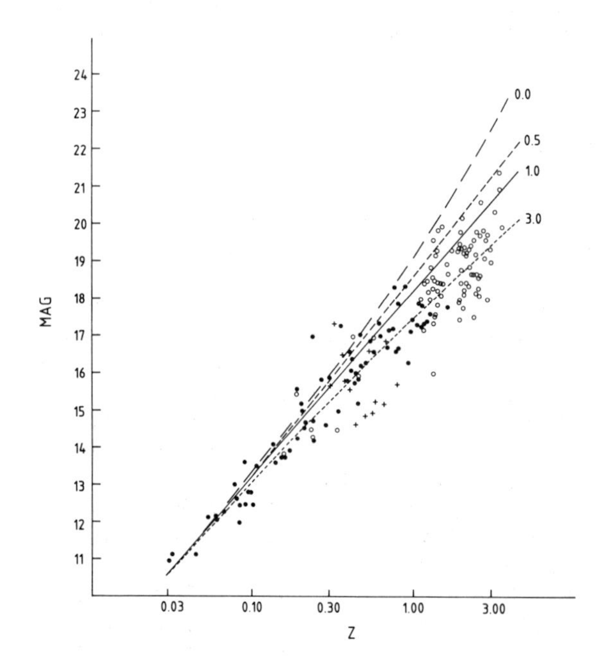

Figure 1. A combined Hubble diagram from Wampler (1987) showing 2.2μ photometry of 3CR galaxies (filled circles) and λ1550 magnitudes of quasars (open circles and crosses). The quasar data has been corrected for luminosity effects and shifted by 6.75 mag relative to the galaxies. The crosses are new data given in Table 2 and may have lower precision than the earlier data.

As described above a series of model universes with no luminosity evolution were calculated as a function of λ and H_oT_o. The locations of these models in the Ω_o,λ plane are given in Fig. 2. By estimating the goodness of fit of the data points to Hubble diagrams calculated from the models it is possible to locate in the Ω_o,λ plane a region in which model calculations give Hubble diagrams that are in reasonable agreement with the observations.

The region of acceptable parameters is shown in Fig. 2 as a stippled area. Also shown is the region of acceptable fits determined from the co-moving galaxy density data of Loh (1986). Fortunately the two types of data delineate regions in the λ,Ω_o diagram that are orthogonal to each other and intersect. Once a model universe has been chosen it is possible to predict the apparent brightness and surface density of standard sources as a function of z. When these figures are combined with a luminosity function one can compare the predictions of the selected model universe with radio source counts. Any model universe that gives a reasonable fit to the observed Hubble diagram is expected to give about the same run in apparent surface density with redshift. Fig. 3 shows a typical observed differential radio source count plot together with the

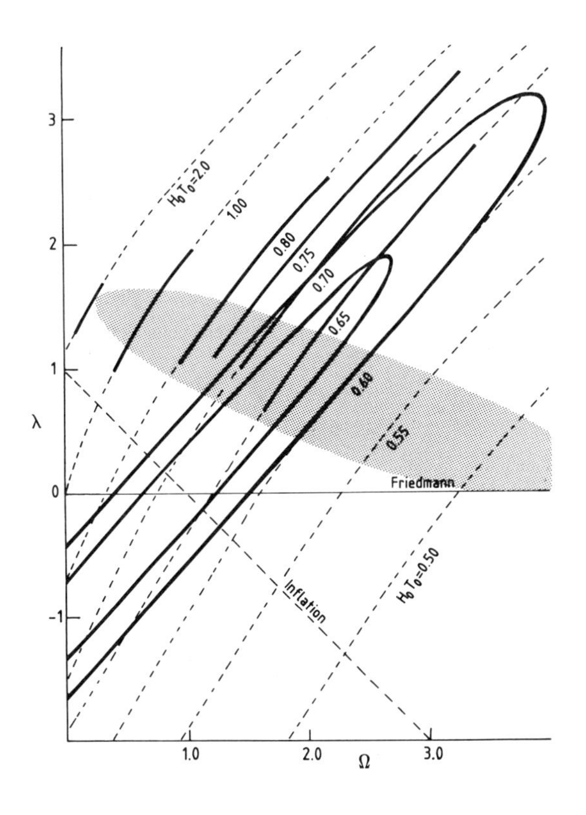

Figure 2. The Ω,λ plane showing the region of acceptable fits to the galaxy counts of Loh (1986) (flattened ellipses) and quasar / galaxy Hubble diagram fits (shaded region). Curves of constant H_oT_o are shown as dashed lines. The regions in this diagram for which model universes were calculated are shown by the solid segments on the H_oT_o curves.

expected relationship for sources of fixed luminosity and redshifts greater than z = 0.1 in an $\Omega = 1$, $H_oT_o = 0.8$ universe. Their apparent flux at 6 cm was taken to be 1 Jy at a distance corresponding to z = 0.1. Because at high z the closed model universes considered here are rapidly running out of volume - that is, a light ray has had time to travel a substantial distance around the universe - there is a sharp decline in the number of distant bright sources relative to the value expected in a static Euclidean universe. This behavior allows a large population of nearby, intrinsically faint sources to be a major contributor to the faint end of the radio source count diagram. And, indeed, it has been found that most of the observed faint sources are in fact intrinsically faint, nearby sources (Weistrop et al. 1987).

Uncertainties in both H_o and T_o lead to a very uncertain value for H_oT_o. Published values for these two quantities (Winget et al. 1987; Giraud 1987; Sandage 1987) give extreme values for the product of the two numbers that lie in the range $0.5 \lesssim H_oT_o \lesssim 2.0$. While this allowed range for H_oT_o is not further restricted by the observed Hubble diagram, closed models will not give acceptable fits to the data for $H_oT_o \gtrsim 2.5$.

Figure 3. Observed source counts at 6 cm normalized to a Euclidian count (Partridge et al. 1986; Kellermann & Wall 1987). The expected locus of a uniform density of sources that have an apparent flux of 1 Jy at z = 0.1 in an $\Omega = 1$, $H_oT_o = 0.8$ Universe is shown as a dashed line.

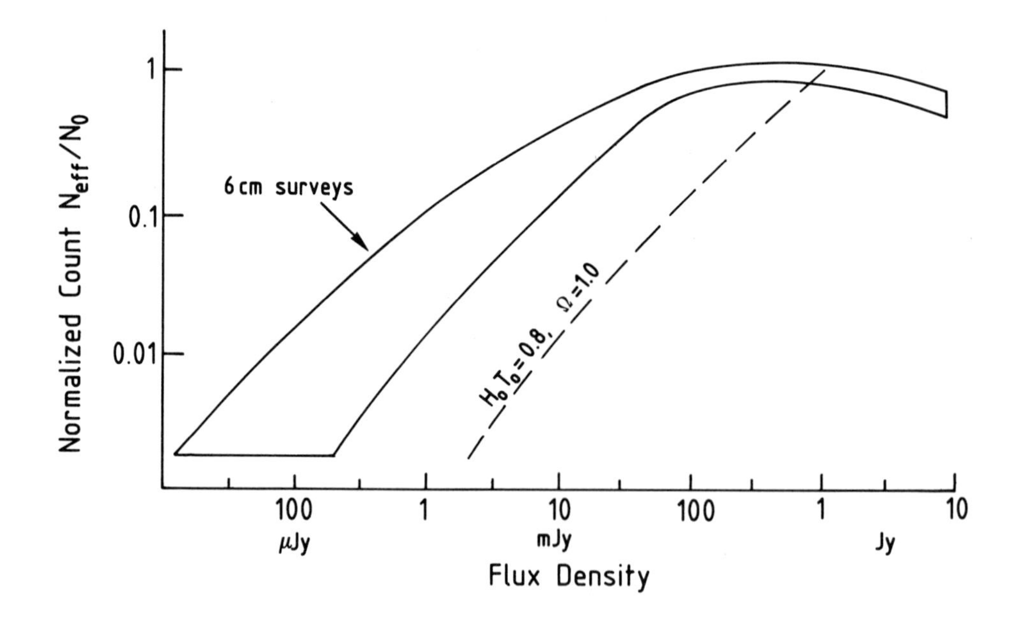

DISCUSSION

Closed Lemaitre universes with $\Lambda > 0$ give Hubble diagrams that are satisfactory fits to the observed galaxy-quasar Hubble diagram without the need for evoking luminosity evolution. The predictions of these closed universes are also compatible with radio source counts and can easily accommodate a large population of nearby, intrinsically faint sources in the counts. The Hubble diagram alone does not constrain H_oT_o to a narrower range than the values popularly quoted in the literature. However if Loh's galaxy count studies are considered H_oT_o is constrained to $0.60 \lesssim H_oT_o \lesssim 0.75$. This implies an upper limit of about 60 km/sec/Mpc for H_o if T_o is $\gtrsim 12$ Gyr (Winget et al. 1987).

We note that unless luminosity evolution is invoked the observed Hubble Diagram requires strong space curvature. Fig. 4A,B gives the angular size-redshift relationship derived from the observed Hubble diagram using the method described above. The solid line gives the mean relationship while the hatched region gives the error estimated directly from the observed Hubble diagram.

Figure 4. θ,z relationships determined from the observed Hubble diagram (see eq. 6 and 7 in the text). The hatched region denotes the estimated uncertainty in determining the departure of the observations from a $z^2/(1+z)$ law. θ_{GR}/θ_E is the ratio of the general relativistic angular size of a distant object to its Euclidian angular size.

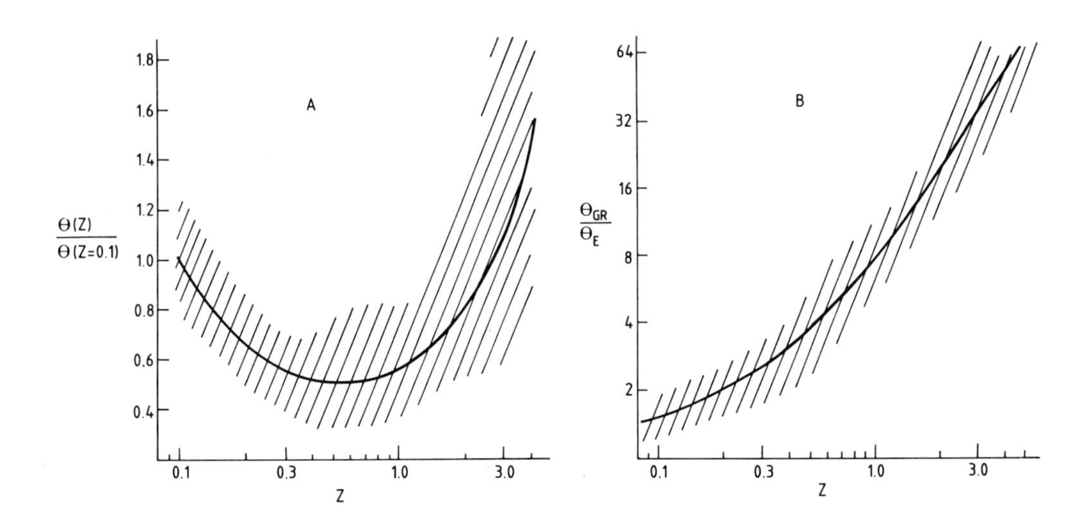

The predicted angular size - redshift relationship seems to be in disagreement with radio observations of distant sources (Kapahi 1987). Kapahi's analysis of faint sources, which include a mixture of measured and estimated redshifts, suggests that the data is best fit by the predictions of a static Euclidean universe. The strongly curved universes hypothesized here would seem to require substantial evolution as a function of redshift. Further evidence for evolution would be a change in the ratio of steep to flat spectrum radio sources as a function of redshift. In fact, at high redshift there is a preponderance of flat spectrum sources (Kellermann & Wall 1987). At first sight this may seem surprising as bright, nearby sources such as Cyg A and Cen A could be seen at very high redshift. Thus the observations might imply that such sources are rare at high redshift. But the interpretation of these observations is not straightforward as the data reflect an interplay of evolution, relativistic effects and the change of volume element on redshift. Fig. 3 shows that the interpretation of source counts in log N - log S diagrams is very model dependent. Because there is a strong luminosity-redshift correlation in the radio data a careful re-examination of the models should prove useful. The reported correlation between the spectral properties of quasars and the radio structure (Wills & Wills 1986) might prove to be a useful tool. In any event, there clearly is intergalactic material in the Universe. This material would be expected to be more dense in the past when the scale of the Universe was smaller. A denser intergalactic medium might be expected to interact with diffuse, extended double radio lobes while nuclear radio sources or galaxies might be less influenced by the comparatively dilute intergalactic medium.

The situation for the optical data seems less certain. Optical observations of high redshift galaxies (T. Tyson, private communication) suggest that the expected increase in the angular diameter of high redshift galaxies is in fact seen. The observations of Loh (1986) of co-moving galaxy counts is consistent with the type of θ-z relationship derived here. The strong space curvature required by the proposed models would modify the interpretation of a number of high redshift observations. For instance, the physical scale of distant absorption clouds, seen in projection against background quasars, would be reduced. The apparent spacing of gravitational lense images would be increased over the separation expected in a Universe with $\Lambda = 0$ and small Ω_0. The apparent velocity of distant superluminal sources would be reduced since space curvature increases the apparent projected separation. This would change the statistics of superluminal sources but not eliminate the class, since nearby sources, such as 3C 120, are too close to us to be significantly magnified by the cosmological space curvature.

Universe models that do not require _ad hoc_ evolution to match observations to the model predictions have the advantage of simplicity. It is difficult to observe evolutionary effects that are independent of the chosen universe model. Color evolution of galaxies is still an uncertain subject (Eisenhardt & Lebofsky 1987). Radio source evolution and quasar number / luminosity evolution is very model dependent.

The purpose of this short note is to point out that this preliminary study shows that there may be conceivable models of the Universe that can fit the observed data without requiring strong evolution of galaxies, quasars or radio sources, at least out to $z \simeq 4$. Further, these ideas can be tested by observation. The extension of Loh's (1986) galaxy studies will be important. The Hubble Space Telescope will be a valuable instrument for determining the optical θ-z relationship at high redshift. And, finally, the projected very large ground based telescopes will be needed to obtain accurate spectra of high redshift galaxies that can be examined for evidence of spectral evolution.

References

Baldwin, J.A. (1977). Astrophys. J., 214, 679-684.
Baldwin, J.A., et al. (1978). Nature, 273, 431-435.
Baldwin, J.A., et al. (1987). In preparation.
Eisenhardt, P.R.M. & Lebofsky, M.J. (1987). Ap.J. 316, 70-83.
Giraud, E. (1987). "New Ideas in Astronomy", Symp. on occasion of 60th birthday of Halton C. Arp (Cambridge University Press: Cambridge).
Kapahi, V.K. (1987). IAU Symp. No. 124, Beijing, China, eds. A. Hewitt, G. Burbidge & L.Z. Fang, D. Reidel, Dordrecht, p. 251-266.
Kellermann, K.I. (1972). Astron. J., 77, 531-542.
Kellermann, K.I. & Wall, J.V. (1987). IAU Symp. No. 124, Beijing, China, eds. A. Hewitt, G. Burbidge & L.Z. Fang, D. Reidel, Dordrecht, p. 545-564.
Kinney, A.L., et al. (1987). Astrophys. J., 314, 145-153.
Loh, E.D. (1986). Phys. Rev. Lett., 57, 2865-2867.
Partridge, R.B., et al. (1986). Astrophys. J., 308, 46-52.
Sandage, A.R. (1987). Astrophys. J., 317, 557-563.
Wampler, E.J. (1987). Astron. Astrophys., 178, 1-6.
Wampler, E.J., et al. (1984). Astrophys. J., 276, 403-412.
Weistrop, D., et al. (1987). Astron. J., 93, 805-810.
Wills, B.J. & Wills, D. (1986). Quasars, IAU Symp. 119, eds. G. Swarup & V.K. Kapahi, D. Reidel, Dordrecht, p. 215-216.
Winget, D.E., et al. (1987). Astrophys. J. Letters, 315, L77-L81.

DISCUSSION

G. Burbidge: In your paper you pointed out the similarity of the Hubble diagram of QSO's and galaxies respectively. Would you like to comment on this business?

J. Wampler: The close agreement between the formal value for q_0 found in the radio galaxy Hubble diagram and that obtained from the quasar Hubble diagram stimulated this present work. We thought it odd that the evolution of giant stars in radio galaxies should so closely match the apparent evolution of the quasar continuum at λ1550Å. So we are investigating the consequences of interpreting the apparent evolution as a consequence of the space-time structure of the universe.

L. Maraschi: What is the status of observations of low surface brightness radio sources at high redshift concerning the θ-z diagram? At least two objects have been found at high z with *large* angular diameters (Nature ~ 1980) (I can provide a complete reference). Has this evidence increased or disappeared?

J. Wampler: I am not a radio astronomer and I can't answer in detail until I have informed myself very much more about the situation. *If* the Hubble diagram is explainable in the terms of these cosmologies then at high z very large angular diameter, very low surface brightness doubles are expected. One usually sees much smaller bent sources. I do not know if these are actually the magnified inner parts of more extended sources.

THE PRICE TO KEEP THE HUBBLE CONSTANT ... CONSTANT

Edmond GIRAUD

European Southern Observatory, Karl Schwarzschild Straße 2, 8046 Garching bei München, Germany.
and
L.A.T. du Collège de France, Institut d'Astrophysique, 98 bis Bd Arago, 75 014 Paris, France.

I. INTRODUCTION.

The Hubble expansion rate measured in the short distance scale varies from a local value of 70-75 to $\sim 90 - 100 kms^{-1}Mpc^{-1}$ as the kinematic distance corrected for infall velocity toward Virgo increases from $D_v = 200 - 400kms^{-1}$ to $D_v \sim 1300kms^{-1}$. Because there is no complete sample of galaxies with luminosity indicators in the kinematic sphere $D_v \leq 5000kms^{-1}$, present studies of the Hubble rate in this range necessarily rely upon biased samples (Sandage and Tammann 1975, Sandage et al. 1979). Therefore it is crucial to be able to derive some estimate of the bias due to the Malmquist effect in the determination of distances to galaxies and clusters. The purpose of this paper is to derive conditions that lead to a constant value of $H_o = H_{loc} \sim 75kms^{-1}Mpc^{-1}$ at $D_v \leq 2500kms^{-1}$ by assuming that the samples are unfair representations of the real world and that the distance indicators are only loosely correlated with absolute magnitudes.

II. THE HUBBLE RATE AT MEDIUM DISTANCE

a) The luminosity index as a distance indicator.

Distances deduced from the luminosity index Λ_c, of 200 spiral galaxies (de Vaucouleurs 1979a, b) have been used to derive a mean Hubble ratio of $96 \pm 10kms^{-1}$ (de Vaucouleurs and Peters 1981) based on de Vaucouleurs' calibrators. The recession velocities of these galaxies corrected for infall toward Virgo can be combined with their distance moduli to determine a Hubble ratio for each object. The resulting values are plotted against the velocities in a log-log diagram ($\log D_v, \log H$) (because the distance moduli are in magnitudes). The diagram (Fig. 1a) reveals an increase of the Hubble ratios with the kinematic distances. The right side of Fig. 1 has qualitatively the shape expected from the Malmquist bias. An implicit evidence for Malmquist effect comes from a trend of Λ_c with kinematic distance indicating that intrinsically faint objects are lost at large distances. It is also probable that only the bright tail of the class of galaxies $\Lambda_c \geq 1$, is observed at large distance.

b) The B- and H-bands Tully-Fisher relations as distance indicators.

The variation of the Hubble rate derived from the B-band Tully-Fisher relation as a function of the velocity (corrected for infall toward Virgo) is illustrated in Fig. 1 of Giraud (1986a). In this sample small spiral galaxies (i.e. objects with de Vaucouleurs' types T = 6, 7, or with low rotation velocity) are progressively lost at large kinematic distance. This is an indication of Malmquist effect.

The variation of the Hubble rate as a function of the velocity, when distances are derived from the infrared Tully-Fisher (IR/HI) relation of Aaronson, Huchra and Mould (1979), is shown in Fig. 1b. In this diagram the bias is much less obvious and it would be more difficult to conclude that the value of log H at medium distance is about 1.8. An evidence for Malmquist effect in this sample comes from the absence of slowly rotating galaxies ($\log W \leq 2.35$) at $\log D_v > 3.2$. An important property of this sample (from Aaronson et al. 1982) is the large range in apparent magnitudes at a given distance ($\sim 5 - 6$mag) compared with the distance depth (~ 5mag).

III. THE SCANNING BIAS MODEL

The correction for Malmquist bias is made by a double scanning on the rotation velocities and on the distances. Let $\phi(M_W)$ the luminosity distribution of galaxies having a rotation velocity W, and σ_{M_W} the dispersion of $\phi(M_W)$. Let $\psi(W_M)$ be the distribution of log W at a given magnitude M, and σ_{W_M} the dispersion of $\psi(W_M)$. The surface density of the Tully-Fisher plane (log W, M)

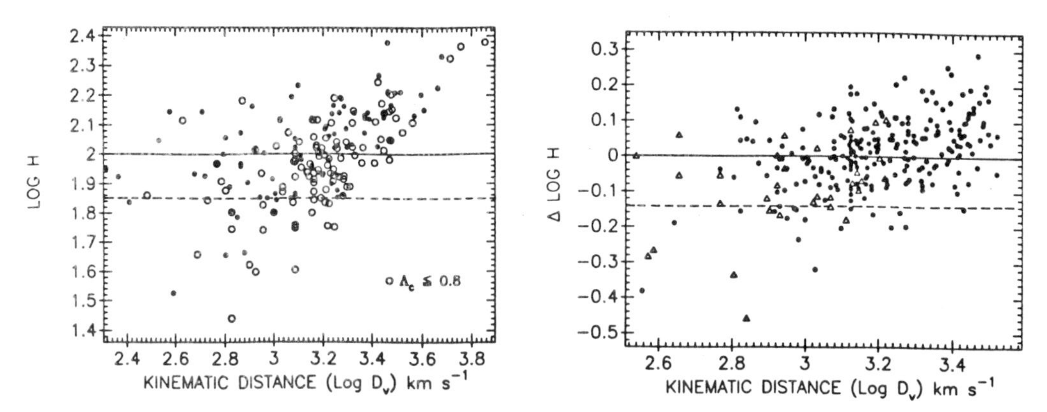

Fig. 1- Individual Hubble ratios H or H/100, plotted against kinematic distances for galaxies having distance moduli derived from a) the luminosity index Λ_c, b) from the infrared Tully-Fisher relation. Both coordinates are in logarithmic scale (1 mag corresponds to 0.2 units in each coordinate). Low rotation galaxies (Log$W(0) \leq 2.35$) are marked by open triangles in Fig. 1b.

is proportional to $\phi(M_W)\psi(W_M)$. Let consider a sharp cut off M_D in absolute magnitude. The bias in the estimate of the luminosity of a galaxy having a rotation velocity W is the difference between the average value of M at constant W and the average of the values of M smaller than M_{D_v} at constant W. If ϕ and ψ are Gaussian and if σ_{M_W} does not depend on W, one finds

$$\triangle M = \frac{1}{2}\frac{\sigma_W \exp - [(M_D - < M_W >)^2/\sigma_W^2]}{\int_{-\infty}^{(M_D - <M_W>)/\sigma_W} \exp - u^2 du}$$

where $\sigma_{M_W} \equiv \sigma_W$. This expression gives the average bias for an object having a rotation velocity W located at a distance D, provided that we know M_D. For a sample populating the space at various distances and observed above various magnitude limits the bias can be scanned as a function of the distance. The main advantage of the model is its flexibility which gives the possibility to have some idea of the bias of objects of various masses drawn from samples observed above variable magnitude limits at various distances. When applied to a biased sample located at a given distance the model predicts that the shift in mean absolute magnitude increases as we go to smaller objects. This leads to a differential Malmquist bias (galaxies with high Λ_c or low rotation velocity give higher Hubble ratios than those with low Λ_c or high rotation).

IV. THE ROLES OF THE INPUT PARAMETERS IN THE DETERMINATION OF THE BIAS.

We want to find a procedure to get $H_o = H_{loc} \sim 75$ from the Λ_c index and from the B- and H-band Tully-Fisher relations by assuming that the samples which lead to 100 are unfair and that the dispersions of the luminosity indicators are systematically underestimated. We drop data and force the parameters of the model until the final result of 75 is obtained.

a) Bias in the B-band Tully-Fisher relation ?

We drop all objects fainter than $B_T^o = 12$ (i.e we reject 35% of the sample) and estimate the individual corrections for bias by assuming that the distances are given by the Hubble velocities corrected for infall. In the case $H_o = 100$ the minimum distance within which there is no significant correction is about 1250 - 1500 kms^{-1}. Then we calculate the absolute magnitudes from the Tully-Fisher relation, make the individual corrections for bias, compute the corrected individual Hubble ratios and plot their mean values in various distance intervals against the kinematic distances (Fig. 2). The main result is that we can get a value of H_o near 75 at $D_v \sim 2200$kms^{-1} where the corrections

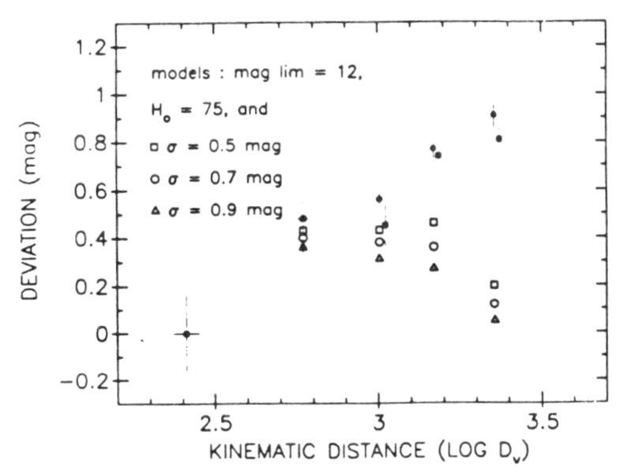

Fig. 2- Difference between the distance moduli calculated from the B-band Tully-Fisher relation (slope 5, zero-point 19.4) and the kinematic distance moduli if $H_o = 75 \mathrm{kms}^{-1}\mathrm{Mpc}^{-1}$, plotted against velocities for galaxies brighter than $B^o_T = 12$ at kinematic distance $\log D_v \leq 3.5$ (filled circles), "corrected" for bias with the following parameters $H_o = 75$, $\sigma = 0.5$mag(open squares), $\sigma = 0.7$mag (open circles), $\sigma = 0.9$mag (open triangles).

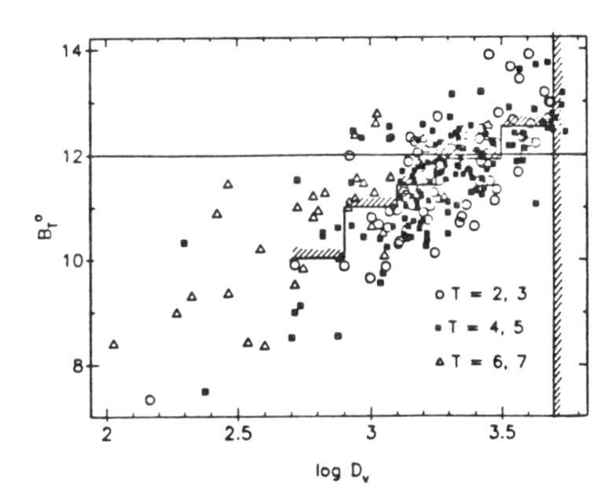

Fig. 3- Apparent magnitudes B^o_T, plotted against kinematic distances in logarithmic scale for galaxies having distances derived from the B-band Tully-Fisher relation. The variable cut off used to get $H_o = 75$ is shown.

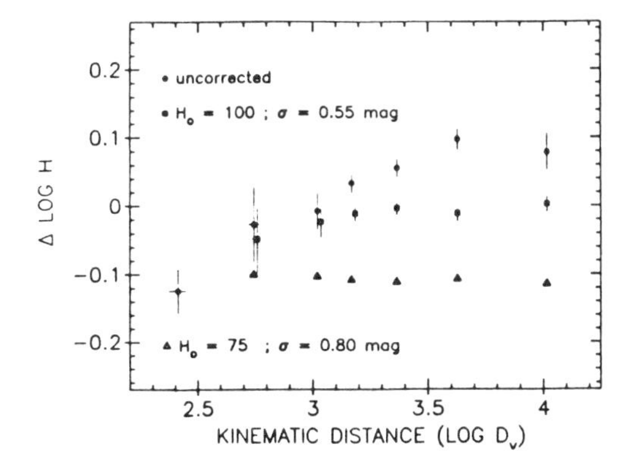

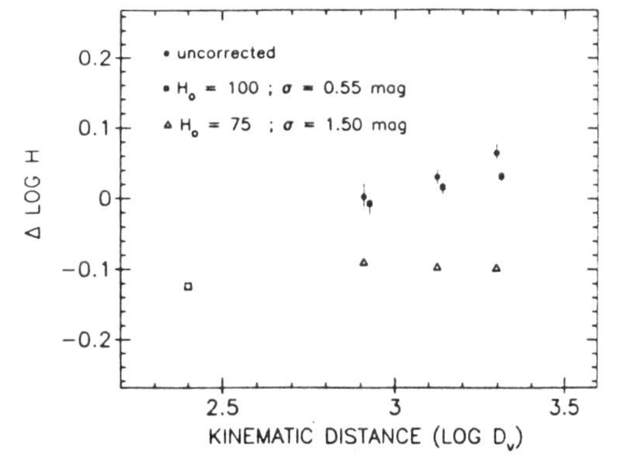

Fig. 4- a) Mean Hubble ratios H/100, derived from the B-band Tully-Fisher relation for galaxies brighter than the magnitude cutoffs shown in Fig. 4, plotted against kinematic distances (a) uncorrected for bias, filled circles, (b) "corrected" for bias in the case of $H_o = 100 \mathrm{kms}^{-1}\mathrm{Mpc}^{-1}$, and a dispersion of the Tully-Fisher relation of $\sigma = 0.55$mag, open squares, (c) in the case of $H_o = 75 \mathrm{kms}^{-1}\mathrm{Mpc}^{-1}$, and $\sigma = 0.8$mag, open triangles. b) Same as a, but for distances derived from the H-band Tully-Fisher relation and variable magnitude cutoffs.

are large but, at small distance, where the corrections are small and the deviation from 75 is already significant, the model cannot flatten the diagram even with a very high dispersion of the indicator. So even after rejection of 35% of the sample this simple model fails.

We now assume that at each distance interval only the brightest objects have been observed. This is to say that some nearby faint galaxies are missing while objects of the same apparent magnitude at larger velocity are observed (i.e. the sample is unfair at each distance interval below a variable magnitude). It is possible to obtain $H_o = 75$ (Fig. 4a) by rejecting approximately 40 – 50% of the sample at each distance interval (Fig. 3) and by assuming that the dispersion of the Tully-Fisher relation is not better than $\sigma = 0.8$mag. If this is true the sample is an extremely poor representation of the real world. On the other hand if we enter $H_o = 100$ and $\sigma = 0.55$mag into the bias model the observed corrected Hubble rate increase from 75 to a constant value of about ... 95 - 100 at medium distance. We conclude that it is not impossible to obtain $H_o = H_{loc} \sim 75$ but we have to assume that the data are very biased and to force $H_o = 75$ in the model.

b) Bias in the infrared Tully-Fisher relation ?

To interpret a high value of H_o derived from the infrared Tully-Fisher relation by the Malmquist effect is a priori more difficult than in the case of the B-band because this relation has a high slope, so at any medium magnitude cut off there are many objects 1 or 2 magnitudes brighter than the limit. It follows that if the high value of H_o is due to the Malmquist bias, the sample is fair only at the extreme bright end of the luminosity distribution and the true dispersion is very large. We reject galaxies fainter than $H_{-0.5} = 10$ at $\log D_v \leq 3.2$ and fainter than $H_{-0.5} = 10.7$ at $3.2 \leq \log D_v \leq 3.4$ (i.e. 35 to 50 % of the sample). In order to obtain 75 at this level of rejection we need to enter an unbelievable dispersion of 1.5 mag in the bias model (Fig. 4b). If we want to reduce the dispersion to 1.2 mag we must reject 57 to 67 % of the sample. An other possibility would be to say that **at each rotation velocity only the brightest objects have been observed** . In that case lower constraints on σ and on the rejections are required. But if at large rotation velocities faint objects are missing, while objects with small rotation velocities and 2 or 3 magnitudes fainter have been observed, we are invoking other unknown sources of bias.

c) Bias in distances from the Λ_c index ?

Since the slope of the relation $\Lambda_c \longrightarrow -M_T^o$ is only 3 , a value of 75 can be easily obtained by using an input value of $\sigma = 0.8$mag and almost the same limits as in the B-band Tully-Fisher relation.

V. CONCLUSION.

It is not impossible to obtain $H_o = H_{loc} \sim 75$ from data in the short distance scale and the scanning bias model. But we need to invoke more and more sophisticated bias as the correlation coefficient of the luminosity indicator is improved. More specifically, because the luminosity range observed in the H-band is large (4 mag or more at a given distance), there is always a non negligible fraction of the sample 1 or 2 mag brighter than any medium magnitude cut off. If the dispersion of the infrared Tully-Fisher relation is ~ 0.5mag or less, the bias correction for these objects is zero and it is not possible to get a value of 75. The possibility that at each rotation velocity, only the brightest objects have been observed, cannot be a priori rejected. This leads to a mixed problem of bias and sampling. To find 75 over the first ~ 2500kms^{-1}, we need in each case to enter the final value of H_o and a large σ into the bias model. If for the same samples, a value of H_o of 100 and a low value of σ are used, the variation of the Hubble rate within ~ 1300kms^{-1} cannot be removed. The results are totally model dependent and imply that all samples are unfair representations of the real world.

REFERENCES.

Aaronson M., Huchra J., and Mould J., 1979 , Ap. J., 229, 1.
Aaronson M., et al., 1982, Ap. J. Suppl., 50, 241.
de Vaucouleurs G., 1979a, Ap. J., 227, 380.
de Vaucouleurs G., 1979b, Ap. J., 227, 729.
de Vaucouleurs G., and Peters W. L., 1981, Ap. J., 248, 395.
Giraud E., 1986a, Ap. J., 301, 7.
Giraud E., 1986b, Astron. Astrophys., 170, 1.
Sandage A., and Tammann G. A., 1975, Ap. J., 196, 313.
Sandage A., Tammann G. A., and Yahil A., 1979, Ap. J., 232, 252.

DISCUSSION

J. P. Vigier: There is certainly a Hubble effect (ie. increase of redshift with distance) but it is difficult to maintain that there is a Hubble constant left. One might interpret an increase of 25% with some hidden mass in our supercluster but how can you interpret the Karoji-Nottale effect (ie. increase of H for sources seen through intervening clusters of galaxies)?

E. Giraud: I don't interpret the Karoji-Nottale effect. I believe that the cheapest hypothesis to test is an overdensity around the Local Group (namely in the cloud of the Sculptor + M81 + M101 + Canes Venatici groups). If the density is larger here we expect a local deceleration due to these masses.

H. Arp: Would you agree with my conclusion that from the present evidence it is extremely unlikely that the Hubble constant is constant? In order to make H_o constant with distance one needs to invoke special and complicated dynamical models.

E. Giraud: We need to invoke more and more complicated bias as we go from the poorer indicators (the Λ index) to the best (actually the infrared Tully-Fisher relation.

THE TESTS OF THE COSMOLOGICAL EXPANSION HYPOTHESIS

T.S. Jaakkola
Observatory and Astrophysics Laboratory
Tähtitorninmäki, SF-00130 Helsinki, Finland

Abstract. The four main presently available groups of tests of the expansion hypothesis are briefly reviewed. A survey of the redshift effect in systems of different scales and levels of hierarchy indicates a higher strength of redshift within the systems than between them. This proves that redshift is an interaction effect, and not due to expansion. The results of various global and local tests are compared, both within the theoretical frame of the expanding models and within that of the static model. In the former frame, the results are internally inconsistent, in the latter frame consistent, giving strong support to the non-expanding alternative. The non-existence of cosmological evolutionary effects, demonstrated by showing that some most forcefully claimed effects are due not to evolution, but to selection effects, also argues against expansion. The powerful surface-brightness test has been applied to four different kinds of data, with the results that contradict the expansion hypothesis.

There is a widespread belief that theories of the Universe are only weakly testable. This is not true at all. There are at least four groups of tests of the cosmological expansion hypothesis available, each containing tens of separate tests along with the classes of objects and the parameters that can be examined.

As the first test, a close enough examination of the properties of the redshift effect in systems of different scales (Jaakkola 1978) will reveal inevitably whether it is a Dopplerian effect or something else. Taking into account the effect of gravitation, the z-effect is in the Dopplerian case fainter within the systems than between the systems. In the case of interaction redshifts, the situation would be the contrary.

There is some evidence for the strength of redshift (h) being higher within the Local supergalaxy than for the homogenous metagalactic distribution. In clusters, groups and pairs of galaxies z appears to depend on, e.g., the type, compactness and status, high z-values usually being connected with features pointing to youth of the galaxy. There are indications that redshift is also a function of position in the systems, witnessing for strong intergalactic z-fields. These results remove the missing mass problem in systems of galaxies.

In individual galaxies there are found redshift gradients from the near to the far sides. A redshift field is found also in the plane of the Milky Way, with h ten times larger than the Hubble constant (H). This field has distorted the structural maps of our galaxy derived from the kinematical maps (Jaakkola et al. 1984). The lines originating in the nucleus of the Galaxy are redshifted by 40-75 km s^{-1}. The O-B type stars in the solar neighborhood, in star clusters and in the Small and Large Magellanic clouds show excess z by 5-15 km s^{-1}. The solar limb redshift is larger than Einstein's prediction by $2 \times 10^{-7} - 10^{-6}$. Significant redshifts of the lines emitted by Taurus A and spacecraft Pioneer 6 are found symmetrically before and after the eclipse by the Sun. These effects, like the related excess light deflections observed for tens of stars, cannot be explained by relativity theory.

Bringing the results together, the strength of redshift appears to depend on the density according to $h \alpha \sqrt{\rho}$; the value of h ranges from the metagalactic value $H \approx 60$ km s^{-1} Mpc^{-1} to 10^{13} km s^{-1} Mpc^{-1} on the surface of the Sun. The results indicate unambiguously the test alternative where h is stronger within the systems than between them, proving that redshift is an interaction effect and not due to expansion. The problem of the QSO redshifts is ignored here; I would only point out that the large intrinsic redshifts in a sub-class of QSO's (Jaakkola et al. 1975; Jaakkola 1982) are also due to one and the same interaction effect as the cosmological redshifts.

Systematization of the global and the local cosmological tests in the expanding and the static theoretical frames forms the second test of expansion (Jaakkola et al. 1979). Within the frame of the standard theory, the empirical value of q_o is dependent on the method: the Hubble diagrams give it the mean value $+0.93 \pm 0.19$, the local tests $+0.03 \pm 0.08$, the (Θ,z)-relation for galaxies $+0.3 \pm 0.2$, for clusters -0.9 ± 0.2, and for radio sources symbolically < -1, meaning inconsistency with all existing relativistic models. The empirical inconsistency of the standard cosmology appears to be of a systematic and stable character.

Concerning the static frame of interpretation, its (m,z)-relation is practically identical with Hubble's linear relation and with that for $q_o = +1$. The value $q_o = \pm 0.93 \pm 0.19$ found above indicates that the data fit the static model. Also the (Θ,z)-diagrams for clusters and radio sources fit the static prediction. The observed convergence of the test results, opposing the status of the standard cosmology, provides strong evidence of the static character of the Universe.

As the third group of tests, if the Metagalaxy expands, it must change also in other respects, if not, its age is infinite and it must not change. There are several kinds of data not indicating evolution. Some other data, such as the counts of bright QSO's (Green & Schmidt 1978), have been claimed to show evolution. It has been shown (Jaakkola 1982) that this result is due to a morphological selection effect, and adding the type 1 Seyfert galaxies identical to the QSO's, the counts fit the static unevolving model closely. The argued spectral evolution of galaxies results from a selection effect arising from the K-term (Laurikainen & Jaakkola 1985). There are several ways to interpret the radio counts without evolution (Jaakkola et al. 1979). The failure of the most strongly urged claims of cosmic evolution demonstrates actual non-existence of such effects, and hence contradicts the expansion hypothesis.

The surface brightness test ($SB \alpha (1+z)^{-a}$; a = 4 for expansion, a = 1 for the static model), first suggested by Hubble & Tolman (1935), forms the fourth test group. It has been applied to four different kinds of data (Jaakkola 1987). Galaxies in six clusters have been argued to support expansion (Crane & Hoffman 1977). However, the data are due to other reasons: procedure of absorption correction, a K-related selection effect and circular reasoning involved in usage of the metric SB. After proper corrections, a ≈ 1. Even without such corrections, 1, 3 and 10 brightest galaxies give a ≦ 1. The SB-profiles of six cD clusters, measured at 0.5 and 1.0 Mpc from the centres, favour the static solution. So do the QSO host galaxy SB-profiles; the contrary conclusion by Gehren (1985) results from inhomogeneity of data and usage of the q_o = +1 metric. The lowest SB-values in the subsequent z-intervals for QSO double radio sources, both using the component sizes and LAS, follow the static model prediction. Hence also the Hubble-Tolman test proves consistently against the expansion hypothesis.

References

Crane, P. & Hoffman, A.W. (1977). In Décalages vers le rouge et expansion de l'Univers, IAU Coll. 37, p. 531.
Gehren, T. (1985). In New Aspects of Galaxy Photometry, ed. J.L. Nieto, p. 227. Berlin: Springer-Verlag.
Green, R.F. & Schmidt, M. (1978). Astrophys. J. Lett., 220, L1.
Hubble, E. & Tolman, R. (1935). Astrophys. J., 82, 302.
Jaakkola, T. (1978). Acta Cosmologica, 7, 17.
Jaakkola, T. (1982). Astrophys. Space Sci., 88, 283.
Jaakkola, T. (1987). Proc. Soviet-Finnish Astron. Meeting, in press.
Jaakkola, T. & Donner, K.J. & Teerikorpi, P. (1975). Astrophys. Space Sci., 37, 301.
Jaakkola, T. & Moles, M. & Vigier, J.P. (1979). Astr. Nachr., 300, 229.
Jaakkola, T. & Holsti, N. & Laurikainen, E. & Teerikorpi, P. (1984). Astrophys. Space Sci., 107, 85.
Laurikainen, E. & Jaakkola, T. (1985). Astrophys. Space Sci., 109, 111.

ON THE CONSTRUCTION OF COSMOLOGICAL MODELS

D. GALLETTO

B. BARBERIS
Istituto di Fisica Matematica "J.-Louis Lagrange"
Via Carlo Alberto 10, 10123 Torino, ITALY.

1. - For the sake of brevity, the procedure which is summarized here refers to the simplest cosmological model and is developed with reference to the case in which the incoherent matter scheme is used to describe the universe. This procedure has been extended by the Authors to the case in which the physical space is a maximally symmetric space. It can be extended to the more general case of homogeneous and anisotropic universes.

The results which are summarized here are based on the following hypotheses:
a. the physical space is the ordinary three-dimensional Euclidean space;
b. the universe, at least on a large scale, is homogeneous and infinite;
c. with regard to our galaxy the behaviour of the universe is radial, in the sense that with respect to the frame of reference with origin in the centre of mass of our galaxy and determined by three other distant galaxies the motion of any typical galaxy is radial.

As we have already said the incoherent matter scheme is used. Let us call natural frame of reference (n.f.r.) every frame of reference determined by four typical and not coplanar galaxies. We have:
I. From the hypotheses made Hubble's law follows, in the sense that, whatever the n.f.r. $\mathcal{R}_0$ is (O being the origin of the frame) and whatever a typical galaxy P is, we have

$$\frac{dOP}{dt} = h(t)OP \quad , \qquad (1)$$

with

$$h(t) = -\frac{1}{3}\frac{\dot{\mu}}{\mu} \quad , \qquad (2)$$

where $\mu(t)$ is the density of the universe.

Therefore we have:
II. The universe has the same kinematical behaviour, which is expressed by (1), with respect to any n.f.r.

The n.f.r. are in translatory motion each with respect to the others.

2. - Let $\mathcal{R}_0$ be an arbitrary n.f.r. By resorting only to the principles of classical mechanics (and precisely to the second law of dynamics and to the principle of superposition of simultaneous forces), from (1) we obtain that the equation of motion of P with respect to $\mathcal{R}_0$, and therefore with respect to any n.f.r., is expressed by

$$\frac{d^2OP}{dt^2} = -\frac{4}{3}\pi k \mu \, OP \quad , \qquad (3)$$

where k is a constant.

If we introduce the deceleration parameter expressed in

terms of h: $q = -(\dot{h} + h^2)/h^2$, we obtain: $k = 3h^2q/4\pi\mu$ and if we assume $0.2 < q < 1$, $10^{-30} < \mu < 10^{-29}$ g cm^{-3} and $50 < h < 100$ km s^{-1} Mpc^{-1} we find that $10^{-8} < k < 10^{-6}$ cm^3 g^{-1} s^{-2}, a result which, bearing in mind that the value of the gravitational constant is $6.7 \cdot 10^{-8}$ cm^3 g^{-1} s^{-2}, permits us to conclude:

III. From the hypotheses made it follows that for the universe there exists a constant k that can be rightly identified with the gravitational constant.

We can therefore say:

IV. From the hypotheses made there follows the explicit equation of motion of P with respect to any n.f.r., given by (3), without resorting to any theory of gravitation.

V. All n.f.r. are equivalent to one another, in the sense that the universe has, from both the kinematical and the dynamical points of view, the same behaviour with respect to them. This behaviour is described by Hubble's law (1) and by the equation of motion (3).

Equation (3) is the same one that would be obtained by assuming the frame $\mathcal{R}_0$ to be inertial, the part of the universe external to the material sphere $\mathcal{S}_{0P}$ with centre at O and radius $|OP|$ to give no contribution to the motion of P, and the forces at a distance (gravitational forces) exerted on P to be expressed by Newton's law of gravitation. A priori, there is nothing, however, to authorise this procedure, which is the one that has been followed up to now in all treatises on cosmology made in Newtonian terms. On the contrary in this context equation (3) is obtained without imposing a priori any limitation on the frame $\mathcal{R}_0$ and above all without resorting to Newton's law of gravitation and resorting instead only to the hypotheses made above.

Moreover, in addition to the previous results, it is possible to obtain the following:

VI. From the hypotheses made it follows that the force exerted between any two typical galaxies is necessarily expressed by Newton's law of gravitation.

From (3) and VI follows:

VII. Whatever the natural frame $\mathcal{R}_0$ is, as far as the motion of P with respect to it is concerned everything happens as if the frame $\mathcal{R}_0$ were inertial and as if the part of the universe external to the material sphere $\mathcal{S}_{0P}$ made no contribution to the motion of P.

Result VII justifies the procedure for the deduction of equation (3) which is followed in all treatises dealing with Newtonian cosmology: this procedure was followed in particular by Milne and McCrea, who in 1934 were the first to make an attempt to give an introduction to cosmology in Newtonian terms. (Of course we do not mention Seeliger's attempt, which dates back to the end of the 19th century, because that attempt - since it was based on the belief that the universe was static - contradicts astronomical observations as well as Newton's law of gravitation).

What has been stated so far, and especially the results summarized in VII, proves that the criticism made by Layzer in 1954 is incorrect: in his criticism Layzer denies the possibility of formulating a Newtonian cosmological theory with an infinite homogeneous universe. In particular, among other things, what we prove to be incorrect is the general conviction that the results summarized in VII can be justified only by resorting to the general theory of relativity.

In general what we have seen so far allows us, at least under the hypotheses b and c, to consider the so-called gravitational paradox to be inconsistent: this paradox in essence states the impossibility of applying Newton's theory of gravitation to an infinite homogeneous fluid.

3. - If the instant t_0 is fixed once for all, and if we define the function (scale factor):

$$R(t) = \exp \int_{t_0}^{t} h(t)dt \quad ,$$

from (1), (2) and (3) there follows

$$\frac{\dot{R}}{R} + \frac{1}{3}\frac{\dot{\mu}}{\mu} = 0 \quad , \tag{4}$$

$$\frac{\ddot{R}}{R} = -\frac{4}{3}\pi k \mu \quad , \tag{5}$$

from which we obtain

$$\frac{\dot{R}^2}{R^2} = \frac{8}{3}\pi k \mu + 2\frac{\alpha}{R^2} \quad , \tag{6}$$

where α is the value of the energy constant which would belong to the element P (which we assume to have a unitary mass), if it were at a distance R from the origin of the n.f.r. with respect to which the motion of P is considered.

From (5), (6) follows

$$\frac{2\ddot{R}}{R} + \frac{\dot{R}^2}{R^2} = 2\frac{\alpha}{R^2} \quad . \tag{7}$$

Equations (4) (continuity equation) and (7) (the equation of evolution), together with (6) (energy integral) are the equations of Newtonian cosmology, obtained without resorting to the Newtonian theory of gravitation.

As Milne and McCrea noticed in 1934, the above-mentioned equations are similar to the equations obtained by working in the corpus of general relativity, but, contrary to what has been observed until now, this is not only a formal analogy, but an essential one, as the results of the next sections prove.

4. - At this stage, if we take into account the property of the velocity of light revealed by the Michelson-Morley experiment, we have:

VIII. The local velocity of light c is the same with respect to any n.f.r.

Let us introduce into the n.f.r. $\mathcal{R}_0$ a cartesian coordinate system x^i (i=1,2,3) with origin in O and let us introduce the coordinates y^i defined by

$$y^i = \frac{x^i}{R(t)} \quad . \tag{8}$$

From VIII and from the Galilean law of addition of velocities follows:

IX. The metric of the space-time manifold is necessarily given by the particular case of the Robertson-Walker metric expressed by

$$ds^2 = R^2(t) \sum_{1}^{3}{}_i (dy^i)^2 - c^2dt^2 \quad , \tag{9}$$

that is by the metric which characterizes the Einstein-de Sitter model of the universe.

At this point we can derive the usual formula for the red-shift and we have:

X. In the present case the formula connecting the red-shift to the expansion of the universe is a consequence of VIII and of the Galilean law of addition of velocities.

Analogous considerations can be made about horizons.

5. - If we impose the condition that the equations of Newtonian cosmology and metric (9) are compatible, in the sense that it would be possible to give an intrinsic form to these equations within the framework of this metric, it follows that the energy constant α, which appears in (6), must be zero. In other words:

XI. The property of light which consists in having the same velocity with respect to any n.f.r. implies that the energy constant α is zero.

This result implies that in the present case for the deceleration parameter we have: $q = 1/2$.

Once it has been proved that necessarily $\alpha=0$, we obtain that the intrinsic form imposed by the Einstein-de Sitter metric on the equations of Newtonian cosmology is precisely the one expressed by Einstein's gravitational equations:

$$G_{\alpha\beta} - \frac{1}{2} G g_{\alpha\beta} = \chi T_{\alpha\beta} .$$

It can therefore be stated that:

XII. The hypotheses that the physical space is Euclidean, that the universe is homogeneous and its behaviour with respect to our galaxy is radial, together with the property of light of having the same velocity with respect to any n.f.r., necessarily lead to Einstein's equations of the general theory of relativity.

What has just been briefly discussed can be therefore considered also a deduction of Einstein's gravitational equations from astronomical observations. This deduction entails, among other things, that Einstein's gravitational constant χ is necessarily expressed by:

$$\chi = \frac{8\pi k}{c^4} ,$$

a result that in all treatises is obtained by resorting to approximation methods.

6. - With appropriate modifications the results summarized in the preceding sections may be extended - as we have already said - to the case that the physical space, instead of being Euclidean, is a three-dimensional maximally symmetric space. Results I,III,IV,VIII,X,XII are maintained, together with equations (4), (5), (6), (7). The energy constant α is no longer zero and determines the curvature of the physical space, whereas the metric we obtain for the space-time is the Robertson-Walker metric. As we have already said, these results can be extended to the more general case of homogeneous and anisotropic universes.

The proofs of the results here summarized, which require very extensive and lengthy considerations, will appear, expressed in a thorough and detailed way, in a series of forthcoming papers.

VIGIER: How can you reconcile your model with Physics? Newton's gravitational law is not compatible with the equations of motion of general relativity. You cannot have them both together ... or you drop the geodetic assumption.

GALLETTO: Vigier's assertion raises and pursues some rather widespread and incorrect convictions which I have mentioned in Sec.2.

As he appears to be unaware of the results which have now been stated, Vigier cannot be acquainted with the fact that in the case of the Einstein-de Sitter model of the universe, characterized by metric (9) besides Einstein's gravitational equations, the results obtained in the relativistic context, i.e. starting from (9) and from the gravitational equations, coincide, and not only formally, with the results obtained in the Newtonian context (of course with the constant $\alpha = 0$: see Sec.5) on condition that in the latter context the procedure be correct and rigorous as is the case of the results I presented in my talk.

Vigier should notice that the sections $t = \text{const.}$ of the space-time characterized by metric (9) are given by the ordinary Euclidean space considered in the instant t and referred to the coordinates y^i. He should notice that in the case of the incoherent matter scheme (which is the case considered in my talk) the explicit form of the Einstein gravitational equations in the case in which the metric is given by (9) coincides with equations (6), (7), from which equation (5) follows. Equation (5) can be written as

$$\ddot{R} = - k \frac{M_R}{R^2} , \qquad (*)$$

where M_R is the mass of the material sphere with radius R.

Equation (*) by itself should already suggest - or at least hint - rather a lot to Vigier.

After having obtained result (*) starting from metric (9) and from Einstein's equations, the natural frames of reference defined in my talk can be introduced into the Euclidean space deriving from metric (9). Moreover, by resorting to the coordinates x^i defined by (8), it is possible to deduce equation (3), from which it is possible to obtain result VII of my talk.

However, for a full answer see the series of forthcoming papers I have just announced in my talk.

Vigier's last objection leaves one rather at a loss: Vigier seems to overlook that in comoving coordinates - as coordinates y^i which appear in metric (9) are - geodetic equations deduced starting from this metric for any element of the universe (i.e. for any galaxy) boil down, as on the other hand they must, to $\dot{y}^i = 0$.

This result, that, in perfect accordance with their definition, states that the comoving coordinates of any galaxy are constant, is also in perfect accordance with the definition (8) that I gave for the coordinates y^i. As I have said in my talk, taking into account the property of the local velocity of light of being the same in every natural frame of reference, this definition allows us to obtain that in the present case the metric of space-time is precisely expressed by (9).

In conclusion, nothing has been dropped.

DARK MATTER AND THE 15 MYR GALACTO-TERRESTRIAL CYCLE

S.V.M. Clube
Department of Astrophysics, South Parks Road, Oxford, UK.

Introduction

To presume a rather specific cosmological scheme (eg the big bang) whilst questions like the origin of spiral structure and the formation of stars have not even been settled in principle is to risk placing the whole of astronomy in an unscientific straitjacket. Thus, a consequence of this approach to astronomy, working from the top down, is that inferred properties of matter tend to acquire a significance that correlates positively with the scale of the astronomical object examined. It is hardly surprising therefore if attempts to understand bodies like comets are often regarded as peripheral to the mainstream of astronomy. Nevertheless, cometary material is arguably the most primitive accessible to scientific examination and the view that comets are the key to progress on a wider front is not without its defenders. Indeed, during the last decade, it has come to be realised that comets may be interacting with the Earth in a far more direct manner than has been thought possible heretofore - at least in modern times - and my purpose in this paper is to address some issues raised by recent discoveries in the fields of cometary and terrestrial science so as to indicate their possibly more general implications.

New evidence

Recent studies of the perihelion distribution of 'new comets' have established that it is strongly depleted in the directions of the poles and the equator of the Galaxy (Delsemme 1987). This finding indicates that the dominant force perturbing the Oort cloud is the tide perpendicular to the galactic plane. Recent theoretical studies (eg Torbett 1986) have also established that weak stellar and molecular cloud encounters are at least as effective as strong encounters over the lifetime of the Solar System in transferring Oort cloud comets into near zero angular momentum orbits, the process whereby these comets come to be observed as the near parabolic flux through the inner planetary system. Taken together, the weak encounters essentially act to produce the vertical galactic tide.

Since a fraction of the parabolic flux is continuously deflected by the major planets (chiefly Jupiter) into Earth-crossing orbits where they or their evolutionary products - disintegrating asteroidal debris and the zodiacal cloud - will interact with the Earth, it is clear in principle that terrestrial evolution will reflect time-variations in the vertical tide T_z and hence $K_z = \int T_z \, dz$ as the Sun moves along its galactic orbit (Clube 1986 and references therein). The question arises therefore whether processes in and on the Earth correlate with changes in K_z.

The approximately steady state number density of selected

stellar species (eg A, F, K stars). expressed as a logarithm ln $\nu(z)$, has long been known to display point(s) of inflexion above the plane at heights of about 50 pc and 150 pc. These have been explained in terms of either (i) a plane stratified disc including dark matter with increasing velocity dispersion $\sigma(z)$ (Oort 1932), or (ii) the additional presence within the plane stratified disc of a discrete mass distribution (eg spiral arms or massive clouds) containing dark matter (Woolley 1957). Only the latter explanation gives the potential $\Phi_z = \int K_z \, dz \propto \ln \nu(z)$ whence it follows that the points of inflexion correspond to a zero galactic tide. Since these points are generally sampled four times per vertical cycle of a typical near circular galactic orbit (period ~ 60 Myr; $z_{max} \leqslant 100$ pc, say), the parabolic comet flux is then in principle modulated by a 15 Myr cycle which may also be reflected in the terrestrial record.

The Earth's H-reversal frequency is in fact modulated by a 15 Myr cycle which its discoverers (Mazaud et al 1983) regard as the most persistent terrestrial cycle ever found. Stochastic variations are more evident in the cratering, climate and vulcanism records though these also correlate with the 15 Myr cycle. The modulation is explicable as a consequence of the comet flux if it is mediated by core-mantle and climatic perturbations brought on by evolving comets in Earth-crossing orbits (Clube and Napier 1984). It is likely therefore that this unmistakable galactic signature in terrestrial processes which is otherwise unexplained reflects a rather fundamental mechanism in which (1) the local dark matter needs to be distributed in molecular clouds, and (2) the molecular cloud system (and hence spiral arms) needs to be a recurring phenomenon (Clube 1987). Indeed it has been argued independently (Bailey et al 1987) that the stochastic and periodic fluctuations in the parabolic flux can only be understood quantitatively if the dark matter is mostly confined to molecular clouds.

The implications

These new findings are prima facie evidence that dark matter in galactic discs largely exists in the generally unobservable form of pre-stellar, sub-stellar condensates (eg brown dwarfs) but concentrated in relatively short-lived GMCs. Such evidence is obviously consistent with a star formation scenario that accounts for the observed planetary mass/angular momentum distribution and in which H_2, CO etc are released during the T Tauri stage from previously formed gravitationally bound aggregates of smaller parent bodies (eg McCrea 1978). Partially differentiated parent bodies are the generally recognized pre-requisite for meteorite formation and seem now to be observed also as giant comets (Clube 1986). Such bodies, in the size range 10^2-10^3 km, appear to be necessary on several counts therefore whilst the continuing failure to detect in situ star forming material in a state of gravitational collapse now leaves open the possibility that these bodies may be the natural Jeans condensations in the rapidly cooling primary medium that might arise in spiral arms periodically injected into the disc from an active galactic nucleus (Bailey et al 1986).

To explore such ideas, one needs to know about the physical and chemical structure of cometary material. The results of the recent fly-by missions to Comet Halley have brought some surprises and may already be altering our general picture of comet structure (eg Wallis and Wickramasinghe 1987). It seems for example that this particular comet may be not unlike Phobos in appearance, with a dead chondritic surface for the most part, but which also includes temporarily (re)activated areas where the closely interwoven volatile/refractory mixture from the interior seems to be released in the form of the large i.d.p. aggregates. Indeed, if the current zodiacal cloud is the principal source of the i.d.p.'s collected from the upper stratosphere, most cometary material could well be initially processed in a steadily less reducing environment as it cools (Fraundorf 1981). The crucial point here is that apparently unaltered submicron crystals embedded within collected i.d.p.'s are not obviously explained as a product of heating during atmospheric entry, and show little indication of a long sojourn in interstellar space. Thus it is not inconceivable that spiral arm condensates pre-exist as dense plasma which undergoes gradually weakening confinement. By locating dark matter in molecular clouds therefore, we seem to be led rather directly to the old idea (eg Ambartsumyan 1965) that spiral arms evolve from short-lived jets.

Of course, it is a corollary of any such picture that the motions of the disc and the local spiral arms are dynamically disjoint, a view that the kinematics of the solar neighbourhood do not at present exclude (Clube 1983a), but if spiral arms are recurrently formed as jets from the nuclear region, the gravitating mass necessary for their temporary containment, averaged over a typical cycle, may also account for the cosmological missing mass (Clube 1983b). Thus the 15 Myr terrestrial cycle provides a basis for regarding spiral structure as an ejection phenomenon, albeit one in which 'local' and 'cosmological' dark matter are fundamentally different.

References

Ambartsumyan V.A. (1965) In The Structure and Evolution of Galaxies (Interscience), 1

Bailey M.E.,Clube S.V.M.&Napier W.M. (1986) Vistas in Astr. 29, 53

Bailey M.E., Wilkinson D.A. & Wolfendale A.W. (1987) Mon.Not.R.astr.Soc. in press

Clube S.V.M. (1983a) ESA - SP 201, 173

Clube S.V.M. (1983b) Liege Coll. 24, 393

Clube S.V.M. (1986) ESA - SP 250, 403

Clube S.V.M. (1987) ESA - SP 278, in press

Clube S.V.M.&Napier W.M. (1984) Mon.Not.R.astr.Soc. 211, 953

Delsemme A.H. (1987) Astr.Astrophys, in press

Fraundorf P. (1981) Geochim.Cosmochim.Acta 45, 915

Mazaud A.,Laj C.,de Seze L.&Verosub K.L. (1983) Nature 304, 328

McCrea W.H. (1978) In Origin of the Solar System (ed Dermott S.F.),75

Oort J.H. (1932) B.A.N. 6, 249

Torbett M. (1986) In The Galaxy and the Solar System (ed Smoluchowski R. et al), 147

Wallis M. & Wickramasinghe N.C. (1987) ESA - SP 278, in press

Woolley R.v.d.R. (1957) Mon.Not.R.astr.Soc 117, 198

LIST OF AUTHORS

INDEX

(References are to first page of relevant article)